Actinobacteria

Loganathan Karthik
Editor

Actinobacteria

Microbiology to Synthetic Biology

Springer

Editor
Loganathan Karthik
Synthetic Biology, R&D Center
Salem Microbes Private Limited
Salem, India

ISBN 978-981-16-5834-1 ISBN 978-981-16-5835-8 (eBook)
https://doi.org/10.1007/978-981-16-5835-8

This Springer imprint is published by the registered company Springer Nature Singapore Pte Ltd.
The registered company address is: 152 Beach Road, #21-01/04 Gateway East, Singapore 189721, Singapore

I dedicate this book to almighty God, My father S. Loganathan, My mother L. Sasikala, My wife K. Ramapriya and my daughter K.R. Dhanya Sri

Foreword

Satya P. Singh

I am delighted to write foreword for this new book entitled *Actinobacteria: Microbiology to Synthetic Biology*, edited by Dr. Karthik Lognathan and published by the Springer Nature. This book includes various upcoming dimensions on the actinomycetes. Being associated with the research on the actinomycetes of the thermophilic and saline habitats for over two decades now, it is my great pleasure to contribute two chapters and write a foreword in this specialized book.

The actinobacteria traditionally explored and investigated for their secondary metabolites are Gram-positive prokaryotes of the bacterial phyla with filamentous morphology and high G+C DNA. More recently, the biocatalytic aspect of these microorganisms has been a point of investigation for the scientific communities. This book includes chapters on various aspects of cultivation, diversity, metabolisms and engineering pathways, CRISPR technology, synthetic biology and regulation of gene expression. The chapters on the secondary metabolites and manoeuvring their synthesis at different levels would provide basic as well as newer dimensions of research. Conventionally, the actinimycetes have been investigated from soil and compost with the limited attention on their investigations from extreme habitats. The contents on the genomics and newer approaches of the cultivation would provide updated account on the role of these microorganisms in the ecosystem and clues on their exploration.

I am confident that this edited book on the actinomycetes with basic, biochemical and molecular dimensions would provide a significant platform for the students, researchers and those who wish to initiate commercial ventures.

Preface

Actinobacteria is an expected wellspring of new bioactive natural products like enzymes, anti-microbials, etc. It is disseminated in far-reaching natural environments. In 1943, Waksman and Henrici initially presented *Streptomyces* sp. From that, several studies were done to uncover its potential in various regions for human welfare. Up until now, two Professors got Nobel Prize in this field, Professor Waksman (1952) and Professor Satoshi Omura (2015).

This book *Microbiology to Synthetic Biology* gives a general picture of actinobacteria. It focuses on diversity, NRPS, sesquiterpenes, lantipeptide, bioinformatics apparatuses, cloning, CRISPR, reverse engineering, FDA-supported medications and marine actinobacteria. The book offers inside and out detail that benefits researchers/analysts and understudies in drug discovery, genetic designing of actinobacteria.

Salem, India

Loganathan Karthik

Contributors

Omnia Mohamed Abdallah Microbiology Department, Faculty of Science, Ain Shams University, Cairo, Egypt

Dina Hatem Amin Microbiology Department, Egyptian Drug Authority (EDA), Giza, Egypt

Palakshi Anitha Madurai Kamaraj University, Madurai, Tamil Nadu, India

Arun A. Bhagwath Yenepoya Research Centre, Yenepoya (Deemed to be University), University Road, Deralakatte, Mangalore, Karnataka, India

Pranami Bharadwaj Microbial Biotechnology Laboratory, Life Sciences Division, Institute of Advanced Study in Science and Technology, Guwahati, Assam, India
Academy of Scientific and Innovative Research (AcSIR), Ghaziabad, Uttar Pradesh, India

Ankita Dobariya UGC CAS, Department of Biosciences, Saurashtra University, Rajkot, Gujarat, India

Alaa Fayez Elsayed Microbiology Department, Faculty of Science, Ain Shams University, Cairo, Egypt

Juan M. Gonzalez Microbial Diversity and Microbiology of Extreme Environments Research Group, Agencia Estatal Consejo Superior de Investigaciones Científicas, IRNAS-CSIC, Sevilla, Spain

Shabiha Nudrat Hazarika Microbial Biotechnology Laboratory, Life Sciences Division, Institute of Advanced Study in Science and Technology, Guwahati, Assam, India
Department of Molecular Biology and Biotechnology, Cotton University, Guwahati, Assam, India

S. Jemimah Naine Department of Life Sciences, Kristu Jayanti College Bengaluru, Bengaluru, Karnataka, India

Loganathan Karthik Synthetic Biology, R&D Center, Salem Microbes Private Limited, Salem, India

Gangarapu Kiran School of Pharmacy, Anurag Group of Institutions, Hyderabad, Telangana, India

Aditya Narayan Konwar Microbial Biotechnology Laboratory, Life Sciences Division, Institute of Advanced Study in Science and Technology, Guwahati, Assam, India
Academy of Scientific and Innovative Research (AcSIR), Ghaziabad, Uttar Pradesh, India

Wen-Jun Li State Key Laboratory of Biocontrol, Guangdong Provincial Key Laboratory of Plant Resources and Southern Marine Science and Engineering Guangdong Laboratory (Zhuhai), School of Life Sciences, Sun Yat-Sen University, Guangzhou, China

Zhiyong Li State Key Laboratory of Microbial Metabolism, School of Life Sciences and Biotechnology, Shanghai Jiao Tong University, Shanghai, P.R. China

Rong Ma Clinical Research Institute of Integrative Medicine, Yueyang Hospital, Shanghai University of Traditional Chinese Medicine, Shanghai, China
Clinical Research Institute of Integrative Medicine, Shanghai Institute of Traditional Chinese Medicine, Shanghai, China

S. Merlyn Keziah School of Biosciences and Technology, Vellore Institute of Technology, Vellore, Tamil Nadu, India

V. Mohanasrinivasan School of Biosciences and Technology, Vellore Institute of Technology, Vellore, Tamil Nadu, India

Kamarudheen Neethu Department of Pharmaceutical Sciences, Oregon State University, Corvallis, OR, USA

Khalid Parwez Yenepoya Research Centre, Yenepoya (Deemed to be University), University Road, Deralakatte, Mangalore, Karnataka, India

Wasu Pathom-aree Research Center of Microbial Diversity and Sustainable Utilization, Department of Biology, Faculty of Science, Chiang Mai University, Chiang Mai, Thailand

Hashti Ramavat UGC CAS, Department of Biosciences, Saurashtra University, Rajkot, Gujarat, India

Manik Prabhu Narsing Rao State Key Laboratory of Biocontrol, Guangdong Provincial Key Laboratory of Plant Resources and Southern Marine Science and Engineering Guangdong Laboratory (Zhuhai), School of Life Sciences, Sun Yat-Sen University, Guangzhou, China

Dalip Singh Rathore UGC CAS, Department of Biosciences, Saurashtra University, Rajkot, Gujarat, India

Bhagwan Narayan Rekadwad Yenepoya Research Centre, Yenepoya (Deemed to be University), University Road, Deralakatte, Mangalore, Karnataka, India

Punchappady Devasya Rekha Yenepoya Research Centre, Yenepoya (Deemed to be University), University Road, Deralakatte, Mangalore, Karnataka, India

Pachara Sattayawat Department of Biology, Faculty of Science, Chiang Mai University, Chiang Mai, Thailand

Amit Kumar Sharma UGC CAS, Department of Biosciences, Saurashtra University, Rajkot, Gujarat, India

Amal Mohamed Shawky Microbiology Department, Egyptian Drug Authority (EDA), Giza, Egypt

Satya P. Singh UGC CAS, Department of Biosciences, Saurashtra University, Rajkot, Gujarat, India

C. Subathra Devi School of Biosciences and Technology, Vellore Institute of Technology, Vellore, Tamil Nadu, India

Debajit Thakur Microbial Biotechnology Laboratory, Life Sciences Division, Institute of Advanced Study in Science and Technology, Guwahati, Assam, India

Jignasha Thumar Department of Microbiology, Government Science College, Gandhinagar, Gujarat, India

Gouthami Thumma Department of Pharmaceutics, University College of Pharmaceutical Sciences, Kakatiya University, Warangal, Telangana, India

Mahendran Ruckmani Vignesh Society of Chemical and Synthetic Biology, Vellore, Tamil Nadu, India

Arivarasan Vishnukirthi Department of Microbiology, Lovely Professional University, Punjab, India

Dabing Zhang School of Life Sciences and Biotechnology, Shanghai Jiao Tong University, Shanghai, China

Fengli Zhang School of Life Sciences and Biotechnology, Shanghai Jiao Tong University, Shanghai, China

Chapter 1
Actinomycetes: Microbiology to Systems Biology

C. Subathra Devi, S. Merlyn Keziah, S. Jemimah Naine, and V. Mohanasrinivasan

Abstract Microbes are the prolific producers of bioactive compounds and their derivatives. More than half the number of versatile bioactive compounds from microbial sources have been produced by actinomycetes. Actinomycetes are remarkable sources of novel antibiotics and compounds which possess medical and industrial importance. This review outlines an introduction to actinobacteria with its diversity and distribution. It also discusses their unique characteristics as prokaryotic organisms, the bioactive molecules, their complex structures, and complicated bioactive molecules. This report summarizes the potential avenues in drug leads research of actinomycetes, we also explore the distinct self-resistant β-lactamase producers, polyketide and non-ribosomal polyketide synthase producers, isoprenoids, and protein inhibitors such as indolocarbazones, prodiginies, antiviral enzymes, antitumor enzymes, tyrosinases, and other hydrolytic enzymes such as amylases, cellulases, and xylanases that break down carbohydrates into sugars; chitinases, proteolytic enzymes like unhairing (nonspecific) and fibrin degrading (specific) proteases, and lipases. Furthermore, the array of actinomycete research has been driven by the break through in genomics and proteomics. Increasing new technical developments such as next-genome sequencing, genome editing, and analyzing secondary metabolite biosynthetic gene clusters (smBGCs) in genome mining which helps us to predict the domain organization of the gene clusters and synthesize new bioactive compounds. This chapter gives us the overall outlook of actinobacteria and its promising future prospects in the industrial and pharmaceutical fields when a hand-in-hand research is carried out with the help of bioinformatics.

Keywords Actinomycetes · Metabolites · Drug leads · Biomolecules · Genome mining

C. Subathra Devi (✉) · S. Merlyn Keziah · V. Mohanasrinivasan
School of Biosciences and Technology, Vellore Institute of Technology, Vellore, Tamil Nadu, India
e-mail: c.subathradevi@vit.ac.in

S. Jemimah Naine
Department of Life Sciences, Kristu Jayanti College Bengaluru, Karnataka, India

L. Karthik (ed.), *Actinobacteria*, https://doi.org/10.1007/978-981-16-5835-8_1

1.1 Introduction to Actinobacteria

Microbes play an essential role in developing the healthcare and pharmaceutical sector as fundamental producers of primary and secondary metabolites and as therapeutic agents. Actinomycetes are the main groups of microorganisms producing various secondary biologically active metabolites along with enzymes of economic importance. They are outstanding sources of bioactive compounds like immunosuppressant, antihelminth, anti-aging, anti-cholesterol, antiprotozoal, antiviral, antifungal, antitumor, and antibiotics agents. The "phylum actinomycetes" are filamentous, Gram-positive, aerobic bacteria generally spread in colonizing, water, and soil plants. Actinobacteria are in nature widespread with high consistency and adhere tightly to agar surface, producing hyphae as well as sporangia/conidia—such as fungi in culture media (Hayakawa et al. 2004; Ilić et al. 2005). This is a distinctive producer of an earthy smell volatile compound geosmin, which is a feature of the actinobacteria (Gerber 1979). Actinobacteria have now been responsible for more than 10,000 antibiotics and 45% of all microbial metabolites (Jackson et al. 2018). The improved industrial production of bioactive metabolites are detailed (Fig. 1.1).

Actinobacteria is considered as an intermediary between fungi and bacteria. Actinobacteria has a radial colony morphology, mycelial shape as that of fungi with a musty odor. Their cell walls are made up of peptidoglycans which is the basic physiological trait of bacteria. Actinobacteria was then designated as Strahlenpilze (ray fungi) by Lieske in 1921 as it exhibited radial colony (Kalakoutskii 2004). The

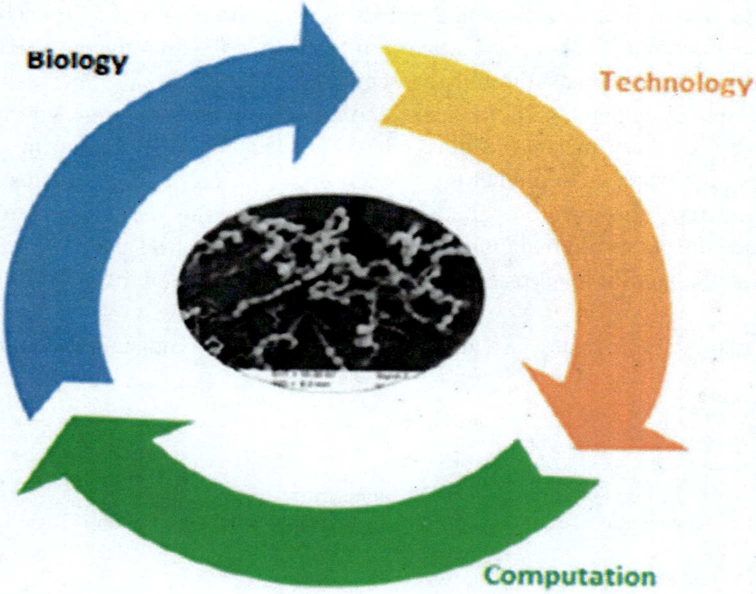

Fig. 1.1 Workflow of synthetic biology to develop improved industrial production of bioactive metabolites

Streptomycetaceae family consists of *Streptacidiphilus, Kitasatospora,* and *Streptomyces* genera which are hard to distinguish with their phenotypic and genotypic characteristics. The feasibility of using site-specific recombination in actinomycetes has been proposed long back but the lack of sufficient genome editing methods restricts the systematic metabolic engineering approaches to discover novel products. Systems biology is essential to recognize that organisms' cells as well as biological entities are complex. Collective behavior of molecular parts called emergent properties is a critical attribute of biological systems. New technological development and approaches aid in studying their biological complexity and predicts the system behavior (Galas et al. 2014).

1.2 Classification and Taxonomy of Actinobacteria

Actinobacteria with the highest taxonomic units and significant lineage within the field of bacteria. "Actinobacteria" are distinct in sporangia and mycelial growth. Actinomycetes are considered to have a content of high G + C in DNA and display high nutrition flexibility as members of actinobacteria. On the basis of branching structures, the phylum actinobacteria divide into six subclasses such as *Thermoleophilales, Rubrobacteraces, Nitriliruptoridae, Coriobacteridae, acidimicrobidae,* and *actinobacteria.* The class of actinobacteria is further classified into *Bifidobacteriales* and *Actinomycetales* in two orders. The order *Actinomycetales* is further classified into orders like *Streptosporangineae, Streptomycineae, Pseudocardineae, Propionibacterineae, Micromonosporineae, Micrococcineae, Kineosporilineae, Jiangellineae, Glycomycineae, Frankineae, Corynebacterineae, Catenulisporineae, Actinopolysporineae, and Actinomycineae* (Ludwig et al. 2012).

1.3 Habitats of Actinobacteria

The genera *Micromonospora* and *Streptomyces* are esteemed prolific producers of various metabolites, immunomodulators, enzyme inhibitors, enzymes, and antibiotics. General actinobacteria habitats are aquatic (both fresh marine and water) and terrestrial (soils). Actinobacteria were isolated from polluted soils, subtropical desert soils, alkaline desert soils, sandy loam soils, Himalayan soils, black alkaline soils, and coal mines. *Actinobacteria* are widely distributed in marine environments like sediments besides seawater, mangroves, molluscs, fish, sponges, and seaweeds (Ward and Bora 2006; Puttaswamygowda et al. 2019).

1.3.1 Actinobacteria from Terrestrial Habitat

Actinosynnema from a glass blade was the first reported endophytic actinobacteria (Matsukuma et al. 1994). The endophytic actinobacteria are widely deemed as non-*Streptomyces* spp. as well as *Streptomyces* spp. The "non-*Streptomyces* spp." is unique and is categorized as rare taxa. Literature shows that the group of the least studied microbes includes plant-associated extremophilic as well as extremotolerant actinobacteria. Huang et al. (2012) recorded actinobacteria from arid plants, which included various genera such as *Amycolatopsis, Nonomuraea, Nocardia, Micromonospora,* and Streptomyces. Drought-tolerant endophytic actinobacteria, *Streptomyces geysiriensis* DE27, *"Streptomyces olivaceus* DE10" as well as *Streptomyces coelicolor* DE07 were identified in several of the endophytic actinobacteria in arid region plants (Yandigeri et al. 2012). Endophytic actinobacteria form a significant portion of the rhizosphere discovered even in plants in which the widely examined species endophytic actinobacteria include genera *Frankia, Streptoverticillium, Nocardia, Micromonospora, Kitasatospora, Pseudonocardia,* and *Streptomyces. "Frankia nitrogen"*-fixing bacteria of nonleguminous plants (Benson and Silvester 1993). The Nocardia species may also be present in the environment such as house dust, garden soil, and beach sand. *N. asteroids, N. cyriacigeorgica, N. nova,* and *N. farcinica* are some of the most general pathogenic species related to human infectious diseases (Wallace et al. 1990; McNeil and Brown 1994; Yassin et al. 2001). The soil actinobacterial isolate which belonged to the *Kitasatospora* genera isolated from European beech (*Fagus sylvatica*) was observed to have antagonist activity against the plant pathogen (*Phytophthora citricola*) (Haesler et al. 2008).

Micromonospora produces about 700 different antibiotics which belong to aminoglycosides, lactones, macrolides, naphthoquinones, peptides, and protylonolides. Microspora antibiotics which belong to antlermicin, calicheamicin, fortimicin, gentamicin, mutamicin, sagamicin, sisomicin, retrocatechin, and verdamicin (Chambers 2006). Some of the other endophytic actinobacteria include *Streptosporangium, Promicromonospora, Polymorphospora, Glycomyces Plantactinospora, Actinomadura, Brevibacterium, Nocardioides, Actinopolyspora, Kibdelosporangium,* and *Microbispora* are also realized in the plants, like *Eucharis cyaneosperma, Siparuna crassifolia, Cantua buxifolia, Croton lechleri, Monstera spruceana, Calycophyllum acreanum,* and *Palicourea longifolia* (Anandan et al. 2016).

Streptomyces siamensis MI22, Streptomyces roietensis MI24, Streptomyces glaucescens MI29, and *Streptomyces sp. MI04* inhabitant of *Madhuca insignis* has been found to have an antimicrobial activity to methicillin-resistant *Candida albicans, Bacillus subtilis, Escherichia coli* as well as *Staphylococcus aureus* (Mondal and Rai 2021).

1.3.2 Actinobacteria from Aquatic Habitat

A study carried out by Cross revealed that certain actinobacteria such as *Streptomyces, Rhodococcus, Micromonospora, Actinoplanes,* and the endospore producing "*Thermoactinomyces*" are freshwater inhabitants (Cross 1981). Any other freshwater inhabitants include *Norcardia, Corynebacterium, Arthrobacter, Mycobacterium kansasii,* and *Actinomadura madurae* species. Marine Actinobacteria evolved to survive at temperatures under 0–8 °C in very high pressures and anaerobic environments in deep-sea level at temperatures over 8–100°C close to hydrothermal vents on the mid-sea ridges. *Verrucosispora, Williamsia maris, Aeromicrobium marinum, Salinibacterium, Solwaraspora, Marinophilus, Salinispora, Streptomyces, Rhodococcus,* and *Dietzia* are the indigenous actinobacteria of the genera (Lam 2006). The Micromonospora are common inhabitants of aquatic environments around the world and were isolated from samples of water from lakes, rivers, and streams, deep marine sediments, littoral sediments, beach sands, river sediments, and lake mud (Cross 1989).

70 % out of 23,000 that produced medicinal metabolites are from actinobacteria. *Salinispora pacifica* has been identified as the most indigenous strain of *Salinispora Arenicola* as well as *Salinispora tropica* shows marine population diversity of bacteria (Buchanan et al. 2005). These actinomycetes prove to be an extremely rich source of secondary metabolites formed in the shapes of organisms that are structurally diverse. *Salinispora* has recently been recorded to produce the salinosporamide A proteasome inhibitor, presently in step I clinical examinations for cancer treatments; and unexpectedly halogenated macrolides—sporolides A and B (Buchanan et al. 2005). Twenty-six different actinobacterial strains characterized under genera *Actinopolyspora, Actinokineospora, Dactylosporangium, Microtetraspora, Nocardiopsis, Sacchyropoluspora, Streptomyces, Streptoverticillium* were isolated from marine sediments of Port Blair, Andaman and Nicobar Islands (Meena et al. 2013).

1.3.3 Actinobacteria as a Symbiont

Actinobacteria have access by host activity to the inaccessible ecological niches. *Acromyrmex octospinosus*, the attine ants were observed to have to share concordance with actinobacteria. "*Pseudocardia* spp." was extracted in 15–20 million old, fossilized tree resin from ants. This mutualism has produced fungicide dentigerumycin, which highly inhibits *Escovopsis weberi.* It could be utilized in weed killer manufacturing (Barke et al. 2010).

1.4 Distribution of Bioactive Compounds in Actinobacteria

It was in the mid-twentieth century that scientists began testing experimental drug molecules in the aquatic sources of the oceans and seas. Sponges are some of the most effective medicines (Magarvey et al. 2004). Different aquatic products show important structural associations with the metabolites that are microbe-originated, which means that microbes are engaged to generate metabolites in their exact biosynthesis source. In addition to key test approaches including the diffusion of tests and testing of substances in solvent samples, the analysis of the metabolic processes of the bacteria is critical to improving production and commercialization. For example, the str/sts and blu genes presented for the streptomycin biosynthesis is more than 30 of the chromosomes of *Streptomyces glaucescens* as well as *Streptomyces griseus*. At the beginning of the 1950s, the Bahamian sea cucumber *Actinopyga agassizii* identified a toxin, holothurin (Allender et al. 2008). As Holothurin has not yet been marketed, the latest observation of "*Iamia majanohamensis*" a new actinobacteria species that was discovered to be 89 % identical to "Acidimicrobium *ferrooxidans* DSM10331 (T)" was obtained from the abdominal cavity of marine cucumber "*Holothuria edulis*" (Kurahashi et al. 2009). Detailed knowledge of *acidimicrobium ferrooxidans*' metabolic processes will help to explore holothurin and compounds with related potential. Most antibiotics and therapeutics available on the market today are based on microbial novel natural goods, which are processed in most pharmaceuticals industries. Until now there has been just a tiny fraction of actinomycete taxon scattered across the globe. 90% of commercial antibiotics are made of bioactive compounds. *Streptomyces* is the main genus of the tropical and subtropic ocean sediments of actinobacteria. Actinobacteria are origins of large antibiotics of spectrum like aminoglycosides (neomycin, gentamicin, tobramycin, kanamycin, and streptomycin,), anthracyclines (doxorubicin), tetracyclines, macrolides (clarithromycin, azithromycin, and erythromycin,), β-lactam (monobactams, carbapenems, cephalosporin, and penicillin), ansamycins (rifamycin) (Laskaris et al. 2010). Antibiotic development and genetic regulation that gives the highest expression are affected by the nature of food and environmental quality factors (Bhattacharyya et al. 1998). Semisynthetic antibiotic rifampicin is developed with "*Amycolatopsis mediterranei*" fermentation and utilized as a major component in anti-tuberculosis treatment (Schulz and Zillig 1981). Actinobacteria secondary metabolites produced contain antitumoral (e.g., bleomycin and doxorubicin), immunosuppressives (for instance, rapamycin and FK-506), antifungals (e.g., nystatin and amphotericin B) (Grasso et al. 2016). Soil actinomycetes isolated from Kavango and Hardap regions in Namibia also have antiplasmodial activity against malarial parasites like *Plasmodium falciparum* and *Plasmodium vivax*. Siderophores are significant as they bypass membrane-associated drug resistance through their ability to deliver drugs into cells by siderophore conjugates as well as antimicrobials via the Trojan Horse method. Drugs that cannot reach the boundary of the bacterial membrane are associated with siderophores (Möllmann et al. 2009).

1.5 Potential Avenues in Actinomycetes Research

1.5.1 Drug Leads Research in Actinobacteria

1.5.1.1 β-Lactamases in Actinobacteria

β-lactamases are the major source of β-lactam resistance in certain pathogenic bacteria as they perform β-lactam antibiotics hydrolysis to develop antibacterially inactive products (Richmond and Sykes 1973). The resistance against β-lactams is owing to extremely low affinity with penicillin-binding proteins in *Streptomyces* (Ogawara 1981). *Actinomadura cremea* subsp. *rifamicini* JCM 3309, *Aeromicrobium erythreum* JCM 8359, *Saccharothrix flava* JCM 3296, *Streptomyces cacaoi* subsp. *cacaoi* JCM 4352, *Saccharothrix aerocolonigenes* subsp. *aerocolonigenes* JCM 4150 are several of the β-lactamases forming actinobacteria (Ogawara et al. 1999). Streptomyces genus is an exceptional natural source of β-lactamase inhibitors like carbepenem, clavam, cephalosporin, monpbactam, penicillin, clavulanic acid, sublactam, and tazobactam have been reported to be produced by *Streptomyces hygroscopicus* D and *Streptomyces clavuligerus* (Viana Marques et al. 2018).

1.5.1.2 Polyketides and Non-ribosomal Peptides

Polyketides are an extensive category of secondary metabolites, which exhibit a broad variety of bioactivities like antifungal (amphotericin), antibacterial (tetracycline), antiviral (balticolid), anticancer (doxorubicin), immunosuppressant (rapamycin), anti-cholesterol (lovaststin), and anti-inflammatory (flavinoids) (Austin et al. 2004; Singh et al. 2011; Risdian et al. 2019, 2021). The biosynthesis of secondary metabolites involves the six different pathways like carbohydrate pathway, β-lactam synthetic pathway, shikimate pathway, hybrid (non-ribosomal polyketide synthetic) pathway, NRPS: "non-ribosomal polypeptide synthase" pathway, PKS: "polyketide synthase" pathway, and peptide pathway. Genes that encode the enzyme of the synthetic pathways are generally organized in the shape of clusters in chromosomal DNA. The polyketides are synthesized with the enzyme PKS.

Actinorhodin (ACT) was first documented in the late 1940s benzoisochromanequinone and derived from *Streptomyces coelicolor* A3(2). This naturally occurring quinone will serve as a bioreductive DNA-alkylating agent, which can possibly clarify the antibacterial activity of ACT. The existence of this compound can easily be identified by the dye as the typical blue-pigment bacteria have litmus-like properties that experience an alkaline blue color shift from blue to red in acid (Brockmann and Hieronymus 1955). Polyketide, phaeochromycins F derived from the marine "*Streptomyces* sp. DSS-18" culture broth with cytotoxic effect was recorded (Cho et al. 2006). Unusual macrodiolides have been shown to inhibit 60 cancer cell lines of NCIs with the IC_{50} of 0.2 to 2.7 μM formed by marinomycins *Marinispora sp.* CNQ-140. Marinomycin displayed considerable selectivity with the IC_{50} of 5.0 nM against human melanoma cell line UACC-62 and Marinomycins B and C has demonstrated powerful "cytotoxic activities by IC_{50} values of 0.9 μM" (Kwon et al. 2006).

"Non-ribosomal peptides" (NRP) are a category of natural peptide metabolites synthesized by NRPS: "non-ribosomal peptide synthetases." "*Thermoactinomyces* sp.YM3–251" is isolated by marine sediment is formed by one such mechercharmycins of NRP. Cytotoxic activity to Jurkat leukemia cells as well as "human 'lung adenocarcnioma A549' by IC_{50} values of 0.04 μM" has been shown to be Mechercharmycin A. No inhibitory potential was identified for Mechercharmycin B (Kanoh et al. 2005). Gene clusters are liable for the development of bioactive compounds PKS I and NRPS. The *Streptomyces nodosus* strain NPS007994 combined polyketide/non-ribosomal peptide lajollamycin derived from marine sediments with an EC_{50} of 9.6 μM was detected to inhibit melanoma cell line (B16-F10) (Zhao et al. 2006).

1.5.1.3 Isoprenoids

The production of isoprenoids by actinobacteria are being relatively large in number and also structurally different from those produced by eukaryotic organisms (Cane and Nachbar 1978). Actinomycete strains are known to manufacture many isoprenoid compounds like squalene-hopene, geosmin, 2-methylisoborneol, etc. (Cane and Ikeda 2012). Streptomyces strains utilize two different pathways, namely MEP (methylerythritol 4-phosphate) and mevalonate pathway to produce isoprenoids (Chappell 1995). Isoprenoids are one of the major natural compounds of isoprene and are graded based on C5 numbering units, like diterpenes (C20), sesquiterpenes (C15), and monoterpenes (C10) (Dairi et al. 2001). Sesquiterpenes and their biological processes have focused on various phytochemical, pharmacological as well as synthetic studies over the past two decades. Terpenoids, amorphane sesquiterpenes originating from marine actinomycetes. *Nocardia brasiliensis,* "*Streptomyces* sp. strain UC5319," as well as "*Streptomyces argenteolus*" form the isoprenoid compounds, *brasilicardin A* (Shigemori et al. 1999), pentalenene (Cane and Ikeda 2012), and KS-505, respectively (Kuzuyama and Seto 2003).

1.5.1.4 Indolocarbazoles

Protein kinase inhibitors development is a promising drug for cancer treatment. Staurosporine, discovered by Omura and his co-workers in 1977, was the first recorded indolocarbazole alkaloid. Staurosporine also with *Micromonospora* sp. are two normal analogs. A synthesize of "5′-hydroxystaurosporine" and "4′-N-methyl-5′-hydroxystaurosporine" was detected and collected by the Fuerteventura Island coast. (Hernandez et al. 2000). Staurosporine contains an indolocarbazole chromophore and has high inhibitory protein kinases activity in mammalian cells

(Salas et al. 2005). Staurosporine from *S. longisporoflavus* and *Streptomyces* sp. TP-A0274 demonstrated a bioactive property with the generation of antitumor derivatives (Wu et al. 2006). Rebeccamycin a DNA topoisomerase I inhibitor contains indocarbazole rings. The difference in rebeccamycin and staurosporine biological activities depending on the pyrrole ring. Rebeccamycin comprises a malcimide skeleton at positions C-5 and C-7 with carbonyl groups. Whereas staurosporine comprises a carbonyl group pyrrolinone skeleton at position C-5. In addition, rebeccamycin also comprises two atoms of chlorine, bound to C-1 and C-11 with the sugar moiety added to aglycone (Nishizawa et al. 2006; Ryan et al. 2007).

1.5.1.5 Prodiginines

The secondary metabolites like spore pigments, melainin, siderophores, and alkyl resorcinol were predominantly produced by Streptomyces sp. Some pigments and vitamins are produced by actinomycetes like *Streptomyces olivaceus,* NRRL B-1125 isolated from Japanese soil in 1949 was found to produce vitamin B12. The strain *Streptomyces fulvissimus* when fortified with cobalt and grown on soybean meal substrate was found to produce vitamin B12. *Streptomyces bellus* MSA1 isolated from marine sediments produced a light pink colored pigment (Srinivasan et al. 2017). A blue pigment from *Streptomyces shaanxiensis* sp. isolated by sewage irrigation soil (Lin et al. 2012) The prodigiosin-like pigments (PdGs) or tripyrrole red-colored prodiginines have attracted attention because of their promising anti-inflammatory, immunosuppressive, and antitumor properties (Montaner and Prez-Toms 2003). Roseophilin, a prodiginine isolated from *Streptomyces griseoviridis* and *Streptomyces* sp. BA18591 was considered to have the antitumor property (Hayakawa et al. 1992; Nakajima et al. 1993). *Actinomadura* spp. isolated from the sediments obtained in the Archipelago of St. Peter and St. Paul, Brazil has been reported to produce the pigments nonylprodigiosin, cyclononylprodigiosin, and methylcyclooctilprodigiosin. The biosynthetic gene clusters presented the unique putative RedJ thioesterase and Red L-like type I PKS involved in the selection of biosynthetic fatty acid acyl precursors and Red G-like Rieske oxygenase suggesting that synthesis of cyclononylprodigiosin and methylcyclooctilprodigiosin is a signature of *Actinomadura* sp. (Silva et al. 2017). High cytotoxicity is recognized as Prodiginine. It is notable that S. *coelicolor* does not secrete the DNA-damaged PdGs, but instead accumulates internally (in the cytoplasm, both in membranes and cell walls) at the period when development is ceased. The well-programmed prodiginines production indicates that these molecules can play a role in PCD progress or/and control (Tenconi and Rigali 2018). Streptomyces producing bioactive compounds are explained (Table 1.1).

Table 1.1 Bioactive compounds produced by Streptomyces sp.

Actinomycetes spp.	Compound	References
Streptomyces lomondensis sp.	Lomofungin	Johnson and Dietz (1969)
Streptomyces oliuoreticulii	Bestatin (aminopeptidase inhibitor)	Wilkes and Prescott (1985)
Saccharopolyspora erythreus	Macrolides (erythromycin)	Kobrehel et al. (1982)
Streptomyces hygroscopicus	Himastatin (antitumor)	Lam et al. (1990)
Streptomyces prunicolor	Benthocyanin A (radical scavenger)	Shinya et al. (1991)
Streptomyces exfoliates	Carquinostatin A (neuronal cell protection)	Shinya et al. (1993)
Streptomyces venezulae ISP5230	Jadomycin B	Han et al. (1994)
Streptomyces nanchangensis	Nachangmycin (insecticidal)	Sun et al. (2002)
Salinispora tropica CNB-392	Salinosporamides	Feling et al. (2003)
Streptomyces echinoruber	Rubrolone (pigment)	Gupta et al. (2004)
Streptomyces sp. 124092	Alkaloids (cyclic dipeptide)	Xie et al. (2008)
Streptomyces griseoruber	Actinomycin D (pigment)	Praveen and Tripathi (2009)
Streptomyces sp. GT2002/1503	Indole alkaloids (Xiamycin)	Ding et al. (2010)
Streptomyces albogriseolus MGR072	Naphthyridines	Li et al. (2010)
Streptomyces sp.	Melanin (Brown-black pigment)	Vasanthabharathi et al. (2011)
Streptomyces sp.	Padanamides A and B	Williams et al. (2011)
Streptomyces sp. 061316	Benzamides	Xu et al. (2012)
Saccharopolyspora sp. RL78	Other alkaloids	He et al. (2012)
"*Streptomyces* sp. 0616208"	Sesquiterpenes	Xie et al. (2012)
Streptomyces griseus HKI0412	Benzene derivatives	Fu et al. (2012)
Streptomyces parvulus	Actinomycin D (Orange pigment)	Shetty et al. (2014)
Streptomyces sp.	N-acetyl N-demethylmayamycin (yellow pigment)	Liang et al. (2016)
Streptomyces sp. 182SMLY	Streptophenazine B (yellow pigment)	Liang et al. (2017)
Streptomyces sp. K30	Functionalized oligoisoprenoids	Rother et al. (2017)
Streptomyces sp.CL190	Isopentyl pyrophosphates and Dimethylallyl pyrophosphates	Clomburg et al. (2019)
Streptomyces nigra	Isoprenoid quinone (LL-diaminopimelic acid)	Chen et al. (2018)

(continued)

Table 1.1 (continued)

Actinomycetes spp.	Compound	References
MAR Streptomyces sp.	Napyradiomycin SF2415B3	Bauermeister et al. (2019)
Streptomyces davaonensis and *Streptomyces cinnabarinus*	Roseoflavin (natural riboflavin)	Mora-Lugo et al. (2019)
Streptomyces sp. 12,045	Bosamycins	Xu et al. (2020a, b)
Streptomyces lydicamycinius	Lydicamycin	Komaki et al. (2020)

1.5.1.6 Antimicrobial Enzymes from Actinobacteria

A flavoprotein oxidase enzyme is also classified as GOD: "Glucose Oxidase" otherwise recognized as "β-D-glucose oxygen 1-oxidoreductase." This enzyme helps in the domain of diagnostics for pyranose oxidase of methanol, choline, cholesterol, amino, and alcohol (Ferri et al. 2011). The enzyme comprises of two similar subunits with mass 80 kDa, acts as an imperative catalyst as well as catalyzes the β-D-glucose oxidation with O_2 atom into "D-glucono-δ-lactone" is transformed into hydrogen peroxide. Peroxidase and amyloglucosidase are two other actinomycetes developed antimicrobial enzymes (Vaijayanthi et al. 2016). Hydrogen peroxide has antimicrobial activity which inhibits dental plaque decreasing halitosis, gingivitis, and dental carries related to oral resident bacteria (Pugliese et al. 2017).

1.5.1.7 Antiviral Compounds from Actinobacteria

Neuraminidase (acylneuraminyl hydrolase) has been added to the broad field of biological and immunological study of biological fluids, animal tissues, viruses, and bacteria. This acylneuraminyl hydrolase breaks into 2,3-, 2,6- 2,8-, and 2,9-glucosidic connections that join terminal nonreducing O- or N-acetylated neuraminyl residues found in the glycoprotein and oligosaccharides (Ray 1977). Neuraminidase inhibitors interrupt the release of host cell influenza and interfere with progeny growth. Three natural NAI compounds, *Streptomyces seoulensis* IFB-A01 which is made by *Penaeus orientalis*, intestinal shrimp were detected, i.e., "streptoseolactone" (1), "limazepines" G (2) as well as H (3). The antiparasitic macrocyclic 22, 23-dihydroavermectin B1b lactone Ivermectin a blend of 80% 22, 23-dihydroavermectin B1a and 20% was discovered by Professor Satoshi Omura from actinobacteria *Streptomyces avermitilis* (González Canga et al. 2008; Campbell et al. 1983). Ivermectin is *found* to exhibit antiviral activity as an inhibitor of nuclear transport regulated by the importin α/β1 heterodimer, accountable for translocating the different species of viral proteins, thereby inhibiting DNA polymerases and considerably affects RNA viruses (Wagstaff et al. 2012). It is found to be effective against Zika virus (Barrows et al. 2016) Pseudorabies virus (inhibiting

Table 1.2 Immunosuppressant and antiviral compounds producing actinobacteria

Actinobacteria spp.	Compounds	Targets	References
Streptomyces hygroscopicus, Streptomyces clavuligerus CKD119	Rapamycin (Sirolimus), hygromycin	Antifungal, active against immunosuppressant and systemic candidosis	Heitman et al. (1991), Kojima et al. (1995), Mo et al. (2009)
Streptomyces chartreusis (termite mounds)	Antiviral compound	Against bovine viral diarrhea virus	Padilla et al. (2015)
Streptomyces spp. AJ8	Antiviral compound	Against killer shrimp virus responsible for white spot syndrome	Jenifer et al. (2015)
Streptomyces sp. SSA 13	Actinomycin D and Iturin A6	Antitumor and antifungal peptides	Sajid and Aftab (2017)
Streptomyces ghanaensis	Clavulanic acid	Antiviral activity against white spot syndrome virus of shrimp	Rajkumar et al. (2018)

DNA polymerase) (Lv et al. 2018) and recently has shown to effectively inhibit the SARS-CoV-2 virus replication utilizing the same system (Caly et al. 2020) Immunosuppressant and antiviral compounds are explained (Table 1.2).

1.5.1.8 Antitumor Enzymes from Actinobacteria

"L-asparaginase" enzyme is a key enzyme utilized in the therapy of acute myelomonocytic leukemia, melanosarcoma, reticulosarbom, lymphosarcoma treatment, acute myelocytic leukemia, acute lymphoblastic leukemia, Hodgkin's disease, acute as well as "chronic lymphocytic leukemia (Verma et al. 2007). L-asparaginase has been noted to be isolated from a variety of actinobacteria including *Pseudonocardiae endophytica* VUK-10 (Kiranmayi et al. 2014), Streptomyces *acrimycini* NGP (Selvam and Vishnupriya 2013), *Streptomyces* sp. (SS7) (Sivasankar et al. 2013), *S. halstedii* (El-Sabbagh et al. 2013), *Streptomyces* sp. WS3/1 (Kumari et al. 2013), and *Streptomyces karnatakensis* (Mostafa 1982). *Streptomyces* spp. AJ8 isolated from solar salt works is found in homology with *Streptomyces* spp. SAUK6068 and *Streptomyces coeruleoprunus* NBRC15400 and have antitumor activity against L929 fibroblast cancer cell lines (Jenifer et al. 2015) Actinomycetes producing antitumor compounds are detailed (Table 1.3).

1.5.1.9 Fibrinolytic Enzymes from Actinobacteria Spp.

Streptomyces sp. has been reported to have thrombolytic and antithrombotic activity (Bono et al. 1996) "*Nocardiopsis dassonvillei* NCIM 5124" with an oil-polluted tropically sea ecosystem near Mumbai, India, which produces two forms of alkaline

Table 1.3 Antitumor compounds producing actinobacteria

Actinobacteria spp.	Compounds	Targets	References
Streptomyces spp. CNQ-583	Deoxybohemamine and Bohemamine	Cell adhesion inhibition	Hernandez et al. (2000)
Streptomyces spp. NTK 937	Caboxamycin	Antiproliferative antitumor	Hohmann et al. (2009)
Salinispora tropica	Salinosporamide A (NPI-0052)	Anticancer	Buchanan et al. (2005)
Streptomyces spp. B692	Himalomycin A and B, fridamycin D	Antitumor	Myhren et al. (2013)
Streptomyces cheonanensis VUK-	2-methyl butyl propyl phthalate	Targets against MCF-7 cell	Mangamuri et al. (2016)
Streptomyces anandii	Ananstrep A (2) Ananstrep B (3) Ananstrep C (4)	Targets against MCF-7 cell	Zhang et al. (2016)
Streptomyces antibioticus	Neoantimycin A, B	Targets against MCF-7 cell	Hu et al. (2017)
Streptomyces artemisiae MCCB 248	Crude extracts from arctic food sediments	Targets against NCI-H460 cells	Dhaneesha et al. (2017)
Streptomyces malaysiense sp. nov.	Bioactive metabolites	Targets against HCT-116 cells	Ser et al. (2016)
Streptomyces sp. MUM265	Bioactive metabolites	Targets against Caco-2 cell line	Tan et al. (2019)
Streptomyces cacaoi subspcacaoi M20	Partially purified compound fractions	Targets against breast cancer (MCF-7) cell lines	Janaki (2019)
Streptomyces monashensis sp. nov	2-methyl butyl propyl phthalate	Targets against colon cancer cell lines HCT-116	Law et al. (2019)
Streptomyces sp. VN1	Unique furan-type compound	Targets against HCT116	Nguyen et al. (2020)

serine endopeptidases like proteases (Dixit and Pant 2000). These proteases verified the activities of fibrinolytic and collagenolytic, serine proteases and chymotrypsin-like serine-type proteases are known to combat thrombotic disorders are prevented atherosclerosis, pulmonary embolism, and other cardiac diseases. These exoproteases are fibrin-specific and act directly on fibrinogen converting them to fibrin-degraded products. The crude protein of *Streptomyces megasporus* SD5 was found to possess 80 IU (Ploug units) of specific activity per milligram (Chitte and Dey 2000). Actinomycetes producing fibrinolytic enzymes are detailed (Table 1.4).

Table 1.4 Fibrinolytic enzyme-producing actinobacteria

Actinobacteria spp.	Homology	Source	References
Streptomyces griseus strain 254	*Streptomyces griseus*	Soil	Chi et al. (1989)
Streptomyces sp. NRC 411	*Streptomyces spp.*	Soil	Abdel-Naby et al. (1992)
Streptomyces megasporus strainSD5	*S. megaspores*	Soil	Chitte and Dey (2000)
Streptomyces sp. R1401	*S. gardneri*	Soil	Ohyama et al. (2002)
Streptomyces sp.CS624	*Streptomyces microflavus*	Soil	Simkhada et al. (2012)
Streptomyces omiyaensis	*S. lividans*	Soil	Uesugi et al. (2011)
Streptomyces venezuelae	*S. venezuelae*	Marine soil	Naveena et al. (2012)
Streptomyces sp.VITSJ4	*Streptomyces* spp.	Marine sediments	Naine et al. (2016)
Streptomyces sp.	*Streptomyces* sp.	Amazonian lichens	Silva et al. (2017)
Streptomyces sp.*CC5*	*Streptomyces* spp.	Soil	Sun et al. (2016)
Streptomyces violaceus VITYGM	*Streptomyces violaceus*	Soil	Mohanasrinivasan et al. (2017)
Streptomyces lusitanus	*Streptomyces lusitanus*	Marine sediment	Warma et al. (2017)
Streptomyces rubiginosus VITPSS1	*Streptomyces rubiginosus*	Marine sediments	Verma et al. (2018)
Streptomyces radiopugnans VITSD8	*Streptomyces radiopugnans*	Marine sponges	Dhamodharan et al. (2019)
Streptomyces faveolus Streptomyces galtieri	*Streptomyces spp.*	Marine soil	Said et al. (2019)
Streptomyces althioticus BN22	*Streptomyces althioticus*	Marine soil	Viswanathan and Rebecca (2019)

1.5.2 Actinobacteria in Treating Skin Diseases

The most common skin disorder worldwide with 85% of teenagers and over 10% of adults is *acne vulgaris* (James 2005). This impacts the oil glands in the trunk, back as well as the face of the body. Gram-positive bacteria colonization of the skin Acne vulgaris is caused by Propionibacterium acnes in combination with *Staphylococcus epidermidis*. This means that cosmetic drugs must be developed to treat acne with strong bactericidal activity. Actinobacterial strains with inhibitory activity against both *S. epidermidis* and *P. acnes* and enzymatic properties like inhibition of tyrosinase, anti-aging, and antioxidant activities (Desbois and Lawlor 2013).

1.6 Industrially Important Biomolecules from Actinobacteria

1.6.1 Hydrolytic Enzymes

1.6.1.1 Amylases

α-Amylases are enzymes that catalyze the α-1,4 hydrolysis—glycosidic links in polysaccharides like starch into glucose, maltose, and maltotriose. α-Amylase has widespread application in the textile industry, distillation, brewing, and food industries. (Gupta et al. 2003).

1.6.1.2 Cellulase, Endoglucanases, and Xylanases

The order Acidothermales (Acidothermaceae) are thermophilic, acidophilic, and cellulolytic bacteria that dwell in hot springs. The cellulose system contains three types of enzymes: endoglucanases, CBH, and β-D glucosidases. Endoglucanases act on amorphous regions of cellulose substrates yielding high degrees of polymerization oligomers. Xylanases are used to degrade xylan the main constituent of hemicelluloses; it is used for biobleaching and paper pulping (Priya et al. 2012).

1.6.1.3 Lipases

Lipases cleave the ester bonds in triglycerides of long-chain fatty acids. It plays an essential role in lipid absorption and assimilation. It is otherwise called fat-splitting enzymes. Lipases have applications in bioremediation, chiral molecules synthesis, detergents, biological pulping of wood, and flavor synthesis (Hasan et al. 2006).

1.6.1.4 Laccase

Laccases are copper-containing oxidases that are reported to involve in the biotransformation of pollutants like PAHs: “polycyclic aromatic hydrocarbons,” synthetic dyes, fertilizers, herbicides, plastics, chlorinated paraffin phthalates, and other contaminants from industrial and hospital effluents (Arregui et al. 2019).

1.6.1.5 Alkaline Proteases and Keratinase

Proteases are proteolytic enzymes that break down proteins into smaller polypeptides and single amino acids. Proteases play important role in biological functions like digestion, pathogenesis, apoptosis, and protein functions. Keratinases are proteases that recycle keratic wastes like chicken feathers, hairs, nails, and wool (Dastager et al. 2009).

1.6.1.6 Pectinases

Pectinases include pectolyase (pectin lyase) and hydrolyze O-glycosyl bonds resulting in α-1,4 polygalacturonic residues. Pectinases are used extensively in processes like fruit juice clarification, juice extraction, wastewater sewage treatment, natural fibers degumming, cocoa, and tobacco industries (OumerO and Abate 2018). Actinomycetes producing enzymes are detailed (Table 1.5).

1.7 Current Trends Explore the Actinomycete Storehouse

Streptomyces species are still gaining the most attention as an origin of new medicinal compounds (Medema et al. 2011a, b). Through inspecting the relations in primary and secondary metabolism components throughout the genome level, new secondary metabolites examination, and biosynthetic gene clusters have become more systemized using high-performance methods. The biodiversity of Streptomyces, which has greatly expanded and enhanced the awareness and development of the evolutionary path, in addition, the comparative genomic analyses carried out in Streptomyces have extended the gene regions towards their possible and special position in bioprospecting.

1.7.1 Genomic Analysis

The genomic research through genetic fingerprinting (identifying and isolating of variable elements within DNA base-pair sequence) (Nübel et al. 1999), DNA–DNA approaches of hybridization (measuring the genetic distances and similarities between pools of DNA sequences) (Pinhassi et al. 1997) and the construction of metagenomics sequencing and library (identifying genes and gene products of interest by direct cloning on determining and altering the respective nucleic acid sequence) (Kisand et al. 2012) was used to determine and characterize the organisms diversity. The growth of NGS: "next-generation sequencing" (Whole-genome sequencing to detect low-frequency variants with ultra-high-throughput) (Weber and Kim 2016) and nanopore sequencing (electrical current to detect biopolymers and proteins to decode the nucleic acids with unique molecular identifiers (UMI) (Deamer et al. 2016) has rendered the procedure easier and time-consuming. The classical method to drug research in natural resources is complementary chemistry and HTS: "High throughput screening." Traditional methods for the exploration of bioactive substances include isolation and testing although the contemporary method

Table 1.5 Recent reports on industrially important enzyme-producing Streptomyces strains

Actinomycetes spp.	Sources	References
Streptomyces sp. SLBA-08 *(α-amylase)*	Semi-arid soil	Santos et al. (2012)
Streptomyces albus, *Streptomyces hygroscopicus* *(xylanase)*	Soil	Priya et al. (2012)
Streptomyces sviceus *(laccase)*	Soil	Gunne and Urlacher (2012)
Streptomyces lonarensis NCL 716 (amylase)	Soil	Sharma and Bhadane (2013)
Streptomyces albus *(keratinase)*	Soil	Nayaka et al. (2013)
Streptomyces sp. *(α-amylase)*	Saltmarsh	Mahmoud (2015)
Streptomyces sp. OC119–7 *(lipase)*	Soil	Ayaz et al. (2015)
Streptomyces aureofaciens K13 *(keratinase)*	Soil	Gong et al. (2015)
Streptomyces griseorubens E44G (protease)	Soil	Al-Askar et al. (2015)
Streptomyces coelicolor A(3) SCO6548 (1,4,β-cellobiosidase)	Soil	Lim et al. (2016)
Streptomyces badius DB-1 *(α-amylase)*	Soil	Shivlata and Satyanarayana (2017)
Streptomyces fragilis DA7–7 *(thermostable α-amylase)*	Desert soil	Nithya et al. (2017)
Streptomyces sp. strains NWU339 and NWU49 (cellulase)	Rhizosphere soil	Adegboye et al. (2018)
Streptomyces gancidicus ASD *(amylase)*	Marine soil	Ashwini and Shanmugam (2019)
Streptomyces macrosporeus *(cellulase)*	Soil	Soeka et al. (2019)
Streptomyces sp. Bse 7–9 (cellulase)	Marine sediments	Ratnakomala (2019)
Acidothermus cellulolyticus (xylanase and Endo-1,4.-β-D glucanase)	Soil	Sun et al. (2016)
Streptomyces sp. TEM 33 *(lipase)*	Soil	Cadirci et al. (2016)
Streptomyces sp.W007 *(lipase)*	Marine sediments	Zhao et al. (2017)
Streptomyces clavuligerus *(lipase)*	Soil	Santos et al. (2017)
Streptomyces ipomoeae CECT 3341 (laccase)	Soil	Blánquez et al. (2017)
Streptomyces cyaneus *(laccase)*	Soil	Ece et al. (2017)
Acidothermus cellulolyticus 11B (endoxylanase, arabinofuranosidase, acetyl xylan esterase)	Soil	Rajkumar et al. (2018)

(continued)

Table 1.5 (continued)

Actinomycetes spp.	Sources	References
Streptomyces griseorubens LH-3 (Extracellulase cellulase-free xylanase)	Soil	Wu et al. (2018)
Streptomyces cellulosae AU-10 *(lipase)*	Soil	Boran (2018)
Streptomyces cyaneus *Streptomyces coelicolor* *Streptomyces bikiniensis* *Streptomyces ipomea* *(laccase)*	Soil	Blánquez et al. (2017)
Streptomyces violascens OC125–8 *(lipase)*	Wastewater	Boran et al. (2019)
Streptomyces thermocarboxydus *(pectinase)*	Soil	Bharadwaj and Udupa (2019)
Streptomyces coelicoflavus GIAL86 *(pectinase)*	Meyghan salt lake	Salehghamari et al. (2019)
*Streptomyces sp. Al-Dhabi-*49 *(lipase)*	Soil	Al-Dhabi et al. (2020a)
Streptomyces sp. HO1518 *(α-glucosidase)*	*Yellow Sea*	Xu et al. (2020a, b)
Streptomyces geysiriensis (xylanases)	Soil	Poornima et al. (2020)
Streptomyces sp. Al-Dhabi-46 *(amylase)*	Soil	Al-Dhabi et al. (2020b)
Streptomyces fumigatiscleroticus *(pectinase)*	Soil	Govindaraji and Vuppu (2020)
Streptomyces sp. isolate SCUT-3 *(keratinase)*	Soil	Li et al. (2020)
Streptomyces thermocarboxydus (xylanases)	Soil	Tran et al. (2021)
Streptomyces sp. *(lipase)*	Soil	Mohamed et al. (2021)
Streptomyces bacillaris *(lipase)*	Soil	Gao et al. (2021)
Streptomyces lopnurensis *(alkaline protease)*	Marine soil	Rathore and Singh (2021)
Streptomyces globisporus *(protease)*	Soil	Blieva et al. (2021)
Streptomyces coelicoflavus *(keratinase)*	Soil	Jadhav et al. (2021)

contains studies in high-resolution metabolomics, metagenomics, genomics, genetics, genome editing, and mining (Jose and Jha 2016). The main issue with automated HTS approaches dependent on targets is the inefficiency of the powerful inhibitors (hits) at the hosts. Furthermore, in both synthetic ($<0.001\%$) and in natural product libraries, there are small hit-to-lead ratios. The majority of the compounds recently identified have little to no therapeutic activity owing to problems in reaching their target, like permeability and efflux, as they cannot pass the cell wall of the target. Awareness of combinatory chemistry alone is however inadequate to identify possible bioactive drugs. It has a complementary function therefore in ensuring differences in the natural product leads. Both random chemical libraries outweigh the benefits of natural product libraries. Symbiotic compositions are known to be a modern "organ" form in which well-structured intra- and inter-communication like quorum sensing occur. Sleep and cryptic biosynthetic pathways of microbes are important for the discovery of inhibitors of biosynthesis in fatty acids leading to promising compounds like platensimycin. Direct fermentation utilizing combinatorial biosynthetic methods is obtained from new drugs like epirubicin as well as erythromycin derivatives (Bérdy 2012). The clusters of cryptic are BGCs: "biosynthetic gene clusters" but there are similar molecular scaffolds in one or more species if single species is likely to be cryptic in the other species; hence, the fraction of the non-discovered secondary metabolites on the basis of genetic capacity should tend to overestimate the number of the pathway that are cryptic, is not secondary to any other species and has to date no systematic study (Fischbach and Walsh 2006).

With this in mind, it is an essential initial move to fully exploit secondary metabolites in bacteria that BGCs can be classified and compared and thereby systematically catalog the depth of natural product diversity. In the early 2000s, genome-led prediction and SMs isolation was established to produce the biosynthetic ability of members of the *Streptomyces avermitilis* and *Streptomyces coelicolor*. Whole-genome sequence analyses of streptomycetes since the twenty-first century indicate that each species can contain more than was predicted of secondary metabolites (Bentley et al. 2002). Genome mining shows a significant number of BGCs indicate that the strain can generate new compounds with biological activity. Advances in sequence and genome mining techniques have revealed that Streptomyces may generate secondary metabolites more than previously expected, as many BGCs were found. Genome mining may be utilized to find gene clusters that contribute to the synthesis and discovery of a vast number of new compounds. To show the full genome review, the entire genomes of the bacteria need to be compiled through a hybrid assembly technique of "short-read sequencing" (Illumina) and "long-read sequencing" (PacBio RS II) this provides an insight into the genome duration and G + C material percentage (Lee et al. 2020).

1.7.2 Genome Mining

The genome mining includes the discovery of previously undistinguished natural product BGCs in the genomes of sequenced species, the sequence analysis, and the experimental detection of gene cluster products of enzymes encoded (Trivella and Felicio 2018). The mining of the genome is solely based on synthetic biology and systems biology. To continue with gene mining, full genome sequencing should be performed. This offers insights into contiguous assemblies, the identification of homologs through annotations of biosynthetic gene clusters, and the potential of construction and redefinition of genetic circuits. Automatic annotation of the genome sequences can be achieved during the RAST tool (Rapid Annotation using Subsystem Technology). A widespread resource for secondary metabolite biosynthetic gene clusters is considered in the anti-SMASH database and comprises gene clusters over 3000 finished genomes (Blin et al. 2019). Numerous bacterial diversity enhancement techniques include innovative cultivations, different culture environments (co-culture or mixed culture), and isolation of uncommon or gradually emerging microbes. Simple engineering of natural product biosynthesis with mutasynthesis (precursor-directed biosynthesis) has demonstrated the high productivity in macrolides, aminoglycosides as well as glycopeptides for extracting hybrid molecules (Bérdy 2012). The pHMMs: "profile hidden Markov models" and the program HMMER were used to find signature enzymes for main groups of secondary metabolites (Eddy 1998).

The pHMMs employed are a combination of the same cut-offs (Medema et al. 2011a, b) listed herein for indolocarbazoles, NRPS, PKS III, PKS II, PKS I, aerobactin, such as β-lactams, aminoglycosides, butyrolactones, and siderophores, include the fatty acid synthase screening affected by the PKS models.

1.7.3 Genome Scanning

An estimated >20 BGCs in this organism are produced by genome scanning of the deep-sea actinomycete "*Verrucosispora maris*" (Cross 1981). Awareness of biosynthetic pathways leads to genetic modification to improve the production of the product. The blue-pigment actinorhodin producing *Streptomyces coeli color* was genetically modified to develop bright yellow polyketide, kalafungin for the use of pigment from anthraquinone. The cell factory for the efficient production of pigments was produced by "heterologous expression" with a biosynthetic pathway of recognized pigment manufacturers (Sankari et al. 2018).

The abundance and diversity of actinobacteria could be explored by signal transduction with the enormous availability sequence data. Actinobacteria are rich in proteins of signal transduction that contain 1CSs: "one-component systems," 2CSs: "two-component systems," as well as ECFs: "extra cytoplasmic-function σ factors" by which they will feel and respond to the changing environment. The genome data obtained could be saved in the MiST: "Microbial Signal Transduction" database for potential usage (Huang et al. 2015). Mechanical bioprospecting across databases and bioinformatics helps one discover the unexplored organisms and bioactive metabolites of actinobacteria and their biosynthesis clusters and others.

In short, actinobacteria appear to be a promising reservoir for bioactive compounds and enzymes of industry importance. The actinobacteria taxonomy was categorized as well as reclassified multiple times as scientists meet novel organisms from time to time; therefore, genome mining in conjunction with traditional screening techniques and HTS expertise on STPs will lead us to new research in the health care sector.

1.7.4 *Systems Biology and Biotechnology of Streptomyces Species*

Streptomyces sp. is one of the key sources of bioactive metabolites (Zhang et al. 2021) bioinformatics analyses and genome sequencing have revealed that the importance of synthesizing secondary metabolites by Streptomyces. Genome mining is an advanced discovery technique focused on gene cluster sequences as well as biosynthetic pathways. Streptomyces, thereby yielding many bioactive molecules with new structures as well as potent activities (Yang et al. 2020). *Streptomyces* genus is one of the most prolific microorganisms generating secondary metabolites. These compounds are made by complex secondary metabolic pathways (Hiltner et al. 2015). They are known to possess many gene clusters encoding bioactive products and it's very clear that approaches based on genomic studies are highly promising for the quest of unique bioactive compounds (Chaudhary et al. 2013). The advancement of effective approaches for heterologous host expression processes has increased the efficiency of combinatorial biosynthesis for bioactive natural products in actinomycetes (Nepal and Wang 2019), hence representing a vast diversity of fascinating molecular architectures (Fig. 1.2).

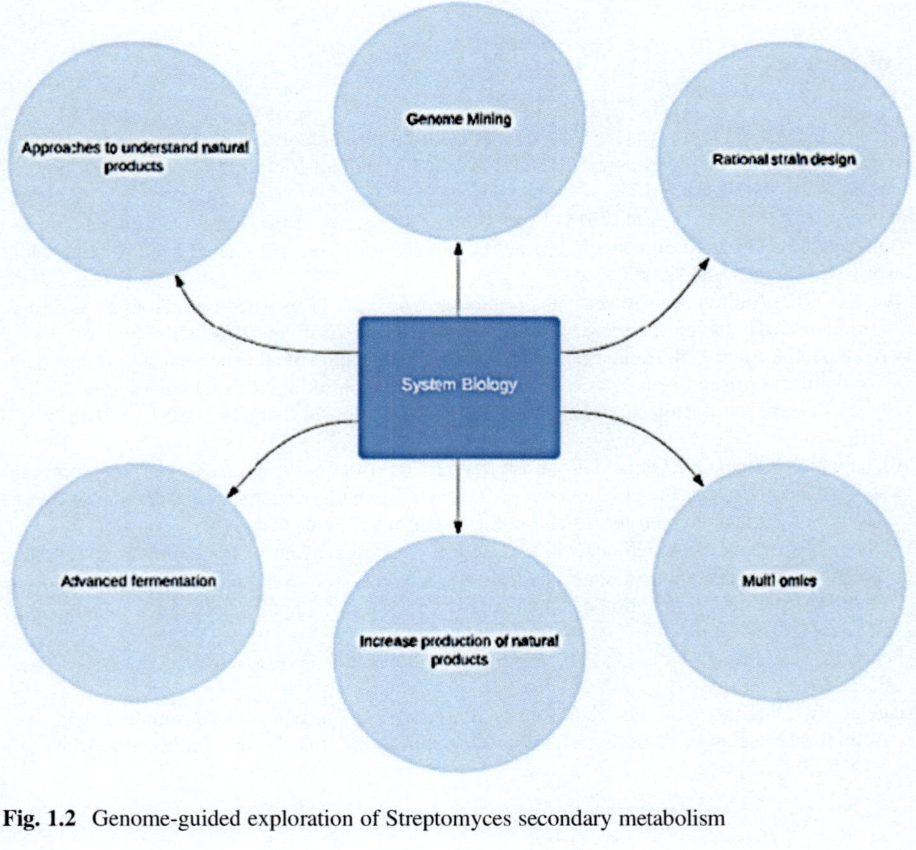

Fig. 1.2 Genome-guided exploration of Streptomyces secondary metabolism

1.8 Conclusion

Streptomyces genus is one of the most prolific microorganisms generating secondary metabolites. Small signaling molecules regulate antibiotic production, and its regulating systems are very complicated, and few specific organisms. These compounds are made by complex secondary metabolic pathways (Hiltner et al. 2015). They are known to possess many gene clusters encoding bioactive products and it is very clear that approaches based on genomic studies are very promising for the investigation of unique bioactive compounds (Chaudhary et al. 2013). The advancement of effective approaches for heterologous host expression processes has increased the efficiency of combinatorial biosynthesis for bioactive natural products in actinomycetes, hence representing a vast diversity of fascinating molecular architectures. For instance, the biosynthetic capacity of *M. aurantiaca* was examined through anti-SMASH in order to obtain the secondary metabolome. The BGCs that encode for domains of PKS, NRPS core genes, as well as accessory, were recognized through ORF ("open reading frames"). About 37 recognized secondary metabolites have been recorded in 42 clusters such as bacteriocin, fatty acid, lantipeptides, siderophores, terpenes, saccharides, oligosaccharides, PKS, NRPS, and putative products (Hu et al. 2020).

References

Abdel-Naby MA, El-Diwany AI, Shaker HM et al (1992) Production and properties of fibrinolytic enzyme from *Streptomyces* sp. NRC 411. World J Microbiol Biotechol 8(3):267–269. https://doi.org/10.1007/BF01201876

Adegboye MF, Briallen L et al (2018) Draft genome sequences of two novel cellulolytic streptomyces strains isolated from south African rhizosphere soil. Genome Announc 6(26). https://doi.org/10.1128/genomeA.00632-18

Aftab SI (2017) Antitumor peptides from *Streptomyces* sp. SSA 13, isolated from Arabian Sea. Int J Peptide Res Therap 23(2):199–211. https://doi.org/10.1007/s10989-016-9552-6

Al-Askar AA, Rashad YM, Hafez EE et al (2015) Characterization of alkaline protease produced by Streptomyces griseorubens E44G and its possibility for controlling Rhizoctonia root rot disease of corn. Biotechnol Biotechnol Equip 29(3):457–462. https://doi.org/10.1080/13102818.2015.1015446

Al-Dhabi NA, Esmail GA, Ghilan AK, Arasu MV (2020a) Isolation and screening of Streptomyces sp. Al-Dhabi-49 from the environment of Saudi Arabia with concomitant production of lipase and protease in submerged fermentation. Saudi J Biol Sci 27(1):474–479

Al-Dhabi NA, Esmail GA, Ghilan AK, Arasu MV, Duraipandiyan V, Ponmurugan K (2020b) Isolation and purification of starch hydrolysing amylase from Streptomyces sp. Al-Dhabi-46 obtained from the Jazan region of Saudi Arabia with industrial applications. J King Saud Univ Sci 32(1):1226–1232

Allender S, Scarborough PV et al (2008) European cardiovascular disease statistics. Eur Heart Netw 3:11–35

Anandan R, Dharumadurai D, Manogaran GP (2016) An introduction to actinobacteria. In: Actinobacteria-Basics And Biotechnological Applications. Intechopen, India. https://doi.org/10.5772/62329

Arregui L, Ayala M, Gómez-Gil X et al (2019) Laccases: structure, function, and potential application in water bioremediation. Microb Cell Factories 18(1):1–33

Ashwini K, Shanmugam S (2019) Enhanced alpha-amylase production using *Streptomyces gancidicus* ASD by process optimization. NISCAIR 48(6):845–852

Austin MB, Bowman ME, Ferrer JL et al (2004) An aldol switch discovered in stilbene synthases mediates cyclization specificity of type III polyketide synthases. Chem Biol 11(9):1179–1194. https://doi.org/10.1016/j.chembiol.2004.05.024

Ayaz B, Ugur A, Boran R (2015) Purification and characterization of organic solvent-tolerant lipase from Streptomyces sp. OC119-7 for biodiesel production. Biocatal Agric Biotechnol 4:103–108

Barke J, Seipke RF, Grüschow S et al (2010) A mixed community of actinomycetes produce multiple antibiotics for the fungus farming ant Acromyrmex octospinosus. BMC Biol 8(1): 1–10. https://doi.org/10.1186/1741-7007-8-109

Barrows NJ, Campos RK, Powell ST et al (2016) A screen of FDA-approved drugs for inhibitors of Zika virus infection. Cell Host Microbe 20(2):259–270. https://doi.org/10.1016/j.chom.2016.07.004

Bauermeister A, Pereira F, Grilo IR et al (2019) Intra-clade metabolomic profiling of MAR4 Streptomyces from the Macaronesia Atlantic region reveals a source of anti-biofilm metabolites. Environ Microbiol 21(3):1099–1112. https://doi.org/10.1111/1462-2920.14529

Benson DR, Silvester (1993) W.B. Biology of Frankia strains, actinomycete symbionts of actinorhizal plants. Microbiol Mol Biol Rev 57(2):293–319

Bentley SD, Chater KF, Cerdeño-Tárraga AM et al (2002) Complete genome sequence of the model actinomycete *Streptomyces coelicolor* A3 (2). Nature 417(6885):141–147

Bérdy J (2012) Thoughts and facts about antibiotics: where we are now and where we are heading. J Antibiot 65(8):385–395. https://doi.org/10.1038/ja.2012.27

Bharadwaj PS, Udupa PM (2019) Isolation, purification and characterization of pectinase enzyme from *Streptomyces thermocarboxydus*. J Clin Microbiol Biochem Technol 5(1):001–006

Bhattacharyya BK, Pal SC, Sen SK (1998) Antibiotic production by *Streptomyces hygroscopicus* D1. 5: cultural effect. Rev Microbiol 29(3). https://doi.org/10.1590/S0001-37141998000300003

Blánquez A, Ball AS, González-Pérez J et al (2017) Laccase SilA from Streptomyces ipomoeae CECT 3341, a key enzyme for the degradation of lignin from agricultural residues. PLoS One 12(11):e0187649. https://doi.org/10.1371/journal.pone.0187649

Blieva RK, Mustafin KG, Akhmetsadykov NN (2021) Optimization of culture medium for enhanced protease biosynthesis in *Streptomyces globisporus*. Rasayan J Chem 14(01): 270–275. https://doi.org/10.31788/RJC.2021.1416123

Blin K, Kim HU, Medema MH, Weber T (2019) Recent development of antiSMASH and other computational approaches to mine secondary metabolite biosynthetic gene clusters. Brief Bioinform 20(4):1103–1113. https://doi.org/10.1093/bib/bbx146

Bono F, Savi P, Tuong A, Maftouh M et al (1996) Purification and characterization of a novel protease from culture filtrates of a *Streptomyces* sp. FEMS Microbiol Lett 141(2–3):213–220

Boran R (2018) Detergent compatible extracellular lipase from streptomyces cellulosae Au-10: a green alternative for the detergent industry. J Surfactant Deterg 21(4):565–573

Boran R, Ugur A, Sarac N, Ceylan O (2019) Characterisation of Streptomyces violascens OC125-8 lipase for oily wastewater treatment. 3 Biotech 9(1):1–7

Breed RS, Murray EG, Kitchens AP (1948) Sergey's manual of determinative bacteriology. Bergey's manual of determinative bacteriology, 6th edn. The Williams & Wilkins Company, Baltimore, Maryland. https://doi.org/10.1002/jps.3030370516

Brockmann H, Hieronymus E (1955) Über Actinomycetenfarbstoffe, V. Mitteil. 1: Zur Konstitution des Actinorhodins, III. Mitteil1. Chem Ber 88(9):1379–1390. https://doi.org/10.1002/cber.19550880908

Buchanan GO, Williams PG, Feling RH et al (2005) Sporolides a and B: structurally unprecedented halogenated macrolides from the marine actinomycete Salinispora tropica. Org Lett 7(13):2731–2734

Bull AT, Stach JE, Ward AC et al (2005) Marine actinobacteria: perspectives, challenges, future directions. Antonie Van Leeuwenhoek 87(1):65–79. https://doi.org/10.1007/s10482-004-6562-8

Cadirci BH, Yasa I, Kocyigit A (2016) Streptomyces sp. TEM 33 possesses high lipolytic activity in solid-state fermentation in comparison with submerged fermentation. Prep Biochem Biotechnol 46(1):23–29

Caly L, Druce JD, Catton MG et al (2020) The FDA-approved drug ivermectin inhibits the replication of SARS-CoV-2 in vitro. Antivir Res 178:104787. https://doi.org/10.1016/j.antiviral.2020.104787

Campbell WC, Fisher MH, Stapley EO et al (1983) Ivermectin: a potent new antiparasitic agent. Science 221(4613):823–828. https://doi.org/10.1126/science.6308762

Cane DE, Nachbar RB (1978) Stereochemical studies of isoprenoid biosynthesis. Biosynthesis of fomannosin from [1, 2-13C2] acetate. J Am Chem Soc 100(10):3208–3212

Cane DE, Ikeda H (2012) Exploration and mining of the bacterial terpenome. Acc Chem Res 45(3): 463–472. https://doi.org/10.1021/ar200198d

Chambers HF (2006) General principles of antimicrobial therapy. Goodman Gilman's the pharmacological basis of therapeutics. McGraw-Hill, New York, NY, pp 1095–1111

Chappell J (1995) Biochemistry and molecular biology of the isoprenoid biosynthetic pathway in plants. Annu Rev Plant Biol 46(1):521–547

Chaudhary AK, Dhakal D, Sohng JK (2013) An insight into the omics based engineering of streptomycetes for secondary metabolite overproduction. BioMed Res Int:968518. https://doi.org/10.1155/2013/968518

Chen C, Ye Y, Wang R et al (2018) Streptomyces nigra sp. nov. is a novel actinobacterium isolated from mangrove soil and exerts a potent antitumor activity in vitro. 9:Front Microbiol, 587. https://doi.org/10.3389/fmicb.2018.01587

Chi CW, Liu HZ, Liu CY, Chibbera BA, Castellino FJ (1989) The inhibition of the enzymic activity of blood coagulation and fibrinolytic serine proteases by a new leupeptin-like inhibitor, and its structural analogues, isolated from *Streptomyces griseus*. J Antibiot 42(10):1506–1512

Chitte RR, Dey S (2000) Potent fibrinolytic enzyme from a thermophilic Streptomyces megasporus strain SD5. Lett Appl Microbiol 31(6):405–410

Cho JY, Kwon HC, Williams PG et al (2006) Actinofuranones A and B, polyketides from a marine-derived bacterium related to the genus Streptomyces (Actinomycetales). J Nat Prod 69(3): 425–428. https://doi.org/10.1021/np050402q

Clomburg JM, Qian S, Tan Z et al (2019) The isoprenoid alcohol pathway, a synthetic route for isoprenoid biosynthesis. Proc Natl Acad Sci U S A 116(26):12810–12815. https://doi.org/10.1073/pnas.1821004116

Cross T (1981) Aquatic actinomycetes: a critical survey of the occurrence, growth and role of actinomycetes in aquatic habitats. J Appl Microbiol 50(3):397–423. https://doi.org/10.1111/j.1365-2672.1981.tb04245.x

Cross T (1989) Growth and examination of actinomycetes-some guidelines. In: Bergey's manual of systematic bacteriology, vol 4. Williams & Wilkins, Baltimore, pp 2340–2343

Dairi T, Hamano Y, Kuzuyama T et al (2001) Eubacterial diterpene cyclase genes essential for production of the isoprenoid antibiotic terpentecin. J Bacteriol 183(20):6085–6094. https://doi.org/10.1128/JB.183.20.6085-6094.2001

Dastager GS, Chan Lee J, Wen Jun L et al (2009) Production, characterization and application of keratinase from *Streptomyces gulbargensis*. Bioresour Technol 100:1868–1871

de Felício R, Ballone P, Bazzano CF, Alves LF et al (2021) Chemical elicitors induce rare bioactive secondary metabolites in deep-sea bacteria under laboratory conditions. Meta 11(2):107

Deamer D, Akeson M, Branton D (2016) Three decades of nanopore sequencing. Nat Biotechnol 34(5):518–524. https://doi.org/10.1038/nbt.3423

Desbois AP, Lawlor KC (2013) Antibacterial activity of long-chain polyunsaturated fatty acids against Propionibacterium acnes and *Staphylococcus aureus*. Mar Drugs 11(11):4544–4557. https://doi.org/10.3390/md11114544

Dhamodharan D, Jemimah Naine S, Merlyn Keziah S et al (2019) Novel fibrinolytic protease producing Streptomyces radiopugnans VITSD8 from marine sponges. Mar Drugs 17(3):164. https://doi.org/10.3390/md17030164

Dhaneesha M, Naman CB, Krishnan KP et al (2017) *Streptomyces artemisiae* MCCB 248 isolated from Arctic fjord sediments has unique PKS and NRPS biosynthetic genes and produces potential new anticancer natural products. 3 Biotech 7(1):32

Ding L, Münch J, Goerls H et al (2010) Xiamycin, a pentacyclic indolosesquiterpene with selective anti-HIV activity from a bacterial mangrove endophyte. Bioorg Med Chem Lett 20(22): 6685–6687. https://doi.org/10.1016/j.bmcl.2010.09.010

Dixit VS, Pant A (2000) Comparative characterization of two serine endopeptidases from Nocardiopsis sp. NCIM 5124. Biochimica et Biophysica Acta (BBA)-General Subjects 1523(2–3):261–268

Dos Santos JB, da Silva Cruz RG, Tardioli PW (2017) Production of whole-cell lipase from *Streptomyces clavuligerus* in a bench-scale bioreactor and its first evaluation as biocatalyst for synthesis in organic medium. Appl Biochem Biotechnol 183:218–240

Du C, van Wezel GP (2018) Mining for microbial gems: integrating proteomics in the postgenomic natural product discovery pipeline. Proteomics 18(18):1700332. https://doi.org/10.1002/pmic.201700332

Ebrahimi-Zarandi M, Bonjar GH, Riseh RS, El-Shetehy M, Saadoun I, Barka EA (2021) Exploring two Streptomyces species to control Rhizoctonia solani in tomato. Agronomy 11(7):1384

Ece S, Lambertz C, Fischer R et al (2017) Heterologous expression of a Streptomyces cyaneus laccase for biomass modification applications. AMB Express 7(1):1–12. https://doi.org/10.1186/s13568-017-0387-0

Eddy SR (1998) Profile hidden Markov models. Bioinformatics 14(9):755–763. https://doi.org/10.1093/bioinformatics/14.9.755

Eliwa EM, Abdel-Razek AS, Frese M et al (2017) New bioactive compounds from the marine-derived actinomycete Nocardiopsis lucentensis sp. ASMR2. Zeitschrift für Naturforschung B 72(5):351–360. https://doi.org/10.1515/znb-2016-0250

El-Sabbagh SM, El-Batanony NH, Salem TA (2013) L-Asparaginase produced by Streptomyces strain isolated from Egyptian soil: purification, characterization and evaluation of its anti-tumor. Afr J Microbiol Res 7(50):5677–5686

Feling RH, Buchanan GO, Mincer TJ et al (2003) Salinosporamide A: a highly cytotoxic proteasome inhibitor from a novel microbial source, a marine bacterium of the new genus Salinospora. Angew Chem 42(3):355–357. https://doi.org/10.1002/anie.200390115

Ferri S, Katsuhiro K, Koji S (2011) Review of glucose oxidases and glucose dehydrogenases: a bird's eye view of glucose sensing enzymes. J Diabetes Sci Technol 5:1068–1076. https://doi.org/10.1177/193229681100500507

Fischbach MA, Walsh CT (2006) Assembly-line enzymology for polyketide and nonribosomal peptide antibiotics: logic, machinery, and mechanisms. Chem Rev 106(8):3468–3496. https://doi.org/10.1021/cr0503097

Fu P, Yang C, Wang Y et al (2012) Streptocarbazoles A and B, two novel indolocarbazoles from the marine-derived actinomycete strain Streptomyces sp. FMA Org Let 14(9):2422–2425. https://doi.org/10.1021/ol3008638

Galas DJ, Sakhanenko NA, Skupin A et al (2014) Describing the complexity of systems: multi-variable "set complexity" and the information basis of systems biology. J Comput Biol 21(2): 118–140. https://doi.org/10.1089/cmb.2013.0039

Gao K, Wang X, Jiang H, Sun J, Mao X (2021) Identification of a GDSL lipase from Streptomyces bacillaris and its application in the preparation of free astaxanthin. J Biotechnol 325:280–287

Gerber NN (1979) Odorous substances from actinomycetes. Dev Ind Microbiol 20:225–238

Gong JS, Wang Y, Zhang DD et al (2015) Biochemical characterization of an extreme alkaline and surfactant-stable keratinase derived from a newly isolated actinomycete Streptomyces aureofaciens K13. RSC Adv 5(31):24691–24699. https://doi.org/10.1039/C4RA16423G

González Canga A, Sahagún PAM, Diez Liebana M et al (2008) The pharmacokinetics and interactions of ivermectin in humans—a mini-review. AAPS J 10(1):42–46. https://doi.org/10.1208/s12248-007-9000-9

Govindaraji PK, Vuppu S (2020) Characterisation of pectin and optimization of pectinase enzyme from novel Streptomyces fumigatiscleroticus VIT-SP4 for drug delivery and concrete crack-healing applications: an eco-friendly approach. Saudi J Biol Sci 27(12):3529–3540. https://doi.org/10.1016/j.sjbs.2020.07.024

Grasso LL, Martino DC, Alduina R (2016) Production of antibacterial compounds from Actinomycetes. Actinobacter Basics Biotechnol Applic 11:177–198. https://doi.org/10.5772/61525

Gunne M, Urlacher VB (2012) Characterization of the alkaline laccase Ssl1 from Streptomyces sviceus with unusual properties discovered by genome mining. PLoS One 7(12):e52360. https://doi.org/10.1371/journal.pone.0052360

Gupta R, Gigras P, Mohapatra H et al (2003) Microbial α-amylases: a biotechnological perspective. Process Biochem 38(11):1599–1616. https://doi.org/10.1016/S0032-9592(03)00053-0

Gupta D, Khare SK, Laha A (2004) Antimicrobial properties of natural dyes against Gram-negative bacteria. Color Technol 120(4):167–171. https://doi.org/10.1111/j.1478-4408.2004.tb00224.x

Haesler F, Hagn A, Frommberger M et al (2008) In vitro antagonism of an actinobacterial Kitasatospora isolate against the plant pathogen Phytophthora citricola as elucidated with ultrahigh resolution mass spectrometry. J Microbiol Methods 75(2):188–195. https://doi.org/10.1016/j.mimet.2008.05.028

Han L, Yang K, Ramalingam E et al (1994) Cloning and characterization of polyketide synthase genes for jadomycin B biosynthesis in Streptomyces venezuelaeISP5230. Microbiology 140(12):3379–3389

Harir M, Bendif H, Bellahcene M et al (2018) Streptomyces secondary metabolites. Basic Biol Appl 6:99–122. https://doi.org/10.5772/intechopen.79890

Hasan F, Shah AA, Hameed A (2006) Industrial applications of microbial lipases. Enzym Microb Technol 39(2):235–251

Hayakawa Y, Kawakami K, Seto H et al (1992) Structure of a new antibiotic, roseophilin. Tetrahedron Let 33(19):2701–2704. https://doi.org/10.1016/S0040-4039(00)79061-7

Hayakawa M, Yoshida Y, Iimura Y (2004) Selective isolation of bioactive soil actinomycetes belonging to the Streptomyces violaceusniger phenotypic cluster. J Appl Microbiol 96(5):973–981. https://doi.org/10.1111/j.1365-2672.2004.02230.x

He J, Magarvey N, Piraee M et al (2001) The gene cluster for chloramphenicol biosynthesis in Streptomyces venezuelae ISP5230 includes novel shikimate pathway homologues and a monomodular non-ribosomal peptide synthetase geneThe GenBank accession number for the sequence reported in this paper is AF262220. Microbiology 147(10):2817–2829

He J, Zhang D, Xu Y et al (2012) Diversity and bioactivities of culturable marine actinobacteria isolated from mangrove sediment in Indian Ocean. Acta Microbiol Sin 52(10):1195–1202

Heitman J, Movva NR, Hiestand PC et al (1991) FK 506-binding protein proline rotamase is a target for the immunosuppressive agent FK 506 in Saccharomyces cerevisiae. Proc Natl Acad Sci U S A 88(5):1948–1952. https://doi.org/10.1073/pnas.88.5.1948

Hernandez LMC, Blanco JADLF, Baz JP et al (2000) 4'-N-Methyl-5′-hydroxystaurosporine and 5'-hydroxy staurosporine, new Indolocarbazole alkaloids from a marine Micromonospora sp. Strain J Antibiot 53(9):895–902. https://doi.org/10.7164/antibiotics.53.895

Hiltner JK, Hunter IS, Hoskisson PA (2015) Tailoring specialized metabolite production in streptomyces. Adv Appl Microbiol 91:237–255. https://doi.org/10.1016/bs.aambs.2015.02.002

Hohmann C, Schneider K, Bruntner C et al (2009) Caboxamycin, a new antibiotic of the benzoxazole family produced by the deep-sea strain Streptomyces sp. NTK 937. J Antibiot 62 (2):99–104

Hu C, Zhou SW, Chen F, Zheng XH, Shen HF, Lin BR, Zhou GX (2017) Neoantimycins a and b, two unusual benzamido nine-membered dilactones from marine-derived *Streptomyces antibioticus* h12-15. Molecules 22(4):557

Hu D, Sun C, Jin T et al (2020) Exploring the potential of antibiotic production from rare actinobacteria by whole-genome sequencing and guided MS/MS analysis. Front Microbiol 11:1540. https://doi.org/10.3389/fmicb.2020.01540

Huang H, Yang T, Ren X, Liu J, Song Y et al (2012) Cytotoxic angucycline class glycosides from the deep sea actinomycete *Streptomyces lusitanus* SCSIO LR32. J Nat Prod 75(2):202–208

Huang X, Pinto D, Fritz G, Mascher T (2015) Environmental sensing in Actinobacteria: a comprehensive survey on the signaling capacity of this phylum. J Bacteriol 197(15): 2517–2535. https://doi.org/10.1128/JB.00176-15

Iikasha AM, Dushimemaria F, Mbewe M et al (2018) Antimicrobial and antiplasmodial activity displayed by actinomycetes isolated from soils from Kavango and Hardap regions in Namibia. Int J Phar Sci Res 9(12):5191–5196. https://doi.org/10.13040/IJPSR.0975-8232.9(12).5191-96

Ilić SB, Konstantinović SS, Todorović ZB (2005) UV/VIS analysis and antimicrobial activity of *Streptomyces* isolates. Facta Univ Med Biol 12:44–46

Jackson SA, Crossman L, Almeida EL et al (2018) Diverse and abundant secondary metabolism biosynthetic gene clusters in the genomes of marine sponge derived Streptomyces spp. isolates. Mar Drugs 16(2):–67. https://doi.org/10.3390/md16020067

Jadhav RS, Oberoi JK, Tejashree R et al (2021) Optimizing the fermentation conditions and enhancing the keratinase production from *Streptomyces coelicoflavus*. Acta sci 4(3): 2581–3226. https://doi.org/10.31080/ASMI.2021.04.0776

James WD (2005) Acne. New Eng J Med 352(14):1463–1472. https://doi.org/10.1056/NEJMcp033487

Janaki T (2019) Anticancer activity of *Streptomyces cacaoi subspcacaoi*. M20 against breast cancer (MCF-7). Cell Lines 12(4):108–116

Jenifer JSCA, Donio MBS, Michaelbabu M et al (2015) Haloalkaliphilic Streptomyces spp. AJ8 isolated from solar salt works and its pharmacological potential. AMB Exp 5(1):1–12. https://doi.org/10.1186/s13568-015-0143-2

Jensen PR, Gontang E, Mafnas C et al (2005) Culturable marine actinomycete diversity from tropical Pacific Ocean sediments. Env Microbiol 7(7):1039–1048. https://doi.org/10.1111/j.1462-2920.2005.00785.x

Johnson LE, Dietz A (1969) Lomofungin, a new antibiotic produced by Streptomyces lomondensis sp. n. Appl Microbiol 17(5):755–759

Jose PA, Jha B (2016) New dimensions of research on actinomycetes: quest for next generation antibiotics. Front Microbiol 19(7):1295

Kalakoutskii LV (2004) Ray fungi and related organisms (Actinomycetales). Microbiology 73(5): 523–530. https://doi.org/10.1023/B:MICI.0000044243.92023.1f

Kanoh K, Matsuo Y, Adachi K et al (2005) Mechercharmycins A and B, cytotoxic substances from marine-derived Thermoactinomyces sp. YM3-251. J Antibiot 58(4):289–292. https://doi.org/10.1038/ja.2005.36\

Karthik L, Kumar G, Venkata K et al (2010) Comparison of methods and screening of biosurfactant producing marine actinobacteria isolated from Nicobar marine sediment. IIOAB J 1(2):221–227

Khalil ZG, Salim AA, Vuong D et al (2017) Amycolatopsins A–C: antimycobacterial glycosylated polyketide macrolides from the Australian soil Amycolatopsis sp. MST-108494. J Antibiot 70(12):1097–1103. https://doi.org/10.1038/ja.2017.119

Kim SK, Chung D, Himmel ME et al (2017) In vivo synergistic activity of a CAZyme cassette from Acidothermus cellulolyticus significantly improves the cellulolytic activity of the C. bescii exoproteome. Biotechnol Bioeng 114(11):2474–2480. https://doi.org/10.1002/bit.26366

Kim MJ, Roh SG, Kim MK et al (2020) Kitasatospora acidiphila sp. nov., isolated from pine grove soil, exhibiting antimicrobial potential. Int J Sys Evol Microbiol 70(10):5567–5575

Kiranmayi MU, Poda S, Vijayalakshmi M (2014) Production and optimization of L-asparaginase by an actinobacterium isolated from Nizampatnam mangrove ecosystem. J Environ Biol 35(5): 799

Kisand V, Valente A, Lahm A et al (2012) Phylogenetic and functional metagenomic profiling for assessing microbial biodiversity in environmental monitoring. PLoS One 7(8):e43630. https://doi.org/10.1371/journal.pone.0043630

Kobrehel G, Radobolja G, Tamburasev Z, Djokic S (1982) 11-aza-10-deozo-10-dihydroerythromycin A and derivatives thereof as well as a process for their preparation. U. S. patent 4,328,334, filed March 28, 1980

Kojima I, Cheng YR, Mohan V, Demain AL (1995) Carbon source nutrition of rapamycin biosynthesis in Streptomyces hygroscopicus. J Ind Microbiol 14(6):436–439

Komaki H, Hosoyama A, Igarashi Y et al (2020) Streptomyces lydicamycinicus sp. nov. and its secondary metabolite biosynthetic gene clusters for polyketide and nonribosomal peptide compounds, Microorganisms. 8(3):370. https://doi.org/10.3390/microorganisms8030370

Kumar P, Kundu A, Kumar M et al (2019) Exploitation of potential bioactive compounds from two soil derived actinomycetes, Streptomyces sp. strain 196 and RI. 24. Microbiol Res 229:126312. https://doi.org/10.1016/j.micres.2019.126312

Kumari PK, Sankar GG, Prabhakar T (2013) 5. L-asparaginase production and molecular identification of marine streptomycete strain WS3/1

Kunimoto S, Aoyagi T, Takeuchi T et al (1974) Purification and characterization of *Streptomyces sialidases*. J Bacteriol 119(2):394–400

Kurahashi M, Fukunaga Y, Sakiyama Y et al (2009) Iamia majanohamensis gen. nov., sp. nov., an actinobacterium isolated from sea cucumber Holothuria edulis, and proposal of Iamiaceae fam. nov. Int J Sys Evol Microbiol 59(4):869–873

Kuzuyama T, Seto H (2003) Diversity of the biosynthesis of the isoprene units. Nat Prod Rep 20(2): 171–183

Kwon HC, Kauffman CA, Jensen PR (2006) Marinomycins A–D, antitumor-antibiotics of a new structure class from a marine actinomycete of the recently discovered genus Marinispora. J Am Chem Soc 128(5):1622–1632. https://doi.org/10.1021/ja0558948

Lacret R, Oves-Costales D, Gómez C et al (2015) New ikarugamycin derivatives with antifungal and antibacterial properties from Streptomyces zhaozhouensis. Mar Drugs 13(1):128–140. https://doi.org/10.3390/md13010128

Lam KS (2006) Discovery of novel metabolites from marine actinomycetes. Curr Opin Microbiol 9(3):245–251. https://doi.org/10.1016/j.mib.2006.03.004

Lam KS, Hesler GA, Mattei JM et al (1990) Himastatin, a new antitumor antibiotic from *Streptomyces hygroscopicus* I. taxonomy of producing organism, fermentation and biological activity. J Antibiot 43(8):956–960

Laskaris P, Tolba S, Calvo-Bado L et al (2010) Coevolution of antibiotic production and counter-resistance in soil bacteria. Environ Microbiol 12(3):783–796. https://doi.org/10.1111/j.1462-2920.2009.02125.x

Law JW, Chan KG, He YW, Khan TM et al (2019) Diversity of *Streptomyces* spp. from mangrove forest of Sarawak (Malaysia) and screening of their antioxidant and cytotoxic activities. Sci Rep 9(1):1–5

Lee N, Kim W, Hwang S et al (2020) Thirty complete Streptomyces genome sequences for mining novel secondary metabolite biosynthetic gene clusters. Sci Data 7(1):1–9. https://doi.org/10.1038/s41597-020-0395-9

Li XL, Xu MJ, Zhao YL, Xu J (2010) A novel benzo [f][1, 7] naphthyridine produced by Streptomyces albogriseolus from mangrove sediments. Molecules 15(12):9298–9307

Li H, Zhang Q, Li S et al (2012) Identification and characterization of xiamycin A and oxiamycin gene cluster reveals an oxidative cyclization strategy tailoring indolosesquiterpene biosynthesis. J American Chem Soc 134(21):8996–9005. https://doi.org/10.1021/ja303004g

Li J, Kim SG, Blenis J (2014) Rapamycin: one drug, many effects. Cell Metab 19(3):373–379. https://doi.org/10.1016/j.cmet.2014.01.001

Li ZW, Liang S, Ke Y et al (2020) The feather degradation mechanisms of a new Streptomyces sp. isolate SCUT-3. Commun Biol 3(1):1–13. https://doi.org/10.1038/s42003-020-0918-0

Li YP, Bu QT, Li JF et al (2021) Genome-based rational engineering of Actinoplanes deccanensis for improving fidaxomicin production and genetic stability. Bioresour Technol 330:124982. https://doi.org/10.1016/j.biortech.2021.124982

Liang Y, Xie X, Chen L et al (2016) Bioactive polycyclic quinones from marine *Streptomyces* sp. 182SMLY. Mar Drugs 14(1):10. https://doi.org/10.3390/md14010010

Liang Y, Chen L, Ye X et al (2017) New streptophenazines from marine Streptomyces sp. 182SMLY. Nat Prod Res 31(4):411–417. https://doi.org/10.1080/14786419.2016.1169419

Lim JH, Lee CR, Dhakshnamoorthy V, Park JS, Hong SK (2016) Molecular characterization of Streptomyces coelicolor A (3) SCO6548 as a cellulose 1,4-β-cellobiosidase. FEMS Microbiol Lett 363(3):1

Lin YB, Wang XY, Fang H et al (2012) Streptomyces shaanxiensis sp. nov., a blue pigment-producing streptomycete from sewage irrigation soil. Int J Sys Evol Microbiol 62(8):1725–1730

Ludwig W, Euzéby J, Schumann P, Busse HJ, Trujillo ME et al (2012) Road map of the phylum Actinobacteria. In: Goodfellow M, Kämpfer P, Buss H-J (eds) Bergey's manual of systematic bacteriology, vol 5, The Actinobacteria

Lv C, Liu W, Wang B et al (2018) Ivermectin inhibits DNA polymerase UL42 of pseudorabies virus entrance into the nucleus and proliferation of the virus in vitro and vivo. Antivir Res 159:55–62. https://doi.org/10.1016/j.antiviral.2018.09.010

Magarvey NA, Keller JM, Bernan V et al (2004) Isolation and characterization of novel marine-derived actinomycete taxa rich in bioactive metabolites. App Environ Microbiol 70(12): 7520–7529. https://doi.org/10.1128/AEM.70.12.7520-7529.2004

Mahmoud K (2015) Statistical optimization of cultural conditions of an halophilic alpha-amylase production by halophilic Streptomyces sp. grown on orange waste powder. Biocatal Agric Biotechnol 4(4):685–693. https://doi.org/10.1016/j.bcab.2015.08.011

Mander P, Cho SS, Simkhada JR, Choi YH, Yoo JC (2011) A low molecular weight chymotrypsin-like novel fibrinolytic enzyme from *Streptomyces* sp. CS624. Process Biochem 46(7): 1449–1455. https://doi.org/10.1016/j.procbio.2011.03.016

Mangamuri U, Muvva V, Poda S et al (2016) Bioactive metabolites produced by Streptomyces Cheonanensis VUK-A from *Coringa* mangrove sediments: isolation, structure elucidation and bioactivity. 3. Biotech 6(1):1–8

Marques AP, Pires C, Moreira H, Rangel AO, Castro PM (2010) Assessment of the plant growth promotion abilities of six bacterial isolates using Zea mays as indicator plant. Soil Biol Biochem 42(8):1229–1235

Matsukuma S, Okuda T, Watanabe J (1994) Isolation of actinomycetes from pine litter layers. Actinomycetologica 8(2):57–65. https://doi.org/10.3209/saj.8_57

McNeil MM, Brown JM (1994) The medically important aerobic actinomycetes: epidemiology and microbiology. Clin Microbiol Rev 7(3):357–417. https://doi.org/10.1128/CMR.7.3.357

Medema MH, Blin K, Cimermancic P et al (2011a) AntiSMASH: rapid identification, annotation and analysis of secondary metabolite biosynthesis gene clusters in bacterial and fungal genome sequences. Nuc Acids Res 39(2):339–346. https://doi.org/10.1093/nar/gkr466

Medema MH, Breitling R, Takano E (2011b) Synthetic biology in Streptomyces bacteria. Meth Enzymol 497:485–502. https://doi.org/10.1016/B978-0-12-385075-1.00021-4

Meena B, Rajan LA, Vinithkumar NV et al (2013) Novel marine actinobacteria from emerald Andaman & Nicobar Islands: a prospective source for industrial and pharmaceutical byproducts. BMC Microbiol 13(1):1–17. https://doi.org/10.1186/1471-2180-13-145

Messina L, Musumeci L, Vaccaro S (2014) Bacteria hyaluronidase and process for its purification. Patent number: WO 2014203133 A1

Messina L, Gavira JA, Pernagallo S et al (2016) Identification and characterization of a bacterial hyaluronidase and its production in recombinant form. FEBS Lett 590(14):2180–2189. https://doi.org/10.1002/1873-3468.12258

Mo S, Ban YH, Park JW et al (2009) Enhanced FK506 production in Streptomyces clavuligerus CKD1119 by engineering the supply of methylmalonyl-CoA precursor. J Ind Microbiol Biotechnol 36(12):1473–1482. https://doi.org/10.1007/s10295-009-0635-7

Mohamed MA, Awad HM, Mostafa ES (2021) Purification, catalytic, kinetic, thermodynamic characterization and stability profile of alkalophilic lipase from Streptomyces sp. SBLWN_MH2. Biocatal Agric Biotechnol 37:102156

Mohanasrinivasan V, Yogesh S, Govindaraj A et al (2017) In vitro thrombolytic potential of actinoprotease from marine Streptomyces violaceus VITYGM. Cardiovasc Hematol Agents Med Chem 14(2):120–124

Möllmann U, Heinisch L, Bauernfeind A et al (2009) Siderophores as drug delivery agents: application of the "Trojan horse" strategy. Biometals 4:615–624

Monaghan RL, Barrett JF (2006) Antibacterial drug discovery—Then, now and the genomics future. Biochem Pharmacol 71(7):901–909. https://doi.org/10.1016/j.bcp.2005.11.023

Mondal S, Rai VR (2021) Molecular profiling and anti-infective potency of endophytic actinomycetes inhabiting Madhuca insignis Radlk., from Western Ghats of India. J Genet Eng Biotechnol 19(1):1–10. https://doi.org/10.1186/s43141-021-00135-0

Montaner B, Prez-Toms R (2003) The prodigiosins: a new family of anticancer drugs. Curr Cancer Drug Targets 3(1):57–65. https://doi.org/10.2174/1568009033333772

Mora-Lugo R, Stegmüller J, Mack M (2019) Metabolic engineering of roseoflavin-overproducing microorganisms. Microb Cell Factories 18(1):1–13. https://doi.org/10.1186/s12934-019-1181-2

Mostafa SA (1982) Properties of L-asparaginase in cell-free extracts of Streptomyces karnatakensis. ZBL Mikrobiol 137(1):63–71. https://doi.org/10.1016/S0232-4393(82)80085-1

Myhren LE, Nygaard G, Gausdal G, Sletta H et al (2013) Iodinin (1, 6-dihydroxyphenazine 5, 10-dioxide) from Streptosporangium sp. induces apoptosis selectively in myeloid leukemia cell lines and patient cells. Mar Drugs 11(2):332–349

Naine SJ, Devi CS, Mohanasrinivasan V (2016) In vitro thrombolytic potential of bioactive compounds from marine Streptomyces sp. VITJS4. Biosci J 32(5):1314–1323

Nakajima S, Kojiri K, Suda H (1993) A new antitumor substance, BE-18591, produced by a streptomycete II. Structure determination. J Antibiot 46(12):1894–1896. https://doi.org/10.7164/antibiotics.46.1894

Naveena B, Gopinath KP, Sakthiselvan P, Partha N (2012) Enhanced production of thrombinase by *Streptomyces venezuelae:* kinetic studies on growth and enzyme production of mutant strain. Bioresour Technol 111:417–424

Nayaka S, Babu KG, Vidyasagar GM (2013) Purification and characterization of keratinase from hair-degrading Streptomyces albus. Int J Bioassays 2(3):599–604

Nepal KK, Wang G (2019) Streptomycetes: surrogate hosts for the genetic manipulation of biosynthetic gene clusters and production of natural products. Biotechnol Adv 37(1):1–20. https://doi.org/10.1016/j.biotechadv.2018.10.003

Nguyen HT, Pokhrel AR, Jodi CT, Dhakal D et al (2020) *Streptomyces* sp. VN1, a producer of diverse metabolites including non-natural furan-type anticancer compound. Sci Rep 10(1):1–4

Nicault M, Tidjani AR, Gauthier A et al (2020) Mining the biosynthetic potential for specialized metabolism of a Streptomyces soil community. Antibiot 9(5):271. https://doi.org/10.3390/antibiotics9050271

Nishizawa T, Grüschow S, Jayamaha DHE et al (2006) Enzymatic assembly of the bis-indole core of rebeccamycin. J Am Chem Soc 128(3):724–725. https://doi.org/10.1021/ja056749x

Nithya K, Muthukumar C, Kadaikunnan S et al (2017) Purification, characterization, and statistical optimization of a thermostable α-amylase from desert actinobacterium *Streptomyces fragilis* DA7-7. 3. Biotech 7(5):1–13. https://doi.org/10.1007/s13205-017-0981-5

Nübel U, Garcia-Pichel F, Kühl M et al (1999) Quantifying microbial diversity: morphotypes, 16S rRNA genes, and carotenoids of oxygenic phototrophs in microbial mats. App Environ Microbiol 65(2):422–430. https://doi.org/10.1128/AEM.65.2.422-430.1999

Ogawara H (1981) Antibiotic resistance in pathogenic and producing bacteria, with special reference to beta-lactam antibiotics. Microbioll Rev 45(4):591. PMCID: PMC281529, PMID: 7035856

Ogawara H, Kawamura N, Kudo T et al (1999) Distribution of β-lactamases in actinomycetes. Antimicrob Agents Chemother 43(12):3014–3017. https://doi.org/10.1128/AAC.43.12.3014

Oh DC, Poulsen M, Currie CR, Clardy J (2009) Dentigerumycin: a bacterial mediator of an ant-fungus symbiosis. Nature Chem Biol 5(6):391–393. https://doi.org/10.1038/nchembio.159

Ohyama S, Wada Y, Hasumi K (2002) Antibiotic A10255 (thioplabin) enhances fibrin binding and activation of plasminogen. J Antibiot 55(1):83–91

Onaka H (2009) Biosynthesis of indolocarbazole and goadsporin, two different heterocyclic antibiotics produced by actinomycetes. Biosci Biotechnol Biochem 73(10):2149–2155. https://doi.org/10.1271/bbb.90263

OumerO J, Abate D (2018) Screening and molecular identification of pectinase producing microbes from coffee pulp. BioMed Res Int Article ID:2961767. https://doi.org/10.1155/2018/2961767

Padilla MA, Rodrigues RAF, Bastos JCS et al (2015) Actinobacteria from termite mounds show antiviral activity against bovine viral diarrhea virus, a surrogate model for hepatitis C virus. J Evid Based Complement Altern Med. https://doi.org/10.1155/2015/745754

Park SR, Yoo YJ, Ban YH et al (2010) Biosynthesis of rapamycin and its regulation: past achievements and recent progress. J Antibiot 63(8):434–441. https://doi.org/10.1038/ja.2010.71

Pinhassi J, Zweifel UL, Hagstroëm A (1997) Dominant marine bacterioplankton species found among colony-forming bacteria. Appl Environ Microbiol 63(9):3359–3366

Pinhassi J, Zweifel UL, Hagstrom A (1998) Dominant marine bacterioplankton species found among colony-forming bacteria. Oceanographic Lit Rev 2(45):299

Poornima S, Divya P, Karmegam N et al (2020) Aqueous two-phase partitioning and characterization of xylanase produced by Streptomyces geysiriensis from low cost lignocellulosic substrates. J Biosci Bioeng 130(6):571–576. https://doi.org/10.1016/j.jbiosc.2020.07.008

Praveen V, Tripathi CKM (2009) Studies on the production of actinomycin-D by Streptomyces griseoruber–a novel source. Lett App Microbiol 49(4):450–455. https://doi.org/10.1111/j.1472-765X.2009.02689.x

Priya BS, Stalin T, Selvam K (2012) Efficient utilization of xylanase and lipase producing thermophilic marine actinomycetes (*Streptomyces albus* and *Streptomyces hygroscopicus*) in the production of ecofriendly alternative energy from waste. Afr J Biotechnol 11(78): 14320–14325. https://doi.org/10.5897/AJB12.835

Pugliese PT, Pugliese SM, Ehrhard JA (2017) U.S. Patent No. 9,540,631. Washington, DC: U.S. Patent and Trademark Office.

Puttaswamygowda GH, Olakkaran S, Antony A, Purayil AK (2019) Present status and future perspectives of marine actinobacterial metabolites. In: Recent developments in applied microbiology and biochemistry. Academic Press, Cambridge, CA, pp 307–319

Rajkumar T, Manimaran M, Taju G et al (2018) Antiviral viral compound from Streptomyces ghanaensis like strain against white spot syndrome virus (WSSV) of shrimp. BioRxiv 340265. https://doi.org/10.1101/340265

Rathore DS, Singh SP (2021) Kinetics of growth and co-production of amylase and protease in novel marine actinomycete, Streptomyces lopnurensis KaM5. Folia Microbiol:1–14. https://doi.org/10.1007/s12223-020-00843-z

Ratnakomala S (2019) Enhancement of Cellulase (CMCase) production from marine actinomycetes Streptomyces sp. Bse 7-9: optimization of fermentation medium by response surface methodology. IOP Conf Series Earth Environ Sci 251(1):012005. https://doi.org/10.1088/1755-1315/251/1/012005

Ray PK (1977) Bacterial neuraminidase and altered immunological behavior of treated mammalian cells. Adv App Microbiol 21:227–267. https://doi.org/10.1016/S0065-2164(08)70043-1

Reading C, ColeM (1977) Clavulanic acid: a beta-lactamase-inhibiting beta-lactam from Streptomyces clavuligerus. Antimicrob Agents Chemother 11(5):852–857. https://doi.org/10.1128/AAC.11.5.852

Reen FJ, Romano S, Dobson AD, O'Gara F (2015) The sound of silence: activating silent biosynthetic gene clusters in marine microorganisms. Mar Drugs 13(8):4754–4783. https://doi.org/10.3390/md13084754

Richmond MH, Sykes RB (1973) The β-lactamases of gram-negative bacteria and their possible physiological role. Adv Microb Physiol 1(9):31–88

Risdian C, Mozef T, Wink J et al (2019) Biosynthesis of polyketides in Streptomyces. Microorganisms 7(5):124. https://doi.org/10.3390/microorganisms7050124

Risdian C, Landwehr W, Rohde M et al (2021) Streptomyces bathyalis sp. nov., an actinobacterium isolated from the sponge in a deep sea. Antonie Van Leeuwenhoek 114(4):425–435. https://doi.org/10.1007/s10482-021-01528-4

Rother W, Birke J, Grond S et al (2017) Production of functionalized oligo-isoprenoids by enzymatic cleavage of rubber. Microbial Biotechnol 10(6):1426–1433. https://doi.org/10.1111/1751-7915.12748

Rudd BA, Hopwood DA (1979) Genetics of actinorhodin biosynthesis by *Streptomyces coelicolor* A3 (2). Microbiology 114(1):35–43

Ryan KS, Howard-Jones AR, Hamill MJ et al (2007) Crystallographic trapping in the rebeccamycin biosynthetic enzyme RebC. Proc Natl Acad Sci 104(39):15311–15316. https://doi.org/10.1073/pnas.0707190104

Said M, Mohamed EO, Shamel Y (2019) Characterization of fibrinolytic enzymes produced by the halophilic Streptomyces Flaveolus and Streptomyces Galtieri isolated from Wady El-Natron area in North Egypt. Med J Cairo Univ 87:2323–2332. https://doi.org/10.21608/mjcu.2019.65659

Salas AP, Zhu L, Sánchez C et al (2005) Deciphering the late steps in the biosynthesis of the anti-tumour indolocarbazole staurosporine: sugar donor substrate flexibility of the StaG glycosyltransferase. Mol Microbiol 58(1):17–27. https://doi.org/10.1111/j.1365-2958.2005.04777.x

Salehghamari E, Nasrollahzadeh Z, Tahmaseb M et al (2019) Pectinase enzyme from Streptomyces coelicoflavus GIAL86 isolated from Meyghan Salt Lake, Arak, Iran. Int J Aquat Biol 7(2):106–111. https://doi.org/10.22034/ijab.v7i2.481

Sankari M, Rao PR, Hemachandran H et al (2018) Prospects and progress in the production of valuable carotenoids: insights from metabolic engineering, synthetic biology, and computational approaches. J Biotechnol 266:89–101. https://doi.org/10.1016/j.jbiotec.2017.12.010

Santos ÉRD, Teles ZNS, Campos NM et al (2012) Production of α-amylase from Streptomyces sp. SLBA-08 strain using agro-industrial by-products. Braz Arch Biol Technol 55(5):793–800. https://doi.org/10.1590/S1516-89132012000500020

Sawa R, Kubota Y, Umekita et al (2018) Quadoctomycin, a 48-membered macrolide antibiotic from Streptomyces sp. MM168-141F8. J Antibiot 71(1) :91–96. doi: https://doi.org/10.1038/ja.2017.140

Schulz W, Zillig W (1981) Rifampicin inhibition of RNA synthesis by destabilisation of DNA-RNA polymerase-oligonucleotide-complexes. Nucleic Acids Res 9(24):6889–6906. https://doi.org/10.1093/nar/9.24.6889

Selvam K, Vishnupriya B (2013) Partial purification and cytotoxic activity of L-asparaginase from Streptomyces acrimycini NGP. Int J Res Pharma Biomed 4:859–869

Ser HL, Palanisamy UD, Yin WF et al (2016) Streptomyces malaysiense sp. nov.: a novel Malaysian mangrove soil actinobacterium with antioxidative activity and cytotoxic potential against human cancer cell lines. Sci Rep 6(1):13

Sharma TK, Bhadane VA, KumarLS et al (2013) Optimization of the production of a maltooligosaccharides producing amylase from the alkalophilic *Streptomyces lonarensis* strain NCL 716 using SVR modeling. Starch-Starke 65(1–2) : 179–185. doi: https://doi.org/10.1002/star.201200094

Shetty PR, Buddana SK, Tatipamula VB et al (2014) Production of polypeptide antibiotic from Streptomyces parvulus and its antibacterial activity. Braz J Microbiol 45(1):303–312. https://doi.org/10.1590/S1517-83822014005000022

Shigemori H, Komaki H, Yazawa K et al (1999) Biosynthesis of diterpenoid moiety of brasilicardin A via non-mevalonate pathway in Nocardia brasiliensis. Tetrahedron Lett 40(23):4353–4354. https://doi.org/10.1016/S0040-4039(99)00689-9

Shinya K, Furihata K, Hayakawa Y et al (1991) The structure of benthocyanin A. A new free radical scavenger of microbial origin. Tetrahedron Lett 32(7):943–946. https://doi.org/10.1016/S0040-4039(00)92126-9

Shinya K, Tanaka M, Furihata K et al (1993) Structure of carquinostatin A, a new neuronal cell protecting substance produced by Streptomyces exfoliatus. Tetrahedron Lett 34(31): 4943–4944. https://doi.org/10.1016/S0040-4039(00)74052-4

Shivlata L, Satyanarayana T (2017) Characteristics of raw starch-digesting α-amylase of Streptomyces badius DB-1 with transglycosylation activity and its applications. Appl Biochem Biotechnol 181(4):1283–1303. https://doi.org/10.1007/s12010-016-2284-4

Silva AE, Guimarães LA, Ferreira EG et al (2017) Bioprospecting anticancer compounds from the marine-derived actinobacteria Actinomadura sp. collected at the Saint Peter and Saint Paul Archipelago (Brazil). J Braz Chem Soc 28(3):465–474. https://doi.org/10.21577/0103-5053.20160297

Simkhada JR, Cho SS, Mander P, Choi YH, Yoo JC (2012) Purification, biochemical properties and antithrombotic effect of a novel Streptomyces enzyme on carrageenan-induced mice tail thrombosis model. Thromb Res 129(2):176–182

Singh R, Shushni MA, Belkheir A (2011) Antibacterial and antioxidant activities of Mentha. https://doi.org/10.1016/j.arabjc.2011.01.019

Siupka P, Piński A, Babicka D et al (2020) Genome mining revealed a high biosynthetic potential for antifungal Streptomyces sp. S-2 isolated from black soot. Int J Mol Sci 21(7):2558. https://doi.org/10.3390/ijms21072558

Sivasankar P, Sugesh S, Vijayan P et al (2013) Efficient production of L-asparaginase by marine Streptomyces sp. isolated from Bay of Bengal, India. Afr J Microbiol Res 7(31):4015–4021

Soeka YS, Suharna N, Triana E, Yulinery T (2019) Characterization of cellulase enzyme produced by two selected strains of *Streptomyces macrosporeus* isolated from soil in Indonesia. Makara J Sci 28:65–71

Srinivasan M, Keziah SM, Hemalatha M, Devi CS (2017) Pigment from *Streptomyces bellus* MSA1 isolated from marine sediments. IOP Conf Ser Mater Sci Eng 263(2):022049. https://doi.org/10.1088/1757-899X/263/2/022049

Stach JE, Maldonado LA, Ward AC et al (2004) Williamsia maris sp. nov., a novel actinomycete isolated from the sea of Japan. Int J Syst Evol 54(1):191–194

Sun Y, Zhou X, Liu J et al (2002) 'Streptomyces nanchangensis', a producer of the insecticidal polyether antibiotic nanchangmycin and the antiparasitic macrolide meilingmycin, contains multiple polyketide gene clusters. Microbiology 148(2):361–371

Sun Z, Liu P, Cheng G et al (2016) A fibrinolytic protease AfeE from Streptomyces sp. CC5, with potent thrombolytic activity in a mouse model. Int J Biol Macromol 85:346–354. https://doi.org/10.1016/j.ijbiomac.2015.12.059

Sun C, Yang Z, Zhang C et al (2019) Genome mining of Streptomyces atratus SCSIO ZH16: discovery of atratumycin and identification of its biosynthetic gene cluster. Org Lett 21(5): 1453–1457. https://doi.org/10.1021/acs.orglett.9b00208

Suthindhiran K, Kannabiran K (2009) Cytotoxic and antimicrobial potential of actinomycete species Saccharopolyspora salina VITSDK4 isolated from the Bay of Bengal coast of India. Am J Infect Dis 5(2):90–98

Tan LT, Mahendra CK, Yow YY et al (2019) *Streptomyces* sp. MUM273b: a mangrove-derived potential source for antioxidant and UVB radiation protectants. Microbiol Open 8(10):e859

Tenconi E, Rigali S (2018) Self-resistance mechanisms to DNA-damaging antitumor antibiotics in actinobacteria. Currt Opin Microbiol 45:100–108. https://doi.org/10.1016/j.mib.2018.03.003

Tran TN, Doan CT, Wang SL (2021) Conversion of wheat bran to xylanases and dye adsorbent by *Streptomyces thermocarboxydus*. Polymers 13(2):287

Trivella DB, de Felicio R (2018) The tripod for bacterial natural product discovery: genome mining, silent pathway induction, and mass spectrometry-based molecular networking. MSys 3(2): e00160–e00117. https://doi.org/10.1128/mSystems.00160-17

Tuan JS, Weber JM, Staver MJ et al (1990) Cloning of genes involved in erythromycin biosynthesis from Saccharopolyspora erythraea using a novel actinomycete-Escherichia coli cosmid. Gene 90(1):21–29. https://doi.org/10.1016/0378-1119(90)90435-T

Uesugi Y, Usuki H, Iwabuchi M, Hatanaka T (2011) Highly potent fibrinolytic serine protease from *Streptomyces*. Enzym Microb Technol 48(1):7–12

Vaijayanthi G, Vijayakumar R, Dhanasekaran D (2016) Actinobacteria—a biofactory of novel enzymes. In: Dhanasekaran D, Jiang Y (eds) Actinobacteria-basics and biotechnological applications. InTechOpen, London, pp 329–352. https://doi.org/10.5772/61436

Vasanthabharathi V, Lakshminarayanan R, Jayalakshmi S (2011) Melanin production from marine Streptomyces. Afr J Biotechnol 10(54):11224–11234. https://doi.org/10.5897/AJB11.296

Verma N, Kumar K, Kaur G, Anand S (2007) L-asparaginase: a promising chemotherapeutic agent. Crit Rev Biotechnol 27(1):45–62. https://doi.org/10.1080/07388550601173926

Verma P, Chatterjee S, Keziah MS et al (2018) Fibrinolytic protease from marine *Streptomyces rubiginosus* VITPSS1. Cardiovasc Hematol Agents Med Chem 16(1):44–55. https://doi.org/10.2174/1871525716666180226141551

Viana Marques DDA, Machado SEF, Ebinuma VCS et al (2018) Production of β-lactamase inhibitors by *Streptomyces* species. Antibiot 7(3):61. https://doi.org/10.3390/antibiotics7030061

Viswanathan K, Rebecca LJ (2019) Optimization of fibrinolytic enzyme from Streptomyces althioticus BN22. Res. J Pharm Technol 12(10):4989–4994. https://doi.org/10.5958/0974-360X.2019.00864.3

Wagstaff KM, Sivakumaran H, Heaton SM et al (2012) Ivermectin is a specific inhibitor of importin α/β-mediated nuclear import able to inhibit replication of HIV-1 and dengue virus. Biochem J 443(3):851–856. https://doi.org/10.1042/BJ20120150

Warma SS, Merlyn Keziah S, Subathra Devi C (2017) Production of fibrinolytic protease from *Streptomyces lusitanus* isolated from marine sediments. In: IOP conference series: materials science and engineering. 263(2): 022048. IOP Publishing

Waksman SA, Reilly HC, Johnstone DB (1946) Isolation of streptomycin-producing strains of Streptomyces griseus. J Bacteriol 52(3):393. PMCID: PMC518198

Wallace RJ, Tsukamura M, Brown BA et al (1990) Cefotaxime-resistant Nocardia asteroides strains are isolates of the controversial species Nocardia farcinica. J Clin Microbiol 28(12):2726–2732

Ward AC, Bora N (2006) Diversity and biogeography of marine actinobacteria. Curr Opin Microbiol 9(3):279–286

Weber T, Kim HU (2016) The secondary metabolite bioinformatics portal: computational tools to facilitate synthetic biology of secondary metabolite production. Syn Sys Biotechnol 1(2):69–79. https://doi.org/10.1016/j.synbio.2015.12.002

Wilkes SH, Prescott JM (1985) The slow, tight binding of bestatin and amastatin to aminopeptidases. J Biol Chem 260(24):13154–13162

Williams PG, Buchanan GO, Feling RH, Kauffman CA, Jensen PR, Fenical W (2005) New cytotoxic Salinosporamides from the marine actinomycete Salinispora t ropica. J Organomet Chem 70(16):6196–6203

Williams DE, Dalisay DS, Patrick BO et al (2011) Padanamides A and B, highly modified linear tetrapeptides produced in culture by a Streptomyces sp. isolated from a marine sediment. Org Lett 13(15):3936–3939. https://doi.org/10.1021/ol2014494

Wu SJ, Fotso S, Li F et al (2006) N-Carboxamido-staurosporine and Selina-4 (14), 7 (11)-diene-8, 9-diol, new metabolites from a marine Streptomyces sp. Antibiot 59(6):331–337. https://doi.org/10.1038/ja.2006.46

Wu H, Cheng X, Zhu Y et al (2018) Purification and characterization of a cellulase-free, thermostable endo-xylanase from Streptomyces griseorubens LH-3 and its use in biobleaching on eucalyptus kraft pulp. J Biosci Bioeng 125(1):46–51. https://doi.org/10.1016/j.jbiosc.2017.08.006

Xie XC, Mei WL, Zeng YB et al (2008) Cytotoxic constituents from marine actinomycete Streptomyces sp. 124092. Chem J Chin Univ 29:2183–2186

Xie QY, Qu Z, Lin HP et al (2012) Micromonospora haikouensis sp. nov., isolated from mangrove soil. Antonie Van Leeuwenhoek 101(3):649–655. https://doi.org/10.1007/s10482-011-9682-y

Xu Z, Baunach M, Ding L, Hertweck C (2012) Bacterial synthesis of diverse indole terpene alkaloids by an unparalleled cyclization sequence. Angew Chem 124(41):10439–10443. https://doi.org/10.1002/ange.201204087

Xu JL, Liu HL, LiuZ F, Ren YH, Wang Y (2020a) Acylated aminooligosaccharides from the yellow sea Streptomyces sp. HO1518 as both α-glucosidase and lipase inhibitors. Mar Drugs 18(11): 576. https://doi.org/10.3390/md18110576

Xu ZF, Bo ST, Wang MJ, Shi J, Jiao RH, SunY GHM (2020b) Discovery and biosynthesis of bosamycins from Streptomyces sp. 120454. Chem Sci 11(34):9237–9245. https://doi.org/10.1039/D0SC03469J

Yandigeri MS, Meena KK, Singh D, Malviya N et al (2012) Drought-tolerant endophytic actinobacteria promote growth of wheat (*Triticum aestivum*) under water stress conditions. Plant Growth Regul 68(3):411–420

Yang Z, He J, Wei X, Ju J, Ma J (2020) Exploration and genome mining of natural products from marine *Streptomyces*. App Microbiol Biotechnol 104(1):67–76. https://doi.org/10.1007/s00253-019-10227-0

Yassin AF, Rainey FA, Steiner U (2001) Nocardia cyriacigeorgici sp. nov. Int J Syst Evol Microbiol 51(4):1419–1423

Zhang L, An R, Wang J, Sun N, Zhang S, Hu J, Kuai J (2005) Exploring novel bioactive compounds from marine microbes. Curr Opin Microbiol 8(3):276–281. https://doi.org/10.1016/j.mib.2005.04.008

Zhang YM, Li HY, Hu C, Sheng HF et al (2016) Ergosterols from the culture broth of marine *Streptomyces anandii* H41-59. Mar Drugs 14(5):84

Zhang Y, Yun K, Huang H et al (2021) Antisense RNA interference-enhanced CRISPR/Cas9 base editing method for improving base editing efficiency in Streptomyces lividans 66. ACS Synth Biol 10(5):1053–1063

Zhao C, Ju J, Christenson SD, Smith WC, Song D, Zhou X, Deng Z (2006) Utilization of the methoxymalonyl-acyl carrier protein biosynthesis locus for cloning the oxazolomycin biosynthetic gene cluster from *Streptomyces albus* JA3453. J Bacteriol 188(11):4142–4147. https://doi.org/10.1128/JB.00173-06

Zhao Z, Hou S, Lan D, Wang X, Liu J, Khan FI, Wang Y (2017) Crystal structure of a lipase from *Streptomyces* sp. strain W007–implications for thermostability and regiospecificity. FEBS J 284(20):3506–3519

Chapter 2
Diversity of *Actinobacteria* in Various Habitats

Manik Prabhu Narsing Rao and Wen-Jun Li

Abstract *Actinobacteria* are ubiquitous, diverse, and one of the biggest lineages in the domain Bacteria. Their existence has been documented in a variety of habitats, including terrestrial, aquatic, and extreme environments. Traditionally, culture-based method has been used to understand the actinobacterial diversity in various habitats, and based on this method many novel strains have been reported. Despite the fact that the culture-based approach has been widely used, the majority of microorganisms in this method were difficult to cultivate. Next-generation sequencing enables culture-free microbial diversity identification based on the molecular phylogeny of the small-subunit ribosomal RNA gene (16S rRNA gene). Actinobacterial diversity in diverse environments has been studied using both culture-dependent and independent methods. In this chapter, we will go through actinobacterial diversity in several ecological niches using both culture-dependent and culture-independent methods.

Keywords *Actinobacteria* · Diversity of actinobacteria · Culture-dependent actinobacterial diversity analysis · Culture-independent actinobacterial diversity analysis · Novel actinobacterial species

2.1 Introduction

Actinobacteria have historically been documented as a contentious form of microorganisms owing to their varied and unique appearances, with some of them resembling the appearance of fungi (Barka et al. 2015). The name is undoubtedly derived from the Greek terms for ray (aktis or aktin) and fungi (mukēs) (van Bergeijk et al. 2020). They have a long evolutionary history, existed on earth around 2.7

M. P. Narsing Rao · W.-J. Li (✉)
State Key Laboratory of Biocontrol, Guangdong Provincial Key Laboratory of Plant Resources and Southern Marine Science and Engineering Guangdong Laboratory (Zhuhai), School of Life Sciences, Sun Yat-Sen University, Guangzhou, China
e-mail: liwenjun3@mail.sysu.edu.cn

L. Karthik (ed.), *Actinobacteria*, https://doi.org/10.1007/978-981-16-5835-8_2

billion years ago, anteceding the great oxidation event that occurred 2.3 billion years ago. This phylum has been associated with the early colonization of land, and they played an important role in assisting Earth's ecosystem's function (Law et al. 2020).

The first hierarchal phylogenetic clustering of members of the phylum "*Actinobacteria*" was provided by Stackebrandt et al. (1997) with the introduction of class *Actinobacteria*. The actinobacterial taxonomy has evolved throughout time, with the most current roadmap dividing the phylum "*Actinobacteria*" into six classes, 46 orders, and 79 families, with 16 new orders and 10 new families included (Salam et al. 2020). They are ubiquitous, diverse, and one of the biggest lineages in the domain Bacteria (Valverde et al. 2012) extensively distributed over both terrestrial (Sharma and Thakur 2020) and aquatic habitats (Kavitha and Savithri 2017), as well as in the microbiomes of higher eukaryotes (Wang et al. 2021).

Actinobacteria are Gram-positive bacteria with a high GC content and a diverse array of morphologies (Whitman et al. 2012). They may be heterotrophic or chemoautotrophic, but the majority are chemoheterotrophic and capable of using a diverse range of nutritional sources, including multiple complex polysaccharides (Barka et al. 2015).

After the discovery of streptomycin, actinobacteria has become a vital resource (Schatz et al. 1944) due to its amazing antimicrobial action. Since then, the golden era of actinobacteria has begun, and till today they are well known for producing antimicrobial agents. *Actinobacteria* produces almost two-thirds of all antibiotics released by microorganisms (Liu et al. 2016). Apart from the development of antibiotics, they play a crucial function in a range of biological processes such as biogeochemical cycles and bioremediation (Alvarez et al. 2017), bio weathering (Abdulla 2009), plant growth promotion (Palaniyandi et al. 2013), and plant growth promotion under abiotic stress (Dong et al. 2019). Due to the wide applications, researchers started exploring actinobacterial diversity and population composition from various environments.

For decades, the conventional culture-dependent approach was used to analyze actinobacterial diversity (Arocha-Garza et al. 2017); however, this method has many drawbacks. The majority of microorganisms in this system are either hidden or difficult to grow (Narsing Rao et al. 2021). Based on the molecular phylogeny of the small-subunit ribosomal RNA gene (16S rRNA gene), next-generation sequencing allows for culture-free microbial diversity detection (Sabat et al. 2017; Myer et al. 2016). In recent years, this approach has been useful in understanding the actinobacterial diversity of different ecological niches (Wei et al. 2021; Rego et al. 2019). In this chapter, we will discuss actinobacterial diversity in different ecological niches using culture-dependent and independent methods.

2.2 Actinobacterial Diversity in Soil

Soil is the soul of infinite life that promotes diverse microflora (Pathma and Sakthivel 2012). *Actinobacteria* are typically soil-dominant microbes, and their presence in soil was documented in 1903 (Zhang et al. 2019; Nalini and Prakash 2020). *Actinobacteria* can be present on the soil's surface as well as at depths of more than 2 m underground. They are usually found in densities ranging from 10^6 to 10^9 cells per gram of soil (Barka et al. 2015). *Streptomyces* accounts for over 95% of the actinobacterial strains isolated from soil (Williams and Vickers 1988). The actinobacterial diversity in the soil also depends on various factors, such as pH, temperature and soil moisture. Guo et al. (2015b) evaluated the actinobacterial diversity of red soil and found *Streptomycetaceae* (57.8%) to be the dominant actinobacterial group followed by members of the families *Pseudonocardiaceae* (10.2%) and *Streptosporangiaceae* (9.5%). Pikoli et al. (2020) evaluated the actinobacterial diversity of acidic soil. They noticed the presence of rare actinobacterial strains, namely *Actinomadura*, *Actinoallomurus*, *Actinospica*, *Streptacidiphilus*, *Aciditerrimonas*, and *Ferrimicrobium*. Temperature was also found to affect the size and structure of actinobacterial populations in soil. *Streptomycetes* were found in greater abundance in grassland during the summer, while *Nocardiae* were found in greater abundance during the winter (Küster 1970; Orchard 1981). pH was also discovered to be an important element in deciding the distribution of soil actinobacteria, with an optimal life occurring around pH 7.0 (Kurtböke 2017). However, Selianin et al. (2005) discovered that alkaline soils with pH values greater than 7 included more *Streptomycetes* capable of thriving at pH 9 than neutral pH soils. They proposed that actinobacterial strains are alkaliphilic if they exhibit high radial rates of colony growth under alkaline circumstances but not at pH 7.0 or below. Soil moisture acts as a major influencing factor for actinobacterial growth. Borowik and Wyszkowska (2016) reported 40% maximum water capacity level for optimum growth of actinobacteria.

Many novel actinobacterial strains have been reported from the soil. *Streptomyces scabichelini* was isolated from a soil sample collected from Hacibektaş, Turkey (Gencbay et al. 2021). A novel *Streptomyces* species (*Streptomyces osmaniensis*) has been isolated from garden soil (Reddy et al. 2010). *Sinomonas mesophila* isolated from ancient fort soil (Prabhu et al. 2015). *Agromyces humi* isolated from farm soil (Lee and Whang 2020). *Phytoactinopolyspora alkaliphila* isolated from saline-alkaline soil (Zhang et al. 2016). A list of some actinobacterial strains isolated from the soil is listed in Table 2.1.

Table 2.1 List of novel actinobacterial members isolated from soil

Sl. No	Novel actinobacterial members	Isolated from	References
1.	*Aeromicrobium alkaliterrae*	Alkaline soil	Yoon et al. (2005)
2.	*Kribbella karoonensis* and *Kribbella swartbergensis*	Soil	Kirby et al. (2006)
3.	*Nocardia acidivorans*	Soil	Kämpfer et al. (2007)
4.	*Nocardioides lentus*	Alkaline soil	Yoon et al. (2006)
5.	*Rhodococcus kyotonensis*	Soil	Li et al. (2007)
6.	*Nonomuraea soli*	Soil	Cao et al. (2012)
7.	*Sphaerisporangium aureirubrum*	Muddy soil	Guo et al. (2015)
8.	*Actinophytocola algeriensis*	Saharan soil	Bouznada et al. (2016)
9.	*Micromonospora fulva*	Forest soil	Lee and Whang (2017)
10.	*Rhodococcus daqingensis*	Petroleum-contaminated soil	Wang et al. (2019b)
11.	*Kribbella jiaozuonensis*	Soil	Zhao et al. (2019)
12.	*Nocardia terrae*	Soil	Kanchanasin et al. (2021)

2.3 Actinobacterial Diversity in Saline Environment

Saline environments have salt concentrations similar to seawater (~ 3.5% (w/v) total dissolved salts) (Díaz-Cárdenas et al. 2017). The diversity of actinobacteria in saline environments has been extensively studied. Meklat et al. (2011) reported isolating 52 actinobacterial strains from 18 saline soil samples that could be affiliated to the genera *Actinopolyspora*, *Nocardiopsis*, *Saccharomonospora*, *Streptomonospora*, and *Saccharopolyspora*. Valenzuela-Encinas et al. (2009) investigated microbial diversity in soils with low, medium, and high salt concentrations. At low and medium salt concentrations, actinobacterial strains (*Streptomyces griseoruber*, *Arthrobacter oxydans*, *Cellulomonas* sp.) and seven clones belonging to the phylum *Actinobacteria* could not be categorized in any order) were found, but not at high salt concentrations. Wu et al. (2009) used culture-dependent and culture-independent approaches to assess the diversity of actinobacterial communities in saline sediments from Yunnan and Xinjiang, China. A total of 163 actinobacterial isolates have been isolated using the culture-dependent technique and were affiliated with the order *Actinomycetales* (distributed into five suborders: *Streptosporangineae*, *Micrococcineae*, *Streptomycineae*, *Pseudonocardineae*, and *Glycomycineae*). Using culture-independent techniques, a total of 748 actinobacterial 16S rRNA gene clones were obtained, and they could be classified into *Actinomycetales*, *Acidimicrobiales*, and unclassified actinobacteria. The *Actinomycetales* sequences were distributed into nine suborders: *Streptosporangineae*, *Glycomycineae*, *Micromonosporineae*, *Pseudonocardineae*, *Corynebacterineae*, *Frankineae*,

Table 2.2 List of novel actinobacterial members isolated from saline habitat

Sl. no	Novel actinobacterial members	Isolated from	References
1.	*Nesterenkonia halotolerans* and *Nesterenkonia xinjiangensis*	Saline soil	Li et al. (2004)
2.	*Zhihengliuella halotolerans*	Saline soil	Zhang et al. (2007)
3.	*Nocardiopsis quinghaiensis*	Saline soil	Chen et al. (2008)
4.	*Haloactinospora alba*	Salt lake	Tang et al. (2008)
5.	*Saccharopolyspora halophila*	Saline lake	Tang et al. (2009)
6.	*Georgenia halophila*	Salt lake	Tang et al. (2010)
7.	*Salinisphaera halophila*	Brine of a salt well	Zhang et al. (2012)
8.	*Isoptericola salitolerans*	Salt lake	Guan et al. (2013)
9.	*Amycolatopsis cihanbeyliensis*	Salt mine	Tatar et al. (2013)
10.	*Amycolatopsis flava*	Dead Sea	Wei et al. (2015)
11.	*Salilacibacter albus*	Dried salt lake	Li et al. (2016)
12.	*Streptomonospora tuzyakensis*	Saline soil	Tatar et al. (2016)
13.	*Streptomyces huasconensis*	High altitude saline wetland	Cortés-Albayay et al. (2019)
14.	*Phytoactinopolyspora halophila*	Saline soil	Ding et al. (2019)

Propionibacterineae, *Streptomycineae*, and *Micrococcineae*. They discovered no important relationship between actinobacterial diversity and salinity. Possible explanations for this inconsistency include the possibility that actinobacteria can withstand a wide variety of salinities or that they are bound to sediment particles. Much, if not all, microorganisms in the sediments might be bound to mineral particle surfaces, so the salinity measured may not reflect the true salinity of the microniches where the actinobacterial sequences were found. In contrast, Jiang et al. (2010) discovered that salinity has a significant impact on actinobacterial community structures. The diversity of the actinobacterial population was shown to be positively correlated with salinity. The actinobacterial diversity of salt lakes was different from freshwater habitats. Saline habitat also reported to harbor many novel actinobacterial strains, and some of them are listed in Table 2.2.

Actinobacteria withstand high salt stress by various mechanisms. We have evaluated the mechanism of *Nocardiopsis gilva* YIM 90087^{T} to withstand salt stress (Han et al. 2018). *Nocardiopsis gilva* YIM 90087^{T} regulates osmotic pressure by the accumulation of ectoine and hydroxyectoine. Further, ABC transporters, glycine, serine, and threonine metabolism played a vital role in overcoming salt stress. *Actinobacteria* isolated from the salt environment also show potential biological activities. *Streptomyces* sp. VITDDK3-a isolated from salt pan soil produced biosurfactant and resist heavy metal (Lakshmipathy et al. 2010).

2.4 Actinobacterial Diversity in Rhizosphere

The rhizosphere is the small zone of the plant roots' surface that is critical for supplying numerous ecosystem services such as nutrient cycling and carbon intake (Kumar and Dubey 2020). Rhizosphere microbiota plays an important role in plant fitness, development, and immunity (Lee et al. 2021). Rhizospheric actinobacteria are a dominating phylum in nature and have a large economic impact on people owing to their contributions to soil systems (Yadav et al. 2018). Oberhofer et al. (2019) investigated the actinobacterial diversity associated with the rhizosphere of plant *Leontopodium nivale* subsp. *alpinum*. A total of 77 actinobacterial strains were isolated, including 43 from the genus *Streptomyces* and 34 from the genera *Micromonospora*, *Asanoa*, *Actinokineospora*, *Mycobacterium*, *Nocardia*, *Leifsonia*, *Microbacterium*, and *Micrococcus*. Khamna et al. (2009) evaluated the actinobacterial diversity of 16 medicinal plant rhizosphere soils. A total of 445 actinobacterial strains were isolated from the rhizosphere soils of 16 medicinal plants. According to taxonomic analysis, 89% of the strains belonged to the genus *Streptomyces*, 11% to the genera were *Actinomadura*, *Microbispora*, *Micromonospora*, *Nocardia*, *Nonomurea*, and three isolates were unclassified. The result suggested that *Curcuma mangga* rhizosphere soil had the greatest number and richness of actinobacterial strains. This finding shows that plants may be one of the most important elements determining rhizosphere community formation. Other variables may also play a role in defining the structure of actinobacterial communities. Nimnoi et al. (2011) evaluated the actinobacterial community in rhizospheres of eaglewood using culture-independent methods of RT-PCR and PCR-DGGE of 16S rRNA gene. The PCR-DGGE 16S rRNA gene profiles specifically showed that the actinobacterial population was separated corresponding to sampling sites, implying that soil characteristics and local climatic conditions. They also discovered that the effect of soil type on the actinobacterial population differed depending on soil properties such as texture, pH, and usable organic matter material.

Rhizosphere-associated actinobacteria produce various secondary metabolites and have multifunctional plant growth-promoting properties. Zhao et al. (2012) isolated 196 actinobacterial strains from the rhizosphere of seven medicinal plant species affiliated to eight suborders (*Corynebacterineae, Glycomycineae, Micrococcineae, Micromonosporineae, Propionibacterineae, Pseudonocardineae, Streptomycineae,* and *Streptosporangineae*) and 13 families (*Actinosynnemataceae, Cellulomonadaceae, Glycomycetaceae, Micrococcaceae, Micromonosporaceae, Mycobacteriaceae, Nocardiaceae, Nocardiopsaceae, Nocardioidaceae, Promicromonosporaceae, Pseudonocardiaceae, Streptomycetaceae, and Streptosporangiaceae)*. Antimicrobial activity was found in all *Streptomyces* isolates. The antimicrobial activities of rare actinobacteria were limited and only can inhibit the growth of *Escherichia coli*, *Verticillium dahliae*, and *Fusarium oxysporum*, and strains related to *Saccharopolyspora shandongensis* and *Streptosporangium roseum* showed wide antimicrobial activity.

Table 2.3 List of novel actinobacterial members isolated from the rhizosphere

Sl. No	Novel actinobacterial members	Isolated from	References
1.	*Sphaerisporangium rhizosphaerae*	Rhizosphere soil of a rubber tree	Mu et al. (2018)
2.	*Rhodococcus kunmingensis*	Rhizosphere of *Taxus chinensis*	Wang et al. (2008)
3.	*Leifsonia soli*	Teak rhizosphere soil	Madhaiyan et al. (2010)
4.	*Streptomyces gramineus*	Bamboo rhizosphere soil	Lee et al. (2012)
5.	*Actinomadura rhizosphaerae*	Rhizosphere soil of *Azadirachta indica*	Malisorn et al. (2018)
6.	*Nonomuraea lactucae*	Rhizosphere soil of lettuce	Cao et al. (2019)
7.	*Cryobacterium tepidiphilum*	Rhizosphere soil of lettuce	Wang et al. (2019a)
8.	*Streptomyces triticagri* and *Streptomyces triticirhizae*	Rhizosphere soil of wheat	Han et al. (2020)
9.	*Streptomyces oryziradicis*	Rhizosphere soil of rice	Li et al. (2020a)
10.	*Glycomyces albidus*	Rhizosphere soil of wheat	Qian et al. (2020)

Han et al. (2018a) isolated two potential *Streptomyces* sp. (*Streptomyces rochei* and *Streptomyces sundarbansensis*) from the rhizosphere of *Mikania micrantha* Kunth. *Streptomyces rochei* and *Streptomyces sundarbansensis* showed antifungal activity against *Fusarium graminearum. The* most abundant compounds produced by these strains were aliphatic ketones, carboxylic acids, and esters, with *n*-hexadecanoic acid being. They can produce plant growth prompter like *indoleacetic acid,* fix nitrogen, and dissolve phosphorus and potassium. Similarly, *Streptomyces* strains isolated for chickpea rhizosphere soil exhibited an increase in nodule number, shoot weight, and yield. The *Streptomyces*-treated plots enhanced total N, available P and organic C (Sreevidya et al. 2016). Rhizosphere has also been reported to harbor many novel actinobacterial strains and a list of some has been tabulated in Table 2.3.

2.5 *Actinobacteria* as Endophytes

The term "endophyte" was introduced by De Bary (1866), who defined it as "any organism that develops inside plant tissues." Endophytic actinobacteria are prevalent in the roots, somewhat plentiful in the branches, and rare in the leaves (Gangwar et al. 2014). It has been documented that actinobacteria can colonize any tissue or organ of the host plant (van der Meij et al. 2018). It has also been observed that various actinobacteria colonize different tissues and organs of the plant, which may be determined by the host–microbe interaction. For instance, *Kitasatospora* frequency was noticed in *Tanacetum sinaicum* and *Artemisia Judaica* but not in *Chiliadenus montanus* and *Echinops spinosu* (El-Shatoury et al. 2013). Within the plant, there was variation in actinobacterial diversity. In *Suaeda glauca* plant, we

Table 2.4 List of novel endophytic actinobacterial members

Sl. No	Novel actinobacterial members	Isolated from	References
1.	*Rhodococcus cercidiphylli*	*Cercidiphyllum japonicum* leaf	Li et al. (2008)
2.	*Micromonospora endophytica*	Leaf of Thai upland rice	Thanaboripat et al. (2015)
3.	*Nocardioides intraradicalis*	Roots of *Psammosilene tunicoides*	Huang et al. (2016)
4.	*Nocardioides zeicaulis*	Stem tissue of healthy maize	Kämpfer et al. (2016)
5.	*Aeromicrobium endophyticum*	Leaves of reed	Li et al. (2019a)

noticed that actinobacteria have been significantly enriched in brown seeds when compared with black seeds. Genera like *Rhodococcus*, *Ralstonia*, *Pelomonas*, and *Bradyrhizobium* have been greatly enriched in brown seeds, while *Marinilactibacillus* was primarily present in black seeds. Furthermore, brown seeds contained a substantial number of bacteria with plant growth-promoting characteristics, while black seeds contained bacteria with enzyme activities (i.e., pectinase, cellulolytic, and xylanolytic activities) (Wang et al. 2021). Such processes, which seem to impart selectivity, may affect endophytic actinobacterial diversity. A novel endophytic actinobacteria *Glutamicibacter halophytocola* isolated from the roots of a coastal halophyte, *Limonium sinense* elevated the salt tolerance of tomato seedlings. *Glutamicibacter halophytocola* boosted tomato development in terms of seedling, root length, and the number of fibrous roots, as well as enhanced osmolyte content (proline), antioxidant defense enzymes, and ion homeostasis management under salt stress (Xiong et al. 2019). Endophytic actinobacteria have also been reported to induce defense mechanisms in plants. Endophytic actinobacteria, isolated from healthy wheat tissue showed induced defense pathways in *Arabidopsis thaliana.* The actinobacteria were reported to activate key genes in the systemic acquired resistance or the jasmonate/ethylene (JA/ET) pathways (Conn et al. 2008). We noticed that actinobacterial strains were also involved in enhancing the salt stress ability of *Arabidopsis thaliana*. Endophytic strains *Arthrobacter endophyticus and Nocardiopsis alba* up-regulated genes encoding chlorophyll a reductase, peptide-methionine (R)-S-oxide reductase, and potassium ion uptake when inoculated into *Arabidopsis thaliana* under salt stress. Carotenoid biosynthesis, phenylalanine metabolism, phenylpropanoid biosynthesis, glycerolipid metabolism, and nitrogen metabolism were all important in improving *Arabidopsis thaliana's* salt stress resistance (Dong et al. 2019). We further reported many novel endophytic actinobacteria species. *Cellulomonas endophytica* (Li et al. 2020b) and *Amycolatopsis alkalitolerans* (Narsing Rao et al. 2020) have been isolated from tubers of *Gastrodia elata* Blume. We were further successful to report a novel genus *Allostreptomyces psammosilenae* isolated from the roots of *Psammosilene tunicoides* (Huang et al. 2017). List of some endophytic novel actinobacterial members listed in Table 2.4.

2.6 Actinobacterial Diversity in Marine Environment

The marine environment includes a wide range of habitats, from the sea surface microlayer to the bulk water column and, eventually, to the ecosystems on and under the seabed. On the seafloor, there are geological sediments, mineral nodule fields, carbonate mounds, cold seeps, hydrocarbon seeps, saturated brines, and hydrothermal vents (Ward and Bora 2006). Epibioses and symbioses are hosted by marine micro- and macrofauna (Ward and Bora 2006). *Actinobacteria* have widely been distributed in the marine environment. Undabarrena et al. (2016) using the culture-dependent method evaluated the actinobacterial diversity of 11 marine sediment samples. The author noticed that the distribution of actinobacterial strains in each sample varied, and the majority of the isolates came from sediments 10 m deep. Only a few isolates were found from deeper sediments or shallow sites. The actinobacterial members isolated belonged to the members of *Micrococcaceae, Dermabacteraceae, Brevibacteriaceae, Corynebacteriaceae, Microbacteriaceae, Dietziaceae, Nocardiaceae*, and *Streptomycetaceae* families. Culture-dependent and independent approaches were employed to evaluate the actinobacterial diversity of Arctic marine sediments. A total of 152 strains and 692 positive clones have been obtained using a culture-dependent and independent method. The actinobacterial strains obtained using the culture-dependent method have been assigned to the members of genera *Actinotalea*, *Arthrobacter*, *Brachybacterium*, *Brevibacterium*, *Kocuria*, *Kytococcus*, *Microbacterium*, *Micrococcus*, *Mycobacterium*, and *Pseudonocardia*. The culture-independent analysis suggests the sample consists of genera *Agrococcus*, *Cellulomonas*, *Demequina*, *Iamia*, *Ilumatobacter*, *Janibacter*, *Kocuria*, *Microbacterium*, *Phycicoccus*, *Propionibacterium*, and *Pseudonocardia*, along with other, unidentified actinobacterial clones. The study suggests that Arctic maritime settings maintain a complex actinobacterial community, some of which appear to be unique yet uncultured species (Zhang et al. 2014).

Hydrothermal vents are seabed locations where hot, anoxic, chemical-rich water is discharged into the cold, oxic, deep ocean (Dick 2019). He and Zhang (2016) reported actinobacterial as the predominant phylum in hydrothermal vents.

Actinobacteria have also been detected or isolated from a wide range of marine creatures such as sponges (Friedrich et al. 2001), fish (Hamada et al. 2009), seaweeds (Lee and Kim 2007), seagrass (Wu et al. 2012), corals (Nithyanand et al. 2011), mollusks (El-Shatoury et al. 2009), and ascidians (Jimenez et al. 2013). Marine environments have also been reported to harbor many novel actinobacterial members and those are listed in Table 2.5.

Table 2.5 List of novel actinobacterial members isolated from the marine environment

Sl. No	Novel actinobacterial members	Isolated from	References
1.	*Kocuria marina*	Marine sediment	Kim et al. (2004)
2.	*Iamia majanohamensis*	Sea cucumber *Holothuria edulis*	Kurahashi et al. (2009)
3.	*Oceanitalea nanhaiensis*	Seawater	Fu et al. (2012)
4.	*Brevibacterium sediminis*	Deep-sea sediments	Chen et al. (2016)
5.	*Microbacterium aureliae*	Moon jellyfish	Kaur et al. (2016)
6.	*Kocuria oceani*	Deep-sea hydrothermal plume	Zhang et al. (2017)
7.	*Rubrobacter indicoceani*	Deep-sea sediment	Chen et al. (2018)
8.	*Williamsia aurantiacus*	Marine sponge	de Menezes et al. (2019)
9.	*Streptomyces marianii*	Subtidal sediment	Iniyan et al. (2021)
10.	*Streptomyces bathyalis*	Sponge	Risdian et al. (2021)

2.7 Actinobacterial Diversity in Terrestrial Hot Springs, Volcanic and Geothermal Soil

Terrestrial hot springs are locations where warm or hot groundwater emerges from the earth. Their water has a high concentration of dissolved elements and minerals, as well as everything needed for life to develop (Narsing Rao et al. 2018, 2021). Volcanic and geothermal environments are characterized by low pH, high temperatures, and gas emissions consisting of mainly CO_2 and varied CH_4, H_2S, and H_2 contents (Picone et al. 2020). Thermophilic and thermotolerant members exist in diverse genera of the phylum *Actinobacteria*, including *Thermopolyspora*, *Thermomonospora*, *Thermotunica*, *Thermotellispora*, *Thermobispora*, *Streptomyces*, *Acidothermus*, *Acidimicrobium*, *Aciditerrimonas*, *Actinomadura*, *Thermoleophilum*, and others (Itoh et al. 2011; Jiao et al. 2015; Shivlata and Satyanarayana 2015; Wu et al. 2018).

Actinobacteria have developed a variety of strategies to deal with thermal stress, including the presence of chaperones (which aid in the refolding of partially denatured proteins), a high GC content, amino acid substitution in proteins, the presence of certain components in the cell wall, and a comparatively higher quantity of charged amino acids (Asp, Glu, Arg, and Lys) than polar amino acids (Asn, Gln, Ser, and Thr) in their proteins (Shivlata and Satyanarayana 2015). Many researchers have reported diversified actinobacterial diversity in a thermal environment. Liu et al. (2016) investigated the presence of culturable actinobacteria in hot springs (40–99 °C) samples from Tengchong County, Yunnan Province, in southwestern China. *Actinomadura*, *Microbispora*, *Micromonospora*, *Micrococcus*, *Nonomuraea*, *Nocardiopsis*, *Promicromonospora*, *Pseudonocardia*, *Streptomyces*, *Thermoactinospora*, *Thermocatellispora*, and *Verrucosispora* were among the 58 thermophilic actinobacterial species recovered. *Streptomyces* was found to be

Table 2.6 List of novel actinobacterial members isolated from thermal environments

Novel actinobacterial members	Isolation source	Optimum temperature	References
Thermoactinospora rubra	Geothermal sandy soil	45–55 °C	Zhou et al. (2012a)
Thermocatellispora tengchongensis	Geothermal sandy soil	45–55 °C	Zhou et al. (2012b)
Planosporangium thailandense	Soil of hot spring	40 °C	Thawai et al. (2013)
Rubrobacter calidifluminis and *Rubrobacter naiadicus*	Hot stream	60 °C	Albuquerque et al. (2014)
Actinomadura amylolytica and *Actinomadura cellulosilytica*	Geothermally heated soil	45 °C	Jiao et al. (2015)
Rubrobacter spartanus	Volcanic soil	50 °C	Norman et al. (2017)
Microbispora soli	Hot spring soil sample	40 °C	Kittisrisopit et al. (2018)
Marmoricola caldifontis	Hot spring sediment	37–45 °C	Habib et al. (2020)

the most prevalent in all sample sites; however, at low temperatures, more diverse actinobacteria were seen.

We used a culture-dependent and independent technique to assess the microbial diversity of seven Indian hot springs. *Actinobacteria* were the dominant phylum in both approaches. We isolated actinobacterial genera such as *Micromonospora*, *Microbacterium*, *Actinocorallia*, and *Micrococcus* using a culture-dependent approach. No strain was isolated at temperatures higher than 55 °C (Narsing Rao et al. 2021). In contrast to our findings, Song et al. (2009) discovered a wide range of actinobacteria in hot springs with temperatures reaching 81 °C. Riquelme et al. (2015) investigated the variety of actinobacterial communities in volcanic caves. The *Actinomycetales* dominated the samples, but there were also numerous newly identified orders present, including *Euzebyales, Gaiellales, Rubrobacterales, Solirubrobacterales,* and *Coriobacteriales*.

The actinobacterial strains isolated from thermal environments possess wide applications. A novel heterotrophic, thermophilic, and extremely acidophilic actinobacteria, *Ferrithrix thermotolerans* reported for iron-oxidization (Johnson et al. 2009). Actinobacterial strains isolated from geothermal springs have been reported for antimicrobial, anti-biofilm, and anticancer activities (Gajanan et al. 2019). Several novel strains that have been reported from the thermal environment are listed in Table 2.6.

2.8 Actinobacterial Diversity in Deserts and Arid Regions

Deserts may be found on every continent and account for 33% of the total land area. Life in these regions is profoundly challenged by harsh stresses like drought, salinity, low or high temperatures, and other environmental extremes (Alsharif et al. 2020). Arid regions are biomes with a mean annual rainfall to mean annual evaporation ratio of less than 0.05 and less than 0.002 for severe hyper-arid environments. Arid zones are the most water-stressed continental ecosystems (Mohammadipanah and Wink 2016).

Ding et al. (2013) evaluated the culturable actinobacterial diversity from the desert ecosystem. A total of 53 actinobacterial strains were isolated and could be affiliated to genera *Streptomyces*, *Micromonospora*, *Saccharothrix*, *Streptosporangium*, and *Cellulomonas*. Similarly, Takahashi et al. (1996) investigated rare actinobacterial diversity from the desert. A total of 335 actinobacterial strains were isolated, among them nine strains were rare actinobacteria belong to genera *Microbispora*, *Nocardia*, *Microtetraspora*, *Amycolatopsis*,

Actinomadura, and *Saccharothrix*. Sun et al. (2018) assessed the diversity of actinobacterial taxa in two desert sands (Badain Jaran and Tengger Deserts). *Actinobacteria* was the predominant phylum, accounting for 35.0 and 29.4% of the communities in the Badain Jaran and Tengger Desert sands, respectively. Members of the *Geodermatophilaceae* were prevalent in both deserts, indicating that they were widespread populations inside the deserts. *Arthrobacter* spp. and *Kocuria* spp. were prevalent at the genus level, accounting for 21.2 and 5.3% of the actinobacterial communities in the Badain Jaran and Tengger Deserts, respectively.

Metagenomic studies of hyper-arid and severe hyper-arid soils in the desert discovered a considerable number of actinobacterial "dark matter," as demonstrated by a 34 % increase in families compared to those that have been legitimately reported. The majority of them belonged to the class *Actinobacteria* but in addition, a few representatives of rare deep lineage actinobacteria belonging to the classes *Acidimicrobia* and *Nitriliruptoria* were also detected (Idris et al. 2017). Similar to other environments mentioned above, deserts and arid regions have also been reported to harbor many novel actinobacterial that are listed in Table 2.7.

2.9 Conclusion

The actinobacterial diversity in various habitats is very diverse. They have been reported from terrestrial and aquatic habitats and also inside the eukaryotes. The diversity of the actinobacterial members depends on various physicochemical factors such as pH and temperature. They can withstand extreme conditions (like high temperature, pH salt concentration) and has various mechanisms to overcome extreme conditions. Although they are important sources for secondary metabolites,

Table 2.7 List of novel actinobacterial members isolated from deserts and arid regions

Sl. No	Novel actinobacterial members	Isolated from	References
1.	*Geodermatophilus siccatus*	Arid sand	del Carmen Montero-Calasanz et al. (2013)
2.	*Streptomyces asenjonii*	Hyper-arid Atacama Desert soils	Goodfellow et al. (2017)
3.	*Pseudonocardia nigra*	Desert rock	Trujillo et al. (2017)
4.	*Desertimonas flava*	Desert soil	Asem et al. (2018)
5.	*Saccharothrix tharensis*	Desert soil	Ibeyaima et al. (2018)
6)	*Micromonospora acroterricola*	High altitude Atacama Desert soil	Carro et al. (2019)
7.	*Streptomyces desertarenae*	Desert sample	Li et al. (2019b)
8.	*Nocardiopsis deserti*	Desert soil	Asem et al. (2020)
9.	*Nonomuraea terrae*	Arid soil	Ay (2020)
10.	*Streptomyces cahuitamycinicus*	Desert soil	Saygin et al. (2020)

only a few genera have been explored and hence it is important to explore various ecological niches to understand their diversity.

Acknowledgments This work was financially supported by the Key Realm R&D Program of Guangdong Province (Grant No. 2018B020206001) and Innovation Group Project of Southern Marine Science and Engineering Guangdong Laboratory (Zhuhai) (No.311021006).

References

Abdulla H (2009) Bioweathering and biotransformation of granitic rock minerals by actinomycetes. Microb Ecol 58(4):753–761

Albuquerque L, Johnson MM, Schumann P, Rainey FA, da Costa MS (2014) Description of two new thermophilic species of the genus *Rubrobacter*, *Rubrobacter calidifluminis* sp. nov. and *Rubrobacter naiadicus* sp. nov., and emended description of the genus *Rubrobacter* and the species *Rubrobacter bracarensis*. Syst Appl Microbiol 37(4):235–243

Alsharif W, Saad MM, Hirt H (2020) Desert microbes for boosting sustainable agriculture in extreme environments. Front Microbiol 11:1666

Alvarez A, Saez JM, Davila Costa JS, Colin VL, Fuentes MS, Cuozzo SA, Benimeli CS, Polti MA, Amoroso MJ (2017) Actinobacteria: current research and perspectives for bioremediation of pesticides and heavy metals. Chemosphere 166:41–62

Arocha-Garza HF, Canales-Del Castillo R, Eguiarte LE, Souza V, De la Torre-Zavala S (2017) High diversity and suggested endemicity of culturable Actinobacteria in an extremely oligotrophic desert oasis. PeerJ 5:e3247

Asem MD, Shi L, Jiao JY, Wang D, Han MX, Dong L, Liu F, Salam N, Li WJ (2018) *Desertimonas flava* gen. nov., sp. nov. isolated from a desert soil, and proposal of *Ilumatobacteraceae* fam. nov. Int J Syst Evol Microbiol 68(11):3593–3599

Asem MD, Salam N, Idris H, Zhang XT, Bull AT, Li WJ, Goodfellow M (2020) *Nocardiopsis deserti* sp. nov., isolated from a high altitude Atacama Desert soil. Int J Syst Evol Microbiol 70 (5):3210–3218

Ay H (2020) *Nonomuraea terrae* sp. nov., isolated from arid soil. Arch Microbiol 202(8):2197–2205

Barka EA, Vatsa P, Sanchez L, Gaveau-Vaillant N, Jacquard C, Meier-Kolthoff JP, Klenk HP, Clément C, Ouhdouch Y, van Wezel GP (2015) Taxonomy, physiology, and natural products of Actinobacteria. Microbiol Mol Biol Rev 80(1):1–43

Borowik A, Wyszkowska J (2016) Soil moisture as a factor affecting the microbiological and biochemical activity of soil. Plant Soil Environ 62(6):250–255

Bouznada K, Bouras N, Schumann P, Spröer C, Sabaou N, Klenk HP (2016) *Actinophytocola algeriensis* sp. nov., an actinobacterium isolated from Saharan soil. Int J Syst Evol Microbiol 66 (7):2760–2765

Cao YR, Jin RX, Jiang Y, He WX, Jiang CL (2012) *Nonomuraea soli* sp. nov., an actinomycete isolated from soil. Int J Syst Evol Microbiol 62(7):1587–1591

Cao P, Wang Y, Sun P, Li C, Zhao J, Jiang S, Zhang Y, Wang X, Xiang W (2019) *Nonomuraea lactucae* sp. nov., a novel actinomycete isolated from rhizosphere soil of lettuce (*Lactuca sativa*). Int J Syst Evol Microbiol 69(2):316–321

Carro L, Golinska P, Nouioui I, Bull AT, Igual JM, Andrews BA, Klenk HP, Goodfellow M (2019) *Micromonospora acroterricola* sp. nov., a novel actinobacterium isolated from a high altitude Atacama Desert soil. Int J Syst Evol Microbiol 69(11):3426–3436

Chen YG, Cui XL, Kroppenstedt RM, Stackebrandt E, Wen ML, Xu LH, Jiang CL (2008) *Nocardiopsis quinghaiensis* sp. nov., isolated from saline soil in China. Int J Syst Evol Microbiol 58(3):699–705

Chen P, Zhang L, Wang J, Ruan J, Han X, Huang Y (2016) *Brevibacterium sediminis* sp. nov., isolated from deep-sea sediments from the Carlsberg and Southwest Indian Ridges. Int J Syst Evol Microbiol 66(12):5268–5274

Chen RW, Wang KX, Wang FZ, He YQ, Long LJ, Tian XP (2018) *Rubrobacter indicoceani* sp. nov., a new marine actinobacterium isolated from Indian Ocean sediment. Int J Syst Evol Microbiol 68(11):3487–3493

Conn VM, Walker AR, Franco CM (2008) Endophytic actinobacteria induce defense pathways in *Arabidopsis thaliana*. Mol Plant-Microbe Interact 21(2):208–218

Cortés-Albayay C, Dorador C, Schumann P, Andrews B, Asenjo J, Nouioui I (2019) *Streptomyces huasconensis* sp. nov., an haloalkalitolerant actinobacterium isolated from a high altitude saline wetland at the Chilean Altiplano. Int J Syst Evol Microbiol 69(8):2315–2322

De Bary A (1866) Morphologie und Physiologie der Pilze, Flechten und Myxomyceten Hofmeister's Handbook of Physiological Botany. Wilhelm Engelmann, Leipzig. https://doi.org/10.5962/bhl.title.120970

de Menezes C, Afonso RS, de Souza WR, Parma MM, de Melo IS, Fugita F, Moraes L, Zucchi TD, Fantinatti-Garboggini F (2019) *Williamsia aurantiacus* sp. nov. a novel actinobacterium producer of antimicrobial compounds isolated from the marine sponge. Arch Microbiol 201 (5):691–698

del Carmen M-CM, Göker M, Rohde M, Schumann P, Pötter G, Spröer C, Gorbushina AA, Klenk HP (2013) *Geodermatophilus siccatus* sp. nov., isolated from arid sand of the Saharan desert in Chad. Antonie Van Leeuwenhoek 103(3):449–456

Díaz-Cárdenas C, Cantillo A, Rojas LY, Sandoval T, Fiorentino S, Robles J, Ramos FA, Zambrano MM, Baena S (2017) Microbial diversity of saline environments: searching for cytotoxic activities. AMB Express 7(1):223

Dick GJ (2019) The microbiomes of deep-sea hydrothermal vents: distributed globally, shaped locally. Nat Rev Microbiol 17(5):271–283

Ding D, Chen G, Wang B, Wang Q, Liu D, Peng M, Shi P (2013) Culturable actinomycetes from desert ecosystem in northeast of Qinghai-Tibet Plateau. Ann Microbiol 63:259–266

Ding ZG, Ji Y, Yin M, Zhao YR, Feng YZ, Chunyu WX, Tang SK (2019) *Phytoactinopolyspora halophila* sp. nov., a halophilic actinomycete isolated from a saline soil. Int J Syst Evol Microbiol 69(2):384–389

Dong ZY, Narsing Rao MP, Wang HF, Fang BZ, Liu YH, Li L, Xiao M, Li WJ (2019) Transcriptomic analysis of two endophytes involved in enhancing salt stress ability of *Arabidopsis thaliana*. Sci Total Environ 686:107–117

El-Shatoury SA, El-Shenawy NS, Abd El-Salam IM (2009) Antimicrobial, antitumour and in vivo cytotoxicity of actinomycetes inhabiting marine shellfish. World J Microbiol Biotechnol 25: 1547–1555

El-Shatoury SA, El-Kraly OA, Trujillo ME, El-Kazzaz WM, El-Din, e., Dewedar, A. (2013) Generic and functional diversity in endophytic actinomycetes from wild compositae plant species at South Sinai-Egypt. Res Microbiol 164(7):761–769

Friedrich AB, Fischer I, Proksch P, Hacker J, Hentschel U (2001) Temporal variation of the microbial community associated with the Mediterranean sponge Aplysina aerophoba. FEMS Microbiol Ecol 38(2–3):105–115

Fu Y, Li Q, Liu K, Xu Y, Wang Y, Jiao N (2012) *Oceanitalea nanhaiensis* gen. nov., sp. nov., an actinobacterium isolated from seawater. Int J Syst Evol Microbiol 62(10):2490–2494

Gajanan T, Vinodh JS, Bhushan BB, Desai D, Santhakumari B, Mahesh SD, Syed GD (2019) Bioactivities and molecular networking-based elucidation of metabolites of potent actinobacterial strains isolated from the Unkeshwar geothermal springs in India. RSC Adv 9: 9850–9859

Gangwar M, Dogra S, Gupta UP, Kharwar RN (2014) Diversity and biopotential of endophytic actinomycetes from three medicinal plants in India. Afr J Microbiol Res 8(2):184–191

Gencbay T, Saygin H, Guven K, Topkara AR, Saricaoglu S, Sahin N, Isik K (2021) *Streptomyces scabichelini* sp. nov., isolated from soil. Int J Syst Evolut Microbiol 71(2). https://doi.org/10.1099/ijsem.0.004639

Goodfellow M, Busarakam K, Idris H, Labeda DP, Nouioui I, Brown R, Kim BY, Del Carmen M-CM, Andrews BA, Bull AT (2017) *Streptomyces asenjonii* sp. nov., isolated from hyper-arid Atacama Desert soils and emended description of *Streptomyces viridosporus* Pridham et al. 1958. Antonie Van Leeuwenhoek 110(9):1133–1148

Guan TW, Teng Y, Yang LL, Zhang XP, Che ZM (2013) *Isoptericola salitolerans* sp. nov., a halotolerant filamentous actinobacterium isolated from a salt lake, China. Extremophiles 17 (3):471–476

Guo L, Zhao J, Liu C, Bai L, Sun P, Han C, Li J, Guo X, Xiang W, Wang X (2015) *Sphaerisporangium aureirubrum* sp. nov., an actinomycete isolated from soil. Int J Syst Evol Microbiol 65(11):4157–4162

Guo X, Liu N, Li X, Ding Y, Shang F, Gao Y, Ruan J, Huang Y (2015b) Red soils harbor diverse culturable actinomycetes that are promising sources of novel secondary metabolites. Appl Environ Microbiol 81(9):3086–3103

Habib N, Khan IU, Xiao M, Li S, Saqib M, Xian WD, Butt S, Li WJ (2020) *Marmoricola caldifontis* sp. nov., a novel actinobacterium isolated from a hot spring. Int J Syst Evol Microbiol 70(3):2053–2058

Hamada M, Iino T, Tamura T, Iwami T, Harayama S, Suzuki K (2009) *Serinibacter salmoneus* gen. nov., sp. nov., an actinobacterium isolated from the intestinal tract of a fish, and emended descriptions of the families *Beutenbergiaceae* and Bogoriellaceae. Int J Syst Evol Microbiol 59 (11):2809–2814

Han D, Wang L, Luo Y (2018a) Isolation, identification, and the growth promoting effects of two antagonistic actinomycete strains from the rhizosphere of Mikania micrantha Kunth. Microbiol Res 208:1–11

Han J, Gao QX, Zhang YG, Li L, Mohamad O, Rao M, Xiao M, Hozzein WN, Alkhalifah D, Tao Y, Li WJ (2018) Transcriptomic and Ectoine analysis of halotolerant *Nocardiopsis gilva* YIM 90087 under salt stress. Front Microbiol 9:618

Han C, Yu Z, Zhao J, Shi H, Hu J, Yu B, Song J, Shen Y, Xiang W, Wang X (2020) *Streptomyces triticagri* sp. nov. and *Streptomyces triticirhizae* sp. nov., two novel *Actinobacteria* isolated from the rhizosphere soil of wheat (*Triticum aestivum* L.). Int J Syst Evol Microbiol 70(1):126–138

He T, Zhang X (2016) Characterization of bacterial communities in Deep-Sea hydrothermal vents from three oceanic regions. Mar Biotechnol 18:232–241

Huang MJ, Huang HQ, Salam N, Xiao M, Duan YQ, Kim CJ, Li QQ, Chen W, Li WJ (2016) *Nocardioides intraradicalis* sp. nov., isolated from the roots of *Psammosilene tunicoides* W. C. Wu et C. Y. Wu. Int J Syst Evol Microbiol 66(10):3841–3847

Huang MJ, Rao M, Salam N, Xiao M, Huang HQ, Li WJ (2017) *Allostreptomyces psammosilenae* gen. Nov., sp. nov., an endophytic actinobacterium isolated from the roots of *Psammosilene tunicoides* and emended description of the family *Streptomycetaceae* [Waksman and Henrici (1943)AL] emend. Rainey et al. 1997, emend. Kim et al. 2003, emend. Zhi et al. 2009. Int J Syst Evol Microbiol 67(2):288–293

Ibeyaima A, Singh AK, Lal R, Gupta S, Goodfellow M, Sarethy IP (2018) *Saccharothrix tharensis* sp. nov., an actinobacterium isolated from the Thar Desert, India. Antonie Van Leeuwenhoek 111(11):2141–2147

Idris H, Goodfellow M, Sanderson R, Asenjo JA, Bull AT (2017) Actinobacterial rare biospheres and dark matter revealed in habitats of the Chilean Atacama Desert. Sci Rep 7:8373

Iniyan AM, Wink J, Landwehr W, Ramprasad E, Sasikala C, Ramana CV, Schumann P, Spröer C, Bunk B, Joseph F, Joshua SA, Shyl E, Kannan RR, Vincent S (2021) *Streptomyces marianii* sp. nov., a novel marine actinomycete from southern coast of India. J Antibiot 74(1):59–69

Itoh T, Yamanoi K, Kudo T, Ohkuma M, Takashina T (2011) *Aciditerrimonas ferrireducens* gen. nov., sp. nov., an iron-reducing thermoacidophilic actinobacterium isolated from a solfataric field. Int J Syst Evol Microbiol 61(6):1281–1285

Jiang H, Huang Q, Deng S, Dong H, Yu B (2010) Planktonic actinobacterial diversity along a salinity gradient of a river and five lakes on the Tibetan Plateau. Extremophiles 14(4):367–376

Jiao JY, Liu L, Zhou EM, Wei DQ, Ming H, Xian WD, Yuan CG, Zhong JM, Li WJ (2015) *Actinomadura amylolytica* sp. nov. and *Actinomadura cellulosilytica* sp. nov., isolated from geothermally heated soil. Antonie Van Leeuwenhoek 108(1):75–83

Jimenez PC, Ferreira EG, Araújo LA, Guimarães LA, Sousa TS, Pessoa ODL, Lotufo TMC, Costa-Lotufo LV (2013) Cytotoxicity of actinomycetes associated with the ascidian *Eudistoma vannamei* (Millar, 1977), endemic of northeastern coast of Brazil. Lat Am J Aquat Res 41:335–343

Johnson DB, Bacelar-Nicolau P, Okibe N, Thomas A, Hallberg KB (2009) *Ferrimicrobium acidiphilum* gen. nov., and *Ferrithrix thermotolerans* gen. nov., sp. nov.: heterotrophic, iron-oxidizing, extremely acidophilic actinobacteria. Int J Syst Evol Microbiol 59:1082–1089

Kämpfer P, Huber B, Buczolits S, Thummes K, Grün-Wollny I, Busse HJ (2007) *Nocardia acidivorans* sp. nov., isolated from soil of the island of Stromboli. Int J Syst Evol Microbiol 57(6):1183–1187

Kämpfer P, Glaeser SP, McInroy JA, Busse HJ (2016) *Nocardioides zeicaulis* sp. nov., an endophyte actinobacterium of maize. Int J Syst Evol Microbiol 66(4):1869–1874

Kanchanasin P, Yuki M, Kudo T, Ohkuma M, Phongsopitanun W, Tanasupawat S (2021) *Nocardia terrae* sp. nov., an actinomycete isolated from soil in Thailand. Arch Microbiol 203(3):1071–1077

Kaur G, Mual P, Kumar N, Verma A, Kumar A, Krishnamurthi S, Mayilraj S (2016) *Microbacterium aureliae* sp. nov., a novel actinobacterium isolated from *Aurelia aurita*, the moon jellyfish. Int J Syst Evol Microbiol 66(11):4665–4670

Kavitha A, Savithri HS (2017) Biological significance of marine actinobacteria of East Coast of Andhra Pradesh, India. Front Microbiol 8:1201

Khamna S, Yokota A, Lumyong S (2009) Actinomycetes isolated from medicinal plant rhizosphere soils: diversity and screening of antifungal compounds, indole-3-acetic acid and siderophore production. World J Microbiol Biotechnol 25:649–655

Kim SB, Nedashkovskaya OI, Mikhailov VV, Han SK, Kim KO, Rhee MS, Bae KS (2004) *Kocuria marina* sp. nov., a novel actinobacterium isolated from marine sediment. Int J Syst Evol Microbiol 54(5):1617–1620

Kirby BM, Roes ML, Meyers PR (2006) *Kribbella karoonensis* sp. nov. and *Kribbella swartbergensis* sp. nov., isolated from soil from the Western Cape, South Africa. Int J Syst Evol Microbiol 56(5):1097–1101

Kittisrisopit S, Pittayakhajonwut P, Tadtong S, Thawai C (2018) *Microbispora soli* sp. nov., isolated from soil of a hot spring. Int J Syst Evol Microbiol 68(12):3863–3868

Kumar A, Dubey A (2020) Rhizosphere microbiome: Engineering bacterial competitiveness for enhancing crop production. J Adv Res 24:337–352

Kurahashi M, Fukunaga Y, Sakiyama Y, Harayama S, Yokota A (2009) *Iamia majanohamensis* gen. nov., sp. nov., an actinobacterium isolated from sea cucumber *Holothuria edulis*, and proposal of *Iamiaceae* fam. nov. Int J Syst Evol Microbiol 59(4):869–873

Kurtböke Dİ (2017) Ecology and habitat distribution of actinobacteria. In: Wink J, Mohammadipanah F, Hamedi J (eds) Biology and biotechnology of actinobacteria. Springer, Cham, pp 123–149

Küster E (1970) Notes on the taxonomy and ecology of S. malachiticus and related species. In: Prauser H (ed) The actinomycetales. VEB Gustav Fisher Verlag, Jena, pp 169–172

Lakshmipathy TD, Prasad AA, Kannabiran K (2010) Production of biosurfactant and heavy metal resistance activity of *Streptomyces* sp. VITDDK3-a novel halo tolerant actinomycetes isolated from saltpan soil. Adv Biol Res 4(2):108–115

Law JWF, Letchumanan V, Tan LTH, Ser HL, Goh BH, Lee LH (2020) The rising of "modern actinobacteria" era. Progress Microbes Mol Biol 3(1):a0000064

Lee SD, Kim SJ (2007) *Aeromicrobium tamlense* sp. nov., isolated from dried seaweed. Int J Syst Evol Microbiol 57(2):337–341

Lee HJ, Whang KS (2017) *Micromonospora fulva* sp. nov., isolated from forest soil. Int J Syst Evol Microbiol 67(6):1746–1751

Lee JC, Whang KS (2020) *Agromyces humi* sp. nov., actinobacterium isolated from farm soil. Int J Syst Evol Microbiol 70(9):5032–5039

Lee HJ, Han SI, Whang KS (2012) *Streptomyces gramineus* sp. nov., an antibiotic-producing actinobacterium isolated from bamboo (*Sasa borealis*) rhizosphere soil. Int J Syst Evol Microbiol 62(4):856–859

Lee SM, Kong HG, Song GC, Ryu CM (2021) Disruption of firmicutes and actinobacteria abundance in tomato rhizosphere causes the incidence of bacterial wilt disease. ISME J 15(1):330–347

Li WJ, Chen HH, Zhang YQ, Schumann P, Stackebrandt E, Xu LH, Jiang CL (2004) *Nesterenkonia halotolerans* sp. nov. and *Nesterenkonia xinjiangensis* sp. nov., actinobacteria from saline soils in the west of China. Int J Syst Evol Microbiol 54(3):837–841

Li B, Furihata K, Ding LX, Yokota A (2007) *Rhodococcus kyotonensis* sp. nov., a novel actinomycete isolated from soil. Int J Syst Evol Microbiol 57(9):1956–1959

Li J, Zhao GZ, Chen HH, Qin S, Xu LH, Jiang CL, Li WJ (2008) *Rhodococcus cercidiphylli* sp. nov., a new endophytic actinobacterium isolated from a *Cercidiphyllum japonicum* leaf. Syst Appl Microbiol 31(2):108–113

Li XJ, Liu JM, Wu Y, Zhang WM, Li J, Liu SW, Wu G, Hu L, Chen L, Huang DL, Li RF, Sun CH (2016) Description of *Salilacibacter albus* gen. nov., sp. nov., isolated from a dried salt lake, and reclassification of *Paraglycomyces xinjiangensis* Luo et al. 2015 as a later heterotypic synonym of *Salininema proteolyticum* Nikou et al. 2015 with emended descriptions of the genus *Salininema* and *Salininema proteolyticum*. Int J Syst Evol Microbiol 66(7):2558–2565

Li FN, Liao SL, Liu SW, Jin T, Sun CH (2019a) *Aeromicrobium endophyticum* sp. nov., an endophytic actinobacterium isolated from reed (*Phragmites australis*). J Microbiol (Seoul, Korea) 57(9):725–731

Li LY, Yang ZW, Asem MD, Fang BZ, Salam N, Alkhalifah D, Hozzein WN, Nie GX, Li WJ (2019b) *Streptomyces desertarenae* sp. nov., a novel actinobacterium isolated from a desert sample. Antonie Van Leeuwenhoek 112(3):367–374

Li C, Cao P, Jiang M, Sun T, Shen Y, Xiang W, Zhao J, Wang X (2020a) *Streptomyces oryziradicis* sp. nov., a novel actinomycete isolated from rhizosphere soil of rice (*Oryza sativa* L.). Int J Syst Evol Microbiol 70(1):465–472

Li YQ, Zhang H, Xiao M, Dong ZY, Zhang JY, Narsing Rao MP, Li WJ (2020b) *Cellulomonas endophytica* sp. nov., isolated from *Gastrodia elata* Blume. Int J Syst Evol Microbiol 70 (5):3091–3095

Liu L, Salam N, Jiao JY, Jiang HC, Zhou EM, Yin YR, Ming H, Li WJ (2016) Diversity of culturable thermophilic actinobacteria in hot springs in Tengchong, China and studies of their biosynthetic gene profiles. Microb Ecol 72(1):150–162

Madhaiyan M, Poonguzhali S, Lee JS, Senthilkumar M, Lee KC, Sundaram S (2010) *Leifsonia soli* sp. nov., a yellow-pigmented actinobacterium isolated from teak rhizosphere soil. Int J Syst Evol Microbiol 60(6):1322–1327

Malisorn K, Kanchanasin P, Phongsopitanun W, Tanasupawat S (2018) *Actinomadura rhizosphaerae* sp. nov., isolated from rhizosphere soil of the plant *Azadirachta indica*. Int J Syst Evol Microbiol 68(9):3012–3016

Meklat A, Sabaou N, Zitouni A, Mathieu F, Lebrihi A (2011) Isolation, taxonomy, and antagonistic properties of halophilic actinomycetes in Saharan soils of Algeria. Appl Environ Microbiol 77(18):6710–6714

Mohammadipanah F, Wink J (2016) Actinobacteria from arid and desert habitats: diversity and biological activity. Front Microbiol 6:1541

Mu S, Zhao Q, Zhao J, Cao T, Zhao X, Guo X, Li Y, Xiang W, Wang X (2018) *Sphaerisporangium rhizosphaerae* sp. nov., an actinomycete isolated from the rhizosphere soil of a rubber tree (*Hevea brasiliensis* Muell. Arg). Int J Syst Evol Microbiol 68(9):2860–2865

Myer PR, Kim M, Freetly HC, Smith TPL (2016) Evaluation of 16S rRNA amplicon sequencing using two next-generation sequencing technologies for phylogenetic analysis of the rumen bacterial community in steers. J Microbiol Methods 127:132–140

Nalini, M.S., Prakash, H.S., 2020. Actinobacteria: diversity, plant interactions and biotechnology applications. In: Yadav, A.N., Singh, J., Rastegari, A.A, Yadav, N., (eds) Plant microbiomes for sustainable agriculture, sustainable development and biodiversity. Springer Nature Switzerland. pp. 199–244

Narsing Rao, M.P., Liu, L., Jiao, J.Y., Xiao, M., Li, W.J., 2018. Hot springs of India: occurrence and microbial diversity. In: Egamberdieva, D., Birkeland, N.K., Panosyan, H., Li, W.J., (eds) Extremophiles in Eurasian ecosystems: ecology, diversity, and applications. Microorg Sustain, 8 Singapore, Springer. Pp. 29–55

Narsing Rao MP, Li YQ, Zhang H, Dong ZY, Dhulappa A, Xiao M, Li WJ (2020) *Amycolatopsis alkalitolerans* sp. nov., isolated from Gastrodia elata Blume. J Antibiot 73(1):35–39

Narsing Rao MP, Dong ZY, Luo ZH, Li MM, Liu BB, Guo SX, Hozzein WN, Xiao M, Li WJ (2021) Physicochemical and microbial diversity analyses of Indian Hot Springs. Front Microbiol 12:627200

Nimnoi P, Pongsilp N, Lumyong S (2011) Actinobacterial community and diversity in rhizosphere soils of *Aquilaria crassna* Pierre ex Lec assessed by RT-PCR and PCR-DGGE. Biochem Syst Ecol 39(4–6):509–519

Nithyanand P, Indhumathi T, Ravi AV, Pandian SK (2011) Culture independent characterization of bacteria associated with the mucus of the coral *Acropora digitifera* from the Gulf of Mannar. World J Microbiol Biotechnol 27(6):1399–1406

Norman JS, King GM, Friesen ML (2017) *Rubrobacter spartanus* sp. nov., a moderately thermophilic oligotrophic bacterium isolated from volcanic soil. Int J Syst Evol Microbiol 67(9):3597–3602

Oberhofer M, Hess J, Leutgeb M, Gössnitzer F, Rattei T, Wawrosch C, Zotchev SB (2019) Exploring Actinobacteria associated with rhizosphere and endosphere of the native alpine medicinal plant Leontopodium nivale subspecies alpinum. Front Microbiol 10:2531

Orchard V (1981) The ecology of Nocardia and related taxa. Zentralblatt für Bakteriologie, Parasitenkunde, Infektionskrankheiten und Hygiene, Abteilung I, Supplement 6:429–434

Palaniyandi SA, Yang SH, Zhang L, Suh JW (2013) Effects of actinobacteria on plant disease suppression and growth promotion. Appl Microbiol Biotechnol 97(22):9621–9636

Pathma J, Sakthivel N (2012) Microbial diversity of vermicompost bacteria that exhibit useful agricultural traits and waste management potential. Springerplus 1:26

Picone N, Hogendoorn C, Cremers G, Poghosyan L, Pol A, van Alen TA, Gagliano AL, D'Alessandro W, Quatrini P, Jetten M, Op den Camp H, Berben T (2020) Geothermal gases shape the microbial community of the volcanic soil of pantelleria. Italy MSyst 5(6):e00517–e00520

Pikoli MR, Sugoro I, Suharti (2020) Diversity analysis of an extremely acidic soil in a layer of coal mine detected the occurrence of rare actinobacteria. Walailak J Sci Technol 17(6):529–542

Prabhu DM, Quadri SR, Cheng J, Liu L, Chen W, Yang Y, Hozzein WN, Lingappa K, Li WJ (2015) *Sinomonas mesophila* sp. nov., isolated from ancient fort soil. J Antibiot 68(5):318–321

Qian L, Duan L, Lin J, Yang Y, Song J, Wang X, Zhao J, Xiang W (2020) *Glycomyces albidus* sp. nov., a novel actinobacterium isolated from rhizosphere soil of wheat (*Triticum aestivum* L.). Int J Syst Evol Microbiol 70(5):3096–3104

Reddy T, Mahmood S, Idris MM, Ciesielski S (2010) *Streptomyces osmaniensis* sp. nov., isolated from soil. Int J Syst Evol Microbiol 60(8):1755–1759

Rego A, Raio F, Martins TP, Ribeiro H, Sousa A, Séneca J, Baptista MS, Lee CK, Cary SC, Ramos V, Carvalho MF, Leão PN, Magalhães C (2019) Actinobacteria and Cyanobacteria diversity in terrestrial antarctic microenvironments evaluated by culturedependent and independent methods. Front Microbiol 10:1018

Riquelme C, Marshall Hathaway JJ, Enes Dapkevicius M, Miller AZ, Kooser A, Northup DE, Jurado V, Fernandez O, Saiz-Jimenez C, Cheeptham N (2015) Actinobacterial diversity in volcanic caves and associated Geomicrobiological interactions. Front Microbiol 6:1342

Risdian C, Landwehr W, Rohde M, Schumann P, Hahnke RL, Spröer C, Bunk B, Kämpfer P, Schupp PJ, Wink J (2021) *Streptomyces bathyalis* sp. nov., an actinobacterium isolated from the sponge in a deep sea. Antonie Van Leeuwenhoek 114(4):425–435

Sabat AJ, van Zanten E, Akkerboom V, Wisselink G, van Slochteren K, de Boer RF, Hendrix R, Friedrich AW, Rossen J, Kooistra-Smid A (2017) Targeted next-generation sequencing of the 16S-23S rRNA region for culture-independent bacterial identification increased discrimination of closely related species. Sci Rep 7(1):3434

Salam N, Jiao JY, Zhang XT, Li WJ (2020) Update on the classification of higher ranks in the phylum *Actinobacteria*. Int J Syst Evol Microbiol 70(2):1331–1355

Saygin H, Ay H, Guven K, Cetin D, Sahin N (2020) *Streptomyces cahuitamycinicus* sp. nov., isolated from desert soil and reclassification of *Streptomyces galilaeus* as a later heterotypic synonym of *Streptomyces bobili*. Int J Syst Evol Microbiol 70(4):2750–2759

Schatz A, Bugie E, Waksman SE (1944) Streptomycin, a substance exhibiting antibiotic activity against gram-positive and gram negative bacteria. Proc Soc Exp Biol Med 55:65–69

Selianin VV, Oborotov GE, Zenova GM, Zviagintsev DG (2005) Alkaliphilic soil actinomycetes. Microbiology 74:729–734

Sharma P, Thakur D (2020) Antimicrobial biosynthetic potential and diversity of culturable soil actinobacteria from forest ecosystems of Northeast India. Sci Rep 10:4104

Shivlata L, Satyanarayana T (2015) Thermophilic and alkaliphilic actinobacteria: biology and potential applications. Front Microbiol 6:1014

Song Z, Zhi X, Li W, Jiang H, Zhang C, Dong H (2009) Actinobacterial diversity in Hot Springs in Tengchong (China), Kamchatka (Russia), and Nevada (USA). Geomicrobiol J 26(4):256–263

Sreevidya M, Gopalakrishnan S, Kudapa H, Varshney RK (2016) Exploring plant growth-promotion actinomycetes from vermicompost and rhizosphere soil for yield enhancement in chickpea. Braz J Microbiol 47(1):85–95

Stackebrandt E, Rainey FA, Ward-Rainey NL (1997) Proposal for a new hierarchic classification system, *Actinobacteria* classis nov. Int J Syst Bacteriol 47:479–491

Sun Y, Shi YL, Wang H, Zhang T, Yu LY, Sun H, Zhang YQ (2018) Diversity of bacteria and the characteristics of actinobacteria community structure in Badain Jaran Desert and Tengger Desert of China. Front Microbiol 9:1068

Takahashi Y, Matsumoto A, Seino A, Iwai Y, Omura S (1996) Rare actinomycetes isolated from desert soils. Actinomycetologica 10(2):91–97

Tang SK, Tian XP, Zhi XY, Cai M, Wu JY, Yang LL, Xu LH, Li WJ (2008) *Haloactinospora alba* gen. nov., sp. nov., a halophilic filamentous actinomycete of the family *Nocardiopsaceae*. Int J Syst Evol Microbiol 58(9):2075–2080

Tang SK, Wang Y, Cai M, Zhi XY, Lou K, Xu LH, Jiang CL, Li WJ (2009) *Saccharopolyspora halophila* sp. nov., a novel halophilic actinomycete isolated from a saline lake in China. Int J Syst Evol Microbiol 59(3):555–558

Tang SK, Wang Y, Lee JC, Lou K, Park DJ, Kim CJ, Li WJ (2010) *Georgenia halophila* sp. nov., a halophilic actinobacterium isolated from a salt lake. Int J Syst Evol Microbiol 60(6):1317–1421

Tatar D, Sazak A, Guven K, Cetin D, Sahin N (2013) *Amycolatopsis cihanbeyliensis* sp. nov., a halotolerant actinomycete isolated from a salt mine. Int J Syst Evol Microbiol 63(10):3739–3743

Tatar D, Guven K, Inan K, Cetin D, Belduz AO, Sahin N (2016) *Streptomonospora tuzyakensis* sp. nov., a halophilic actinomycete isolated from saline soil. Antonie Van Leeuwenhoek 109(1):35–41

Thanaboripat D, Thawai C, Kittiwongwattana C, Laosinwattana C, Koohakan P, Parinthawong N (2015) *Micromonospora endophytica* sp. nov., an endophytic actinobacteria of Thai upland rice (*Oryza sativa*). J Antibiot 68(11):680–684

Thawai C, Thamsathit W, Kudo T (2013) *Planosporangium thailandense* sp. nov., isolated from soil from a Thai hot spring. Int J Syst Evol Microbiol 63:1051–1055

Trujillo ME, Idris H, Riesco R, Nouioui I, Igual JM, Bull AT, Goodfellow M (2017) *Pseudonocardia nigra* sp. nov., isolated from Atacama Desert rock. Int J Syst Evol Microbiol 67(8):2980–2985

Undabarrena A, Beltrametti F, Claverías FP, González M, Moore ER, Seeger M, Cámara B (2016) Exploring the diversity and antimicrobial potential of marine Actinobacteria from the Comau Fjord in Northern Patagonia, Chile. Front Microbiol 7:1135

Valenzuela-Encinas C, Neria-González I, Alcántara-Hernández RJ, Estrada-Alvarado I, Zavala-Díaz de la Serna FJ, Dendooven L, Marsch R (2009) Changes in the bacterial populations of the highly alkaline saline soil of the former lake Texcoco (Mexico) following flooding. Extremophiles 13(4):609–621

Valverde A, Tuffin M, Cowan DA (2012) Biogeography of bacterial communities in hot springs: a focus on the actinobacteria. Extremophiles 16:669–679

van Bergeijk DA, Terlouw BR, Medema MH, van Wezel GP (2020) Ecology and genomics of Actinobacteria: new concepts for natural product discovery. Nat Rev Microbiol 18(10):546–558

van der Meij A, Willemse J, Schneijderberg MA, Geurts R, Raaijmakers JM, van Wezel GP (2018) Inter- and intracellular colonization of *Arabidopsis* roots by endophytic actinobacteria and the impact of plant hormones on their antimicrobial activity. Antonie Van Leeuwenhoek 111 (5):679–690

Wang YX, Wang HB, Zhang YQ, Xu LH, Jiang CL, Li WJ (2008) *Rhodococcus kunmingensis* sp. nov., an actinobacterium isolated from a rhizosphere soil. Int J Syst Evol Microbiol 58(6):1467–1471

Wang Y, Cao P, Sun P, Zhao J, Sun X, Zhang J, Li C, Xiang W, Wang X (2019a) *Cryobacterium tepidiphilum* sp. nov., isolated from rhizosphere soil of lettuce (var. ramosa Hort.). Antonie Van Leeuwenhoek 112(11):1611–1621

Wang L, Zhang L, Zhang X, Zhang S, Yang L, Yuan H, Chen J, Liang C, Huang W, Liu J, Zhao Y, Yang Q (2019b) *Rhodococcus daqingensis* sp. nov., isolated from petroleum-contaminated soil. Antonie Van Leeuwenhoek 112(5):695–702

Wang H, Narsing Rao MP, Gao Y, Li X, Gao R, Xie Y, Li Q, Li W (2021) Insights into the endophytic bacterial community comparison and their potential role in the dimorphic seeds of halophyte *Suaeda glauca*. BMC Microbiol 21(1):143

Ward AC, Bora N (2006) Diversity and biogeography of marine actinobacteria. Curr Opin Microbiol 9(3):279–286

Wei X, Jiang Y, Chen X, Jiang Y, Lai H (2015) *Amycolatopsis flava* sp. nov., a halophilic actinomycete isolated from Dead Sea. Antonie Van Leeuwenhoek 108(4):879–885

Wei Y, Wang F, Gao J, Huang Y, Ren W, Sheng H (2021) Culture-dependent and culture-independent characterization of bacterial community diversity in different types of sandy lands: the case of Minqin County, China. BMC Microbiol 21:87

Whitman W, Goodfellow M, Kämpfer P, Busse HJ, Trujillo M, Ludwig W, Suzuki K (2012) Bergey's manual of systematic bacteriology: volume 5: the Actinobacteria, 2nd edn. Springer

Williams ST, Vickers JC (1988) Detection of actinomycetes in the natural environment: problems and perspectives. In: Okami Y, Beppu T, Ogawara H (eds) Biology of actinomycetes. Japan Scientific Societies Press, Tokyo, pp 165–270

Wu J, Guan T, Jiang H, Zhi X, Tang S, Dong H, Zhang L, Li W (2009) Diversity of Actinobacterial community in saline sediments from Yunnan and Xinjiang, China. Extremophiles 13(4):623–632

Wu H, Chen W, Wang G, Dai S, Zhou D, Zhao H, Guo Y, Ouyang Y, Li X (2012) Culture-dependent diversity of actinobacteria associated with seagrass (*Thalassia hemprichii*). Afr J Microbiol Res 6(1):87–94

Wu H, Liu B, Shao Y, Ou X, Huang F (2018) *Thermostaphylospora grisealba* gen. nov., sp. nov., isolated from mushroom compost and transfer of *Thermomonospora chromogena* Zhang et al. 1998 to *Thermostaphylospora chromogena* comb. nov. Int J Syst Evol Microbiol 68:602–608

Xiong YW, Gong Y, Li XW, Chen P, Ju XY, Zhang CM, Yuan B, Lv ZP, Xing K, Qin S (2019) Enhancement of growth and salt tolerance of tomato seedlings by a natural halotolerant actinobacterium *Glutamicibacter halophytocola* KLBMP 5180 isolated from a coastal halophyte. Plant Soil 445:307–322

Yadav AN, Verma P, Kumar S, Kumar V, Kumar M, Sugitha TCK, Singh BP, Saxena AK, Dhaliwal HS (2018) Actinobacteria from rhizosphere: molecular diversity, distributions, and potential biotechnological applications. In: Singh BP, Gupta VK, Passari AK (eds) New and future developments in microbial biotechnology and bioengineering. Elsevier, Amsterdam, pp 13–41

Yoon JH, Lee CH, Oh TK (2005) *Aeromicrobium alkaliterrae* sp. nov., isolated from an alkaline soil, and emended description of the genus *Aeromicrobium*. Int J Syst Evol Microbiol 55 (5):2171–2175

Yoon JH, Lee CH, Oh TK (2006) *Nocardioides lentus* sp. nov., isolated from an alkaline soil. Int J Syst Evol Microbiol 56(1):271–275

Zhang YQ, Schumann P, Yu LY, Liu HY, Zhang YQ, Xu LH, Stackebrandt E, Jiang CL, Li WJ (2007) *Zhihengliuella halotolerans* gen. nov., sp. nov., a novel member of the family *Micrococcaceae*. Int J Syst Evol Microbiol 57(5):1018–1023

Zhang YJ, Tang SK, Shi R, Klenk HP, Chen C, Yang LL, Zhou Y, Li WJ (2012) *Salinisphaera halophila* sp. nov., a moderately halophilic bacterium isolated from brine of a salt well. Int J Syst Evol Microbiol 62(9):2174–2179

Zhang G, Cao T, Ying J, Yang Y, Ma L (2014) Diversity and novelty of actinobacteria in Arctic marine sediments. Antonie Van Leeuwenhoek 105(4):743–754

Zhang YG, Lu XH, Ding YB, Zhou XK, Li L, Guo JW, Wang HF, Duan YQ, Li WJ (2016) *Phytoactinopolyspora alkaliphila* sp. nov., an alkaliphilic actinomycete isolated from a saline-alkaline soil. Int J Syst Evol Microbiol 66:2058–2063

Zhang L, Xi L, Ruan J, Huang Y (2017) *Kocuria oceani* sp. nov., isolated from a deep-sea hydrothermal plume. Int J Syst Evol Microbiol 67(1):164–169

Zhang B, Wu X, Tai X, Sun L, Wu M, Zhang W, Chen X, Zhang G, Chen T, Liu G, Dyson P (2019) Variation in actinobacterial community composition and potential function in different soil ecosystems belonging to the arid Heihe River Basin of Northwest China. Front Microbiol 10: 2209

Zhao K, Penttinen P, Chen Q, Guan T, Lindström K, Ao X, Zhang L, Zhang X (2012) The rhizospheres of traditional medicinal plants in Panxi, China, host a diverse selection of actinobacteria with antimicrobial properties. Appl Microbiol Biotechnol 94(5):1321–1335

Zhao J, Duan L, Qian L, Cao P, Tian Y, Ju H, Xiang W, Wang X (2019) *Kribbella jiaozuonensis* sp. nov., a novel actinomycete isolated from soil. Int J Syst Evol Microbiol 69(11):3500–3507

Zhou EM, Tang SK, Sjøholm C, Song ZQ, Yu TT, Yang LL, Ming H, Nie GX, Li WJ (2012a) *Thermoactinospora rubra* gen. nov., sp. nov., a thermophilic actinomycete isolated from Tengchong, Yunnan province, south-West China. Antonie Van Leeuwenhoek 102(1):177–185

Zhou EM, Yang LL, Song ZQ, Yu TT, Nie GX, Ming H, Zhou Y, Tang SK, Li WJ (2012b) *Thermocatellispora tengchongensis* gen. nov., sp. nov., a new member of the family *Streptosporangiaceae*. Int J Syst Evol Microbiol 62(10):2417–2423

Chapter 3
Traditional Screening and Genome-Guided Screening of Natural Products from *Actinobacteria*

Amal Mohamed Shawky, Omnia Mohamed Abdallah, and Dina Hatem Amin

Abstract Since the golden era of antibiotics, around one half of the antibiotics that are available in the market were discovered. From 1960, a drop in the discovery of new antibiotics has been reported which paid attention to the necessity to develop new screening techniques, especially from high potential microorganisms such as Actinobacteria. In this review, we reported some traditional screening approaches to achieve this goal such as exploring rare Actinobacteria taxa in extreme habitats and in a combined culture to induce secondary metabolite production. We also highlight several genomic approaches including, whole genome sequencing, metagenomics, and CRISPR/Cas to provide more knowledge regarding biosynthetic gene clusters and predict natural compounds secreted by Actinobacteria. This will open the door for gene regulation, expression, and editing gene sequences in Actinobacteria, which will guide the discovery of novel antibiotics and new chemical entities to combat antimicrobial resistance problem.

Keywords Bioactive compounds · Actinobacteria · Whole genome sequence · NGS · Metagenomics

3.1 Introduction

Actinobacteria are the most prolific and potent source of a lot of bioactive molecules. Actinobacteria are a group of Gram-positives bacteria whose genetic material DNA is rich in GC content (70%) relative to other bacteria such as *Escherichia coli* (GC content = 50%) (Bentley et al. 2002). It is estimated that approximately three-

A. M. Shawky · D. H. Amin (✉)
Microbiology Department, Egyptian Drug Authority (EDA), Giza, Egypt
e-mail: dina.hatem@sci.asu.edu.eg

O. M. Abdallah
Microbiology Department, Faculty of Science, Ain Shams University, Cairo, Egypt

L. Karthik (ed.), *Actinobacteria*, https://doi.org/10.1007/978-981-16-5835-8_3

quarters of the microbial-produced bioactive compounds are generated from actinobacteria. The most widely identified group is Streptomycetes, which produce a broad variety of bioactive compounds, mainly antibiotics, antifungal substances, antioxidants, anticancer compounds, enzymes, pesticides, etc. These valuable molecules have been developed to products including human medicine, veterinary treatment, agrochemicals, and other industrial products (Abdel-Razek et al. 2020; Edelvio et al. 2018; Kämpfer 2006 Salwan and Sharma 2020; Tiwari and Gupta 2012).

Phylogenetically, *Streptomyces* spp. are the most common genus of Actinobacteria; furthermore, 70% discovered antibiotics are produced by *Streptomyces* spp. alone (Tishkov 2001). In the health care facilities and hospitals, most of used drugs were of *Streptomyces* origin or their synthetic forms. However, clinical studies revealed that there were more than 40% drop of the discovery of new based actinomycetes secondary metabolites drugs from 2001 to 2008 compared to their sharing during the 1960s and 1980s (Tishkov 2001; Li and Vederas 2009; Tiwari and Gupta 2012). Considering the emergence of resistance of different pathogens towards antibiotics such as multidrug-resistant superbugs, especially *Enterobacter* spp., *Staphylococcus* spp., *Klebsiella* spp., *Acinetobacter* spp., *Pseudomonas* spp., and *Escherichia coli* are increasingly resistant to common antibacterial agents (Mulani et al. 2019; Wright 2013). In addition, the development of new drugs by the pharmaceutical industry became difficult due to economic perspectives and challenging regulatory requirements. Accordingly, attention has turned to develop and innovate new screening tools and techniques for new bioactive compounds. This review highlights the traditional screening approaches and genome-guided screening strategies for new bioactive compounds discovery.

3.2 Traditional Approaches

Classical methods have been used for effective drug discovery and new bioactive molecules which are:

3.2.1 Phenotypic Assay

Biological activity of the substances, crude extracts or purified chemicals were screened phenotypically through whole cell assays. This technique does not require knowledge of the identity or the chemical structure of the drug or the mode of action. Instead, the bioactive compound is selected on the basis of their empirical activity. This strategy is known as bioactive-guided screening. One of the main disadvantages of this method is the high probability of repeated rediscovery same bioactive compounds and time consuming (Lee et al. 2012; Wohlleben et al. 2016; Moffat et al. 2017). The extract that shows biological activity is fractionated then purified fractions are obtained. These purified fractions were rescreened and their activities were confirmed. The chemical structure of the relatively pure compound was elucidated using NMR and high-resolution mass spectrometry (HRMS) (Henke and Kelleher 2016).

By this approach, the screening of bioactive compound that is produced by actinobacteria is usually based on agar plate or broth assay. Any inhibition for the growth or a colorimetric reaction indicating specific activity is detected. This technique can still be very effective to find new natural products if it is conjugated with selective isolation of rare actinomycetes, isolation of previously uncultivable bacteria, modification of nutrients conditions in medium, co-cultivation with other microorganism (Scherlach and Hertweck 2009; Zarins-Tutt et al. 2016; Ochi 2017; Onaka 2017).

3.2.1.1 Rare Actinomycetes as a Potential Source for New Bioactive Compounds

"Rare actinomycetes" are known as the actinomycete strains less frequently isolated than that of the "commonly" isolated *Streptomyces* spp., even though they may not necessarily be rare in the environment (Tiwari and Gupta 2014). Rare Actinomycetes are usually considered as non-streptomycetes strains which require for their isolation non-conventional methods. Pretreatment of soil samples through adding chemicals (phenol, chloramine, and antibiotics), radiation (ultraviolet, microwave, ultrasonic and super high frequency radiation), drying and thermal treatment are required for selective isolation of rare actinomycetes (Subramani and Aalbersberg 2013). The development and application of selective isolation makes some genera like *Actinomadura*, *Actinoplanes, Micromonospora,* and *Microtetraspora*, not rare at all and can be obtained from many soil samples (Lazzarini et al. 2001; Baltz 2006; Tiwari and Gupta 2014). In addition, the isolation of rare actinomycetes from underexplored diverse natural habitats have contributed in the isolation of about 220 rare actinomycete genera of which more than 50 taxa. These genera were demonstrated as producers of 2500 bioactive compounds which represents more than 25% of all actinomycetes metabolite (Bérdy 2005; Subramani and Aalbersberg 2013). Currently, a lot of bioactive molecules which produced by rare actinomycetes are available and exhibit high potent activity such as vancomycin by *Amycolatopsis orientalis* (Padma et al. 2002), rifamycins by *Amycolatopsis mediterranei* (Solanki et al. 2008), and telithromycin (derivative of erythromycin) by *Saccharopolyspora erythrae* (Butler et al. 2017).

3.2.1.2 Exploring for New Isolates in Diverse Natural Habitats

As a result of over exploitation of terrestrial streptomycetes, the unexplored and underexplored extreme environments are promising sources of novel actinomycetes that are believed to be rich sources of interestingly new compounds (Qin et al. 2016; Subramani and Sipkema 2019).The high incidence of actinobacterial species was reported in all extreme environments which had broken the conventional model of restricted predominance of actinobacteria in soil and fresh water habitats (Baltz 2005). Actinobacteria were recovered from extreme environments conditions, which are characterized by high acidity or alkalinity, low or high temperatures, salinity, high radiation, low levels of available moisture, and nutrients (Zenova et al. 2011). Moreover, extremophilic actinobacteria can survive in distinct extreme geographical locations such as desert, deep sea, volcanic environment, and cryoenvironment (Chitte and Dey 2002; Kurapova et al. 2012; Shivlata and Tulasi 2015; Sivalingam et al. 2019).These extremophilic actinobacteria had various adaptation strategies to overcome and survive in extreme conditions which makes them prolific source for distinct new bioactive molecules. Thus, it is of great significance to reveal the diversity of the Actinobacteria and their survival mechanisms, and subsequently expound the application of their secondary metabolic products (Sivalingam et al. 2019).

The ocean covers more than 70% of the Earth's surface and represents more than 95% of the total biosphere. Accordingly, marine ecosystem considered a rich habitat where marine actinobacteria with high potential productivity of new active compounds is estimated. Different genera of Actinobacteria have been recovered from different marine ecological niches include *Actinomadura, Actinosynnema, Amycolatopsis, Arthrobacter, Brachybacterium, Frankia, Gordonia, Kitasatospora, Micromonospora, Micrococcus, Microbacterium, Mycobacterium, Nocardioides, Nocardiopsis, Nonomurea, Pseudonocardia, Rhodococcus, Saccharopolyspora, Salinispora, Serinicoccus, Solwaraspora, Streptomyces, Streptosporangium, Tsukamurella, Turicella, and Verrucosispora* (Ward and Bora 2006). In addition, novel bioactive products were detected with variable application such as Salinipostins (antimalarial) by *Salinispora* sp., Cyclomarine and Cyclomarazine (anti-inflammatory agents) by *S. arenicola,* Saccharothrixones (cytotoxic activity) by *Saccharothrix* sp., Fluostatin (antimicrobial substance) by *Micromonospora rosaria,* and Juvenimicin C (cancer chemo-preventive agent) by *Micromonospora* sp. (CNJ-878) (Carlson et al. 2013; Gan et al. 2015; Schultz et al. 2008; Schulze et al. 2015; Zhang et al. 2012).

3.2.1.3 Combined Culture (Co-Cultivation) Technique to Induce Secondary Metabolite Production

In nature, bacteria exist in a community where they compete with other microbes by producing secondary metabolites to survive in the environment. Consequently, co-cultivation technique is a trial to simulate the environmental conditions which may induce production potentially new compounds. The co-cultivation technique involves cultivation of two or more microorganisms which may stimulate the silent genes or gene clusters of one partner or increase the yields of previously described metabolites (Tan et al. 2019; Yu et al. 2019).

Co-cultivation of *Streptomyces rochei* MB037 and *Rhinocladiella similis* 35 stimulates the production of five metabolites which their chemical structure and biological activity were detected. The study revealed that the fungus *Rhinocladiella similis* 35 stimulate the actinobacteria *Streptomyces rochei* MB037 for producing new metabolites. One of these metabolites has antibacterial effect against methicillin-resistant *Staphylococcus aureus* (Yu et al. 2019). Similarly, biosynthesis of three natural products, namely N-(2-hydroxyphenyl)-acetamide, 1,6-dihydroxyphenazine, and 5a,6,11a,12-tetrahydro-5a,11a-dimethyl[1,4]benzoxazino[3,2-b][1,4] benzoxazine were detected by co-cultivation of *Actinokineospora* sp. EG49 and *Nocardiopsis* sp. RV163 (Dashti et al. 2014). Novel tropolone alkaloids named rubterolones and the biosynthetic pathway of rubterolones were reported by co-cultivation of *Actinomadura* sp. 5–2 (Guo et al. 2017) .

Recently, several analytical techniques facilitate the separation, identification, and elucidation chemical structures of secondary metabolites produced in co-cultivation method which allows studying a wide range of components (Tan et al. 2019).

3.2.2 Chemical Structure-Based Screening

This technique is used to identify new molecules with different chemical structures and properties either from biological sources or chemical libraries. In this method, high-performance liquid chromatography (HPLC), mass spectrometry (MS), or nuclear magnetic resonance spectroscopy (NMR) are applied for chemical structure elucidation (Moffat et al. 2017). Recently, the advances of combinatorial chemistry and computational chemistry help the researchers to synthesize a huge number and structurally diverse compounds, detect the molecules identity, screening and encoding combinatorial libraries (Liu et al. 2017).

In 2009, Esquenazi et al. demonstrated techniques that enable researcher to detect the production of secondary metabolites in microbes during their growth phase using imaging mass spectrometry (IMS). By this method, researchers can monitor secondary metabolite production in bacterial cells by using the mass spectrum of the chosen metabolite as the sensor. They also able to address the functional roles of natural products in metabolic exchange, including their role in communication, defense, and as entities that control morphological changes in organisms and how they may aid in the spatial analysis of natural products from heterogeneous samples. One of the outstanding imaging mass spectrometry (IMS) tool is Maldi-TOF which provides spatial and temporal resolution of secondary metabolites from marine sponges, cyanobacteria, zoanthids, bacteria, and plants (Esquenazi et al. 2009). Moreover, Maldi-TOF MS is used as screening tool for taxonomic dereplication of large actinobacteria collections through similarity protein profile to facilitate detection of low abundance actinobacteria with potential as a source of antimicrobial agents (Arango et al. 2018).

3.3 Genetic Tools for Screening for Novel Bioactive Compounds

Novel molecular tools and methods of genome analysis have been developed which provide more knowledge of natural product biosynthesis in many organisms. Biosynthetic gene clusters (BGS) are organized group of genes which contains whole pathway necessary for secondary metabolites production. The secondary metabolite biosynthetic gene clusters expression generally led to precursor biosynthesis, assembly, modification, resistance, and regulation of their product. The expression of these clusters is extremely controlled by complex regulatory networks that affected by biotic and abiotic stresses found in the bacteria's natural habitat (Craney et al. 2013). Presence of more than 30 secondary metabolite biosynthetic gene clusters (BGC) in *Streptomyces* spp. genome suggests that they can produce over 30 secondary metabolites in each strain. However, only a few metabolites are detected under laboratory conditions because of conditional or low production of these metabolites. Thus, new techniques are needed to get the biosynthetic potential of these compounds, including the genetic recombination of biosynthetic gene clusters (Albarano et al. 2020; Ochi 2017; Onaka 2017; Scherlach and Hertweck 2009; Zarins-Tutt et al. 2016).

Currently, a significant improvement in genetic engineering strategies for activation of BGC-expression and production of the respective compounds in actinobacteria were developed. These approaches include the expression of multiple copies of the whole BGC or factors that are limiting the production, expression of (Xuan et al. 2013) activator genes, deletion of genes encoding repressors of the BGC, substitution or modification of native regulatory elements (e.g., promoters), and/or expression of the BGC in optimized (e.g., genome-minimized, precursor-optimized), native or heterologous hosts. Interestingly, a variety of technologies and protocols for engineering of actinomycetes genomes have been established which will contribute to manipulate actinomycetes genetic materials and their novel natural products, and to access silent biosynthetic pathway (Pham et al. 2019) as listed in Table 3.1.

3.3.1 Whole Genomic Sequencing

Whole genome sequencing (WGS) is a powerful modern tool for the discovery of natural products. Complete genome sequences of several actinobacteria have demonstrated that they also contain many cryptic BGCs that probably encode novel metabolites. Biosynthetic gene clusters (BGCs) are operons of genes, which their expression led to proteins production. These proteins are responsible for the metabolism (including production, transport, resistance, and regulation of expression) of a natural product. It was demonstrated that most potent natural products belong to the chemical families of Type I polyketides and non-ribosomal peptides. The backbone of these compounds is synthesized by large enzymes, polyketide synthases (PKS),

Table 3.1 Selected examples of bioactive molecules identified by new genomic approaches besides traditional techniques

Bioactive compound	Activity	Source	Approach	Reference
Quinomycin G	Antibacterial and anti-tumor	*Streptomyces* sp.	Metagenomics	Andrade et al. (2000)
Isopimara-2-one-3-ol-8,15-diene	Cytotoxic	*Micromonospora* sp.	Metagenomics	Gomez-Escribano and Bibb (2014)
2-methyl butyl propyl phthalate	Antibacterial, antifungal, cytotoxic	*Streptomyces cheonanensis*	Metagenomics to discover marine isolate	Zhou et al. (2015)
Angucyclinone	Anticancer	*Streptomyces* sp.	Metagenomics to discover marine isolate	Chao et al. (2015))
N-(4-minocyclooctyl)-3,5-dinitrobenzamide	Antibacterial, cytotoxic	*Pseudonocardia endophytica*	Metagenomics	Zhen et al. (2015)
Salinipostins	Antimalarial	*Salinispora* sp.	Metagenomics	Schulze et al. (2015)
Antimycin	Cytotoxic	*Streptomyces* sp.	Metagenomics	Yang et al. (2017)
Formicamycins	Antibacterial	*S. formicae*	CRISPR/Cas9	Qin et al. (2017)
Nesterenkoniane	Antiallergic	*Nesterenkonia flava*	Metagenomics	Sarmiento-Vizcaíno et al. (2017)
New spiroindimicins E and F and lagunapyrones D and E	Antimicrobial	*Streptomyces* sp. MP131–18	antiSMASH, spectroscopic method (MS and NMR)	Paulus et al. (2017)
Alpiniamide A and its new derivatives B-D	Antibacterial	*Streptomyces* sp. IB2014/011–12	Genome mining and NMR	Paulus et al. (2018)
Antibiotic clusters: Ectoine, paenibactin, albachelin, erythrochellin, and labrinthopeptin	Antimicrobial	*Streptomyces* sp. Ru87	Bioassay and Illumina sequencing	Amin et al. (2019)
Antibiotic clusters: Ioxanthin, SapB, desferrioxamine B, methoxyhydroquinones, and tetrocarcin	Antimicrobial	*Micromonospora* sp. Rc5	Bioassay and Illumina sequencing	Amin et al. (2019)
Gacamide A	Antibacterial	*Pseudomonas fluorescens* Pf0–1	Genome mining	Jahanshah et al. (2019)

(continued)

Table 3.1 (continued)

Bioactive compound	Activity	Source	Approach	Reference
P_24306 (C10H13N2) and N_12799 (C18H32O3)	Antimicrobial	*Streptomyces* sp.	Bioassay and metabolomics approach	Sebak et al. (2019)
Scleric acid	Antimicrobial and anticancer	*Streptomyces sclerotialus*	CRISPR/Cas9	Alberti et al. (2019)
Empedopeptins	Antibacterial	*Massilia* sp. YMA4	Genome mining	Ho et al. (2021)
Nosiheptide and Siomycin	Antibacterial	*Streptomyces. actuosus* and *S. sioyaensis*	PacBio RSII	Majer et al. (2021)

and non-ribosomal peptide synthetases (NRPS)(Gomez-Escribano et al. 2016). Polyketide synthases (PKS) and non-ribosomal peptide synthetases (NRPS) have modular and repetitive sequences, which allow to mine genomes for their BGCs (Beld et al. 2014; Walsh 2004; Williams 2013). Consequently, the development of a range of experimental methodologies for identifying the metabolic products of these clusters was applied.

The development of next-generation sequencing (NGS) technologies has greatly accelerated the rate and reduced the cost of genomic data acquisition compared to Sanger sequencing technique (Metzker 2010; Nett et al. 2009; Zerikly and Challis 2009).Various platform families were developed as NGS tools such as Illumina, PacBio, SOLiD, and Ion Torrent which provide different read length and based on different sequencing techniques (Hodkinson and Grice 2015). According to the RefSeq database, a total of 1749 *Streptomyces* genomes had been deposited and more than 73% of the genomes were sequenced by NGS techniques, such as Illumina, PacBio, 454, and MinION (Lee et al. 2020a, b). Illumina technology was first introduced in 2006 which allow researchers to sequence DNA and RNA generating large amount of data. Despite the great output of this technology and economic benefits, the short reads produced by Illumina platform remain one of major challenges. On the other hand, PacBio provides long reads which enable the assembly of high-quality genomes with few contigs and bypasses the inaccuracies generated by the high G + C content of *Streptomyces*' genomes (Donkor 2013; Rhoads and Au 2015).

Next-generation sequencing technologies (NGS) provide huge sequencing data of bacterial genomes, but this raw sequencing data need more analysis using variable bioinformatics tools. Advances in computational technology and computing algorisms facilitate genome assembly and annotation. Comprehensive bioinformatic tools such as the web tool antiSMASH 3.0, the BGC repository, PRISM, ClustScan, CLUSEAN, BAGEL, and NP searcher enable detection and annotation of secondary metabolite biosynthetic gene clusters which resulting unprecedented rate of new natural product biosynthetic pathway discovery (Blin et al. 2019; de Jong et al. 2006; Li et al. 2009; Medema et al. 2015; Starcevic et al. 2008; Weber et al. 2009; Tilmann

Weber et al. 2015). A study in 2021 revealed secondary metabolomes with antibiotic potential using WGS of *Streptomyces actuosus* ISP-5337, *Streptomyces sioyaensis* B-5408, and *Actinospica acidiphila* B-2296 (Majer et al. 2021). In another study, whole genome sequencing for 30 isolates of streptomycetes using both present WGS using both Illumina and PacBio were applied. Based on assembly and annotation, the study predicted a total of 922 secondary metabolites biosynthetic gene clusters (smBGC) which may be resource for new secondary metabolites (Lee et al. 2020a, 2020b).

3.3.2 Metagenomics Technique

Metagenomics is a culture-independent technique which based on extracting DNA from environmental sample and initially involved the cloning of either total or enriched DNA (eDNA) directly from the environment into a host that can be easily cultivated (Handelsman 2004). Then, sequencing and analysis of total metagenomic DNA are carried out which enable the study of the identity of species, the metabolic activities, functional roles of the microbes present in a given population and characterization the microbial life in the community (Langille et al. 2013).

The heterologous expression of natural product gene clusters captured on individual clones or on small numbers of overlapping clones should provide a means of obtaining previously unidentified bioactive small molecules (Feng et al. 2011). Accordingly, metagenomics enable the discovery of novel natural products by increasing the recovery rates up to 40% compared to traditional methods (Jakubiec-Krzesniak et al. 2018; Mahapatra et al. 2020).

Two types of analysis have been applied using metagenomic libraries: a functional-based approach and sequence-based approach (Fig. 3.1). Functional-based approach depends on identification of clones that express desired trait then characterization of active clones by sequencing and biochemical analysis. This approach is highly recommended for identification of clones with high potential applications (enzymes, antibacterial, antifungal, antitumors, natural products). Thus, functional-based metagenomics is applied to investigate novel gene sequence with desired function. On the other hand, sequence-based approach relies on designing PCR primers or hybridization probes by using conserved DNA sequences to screen target sequence through metagenomic libraries (Schloss and Handelsman 2003). In most metagenomics studies, *Escherichia coli* has been used as the cloning host. The vector is selected according to the size of DNA fragment that needs to be inserted. For small fragments, plasmids <15 kb, for larger fragments cosmids (15–40 kb), fosmids (25–45 kb), and/or bacterial artificial chromosomes (BACs) (100–200 kb) have been successfully used (Angelov et al. 2009; Kakirde et al. 2011; Uchiyama and Miyazaki 2009; van Elsas et al. 2008). Despite metagenomics approach had a great impact on the discovery of compounds like polyketide synthase (PKS),

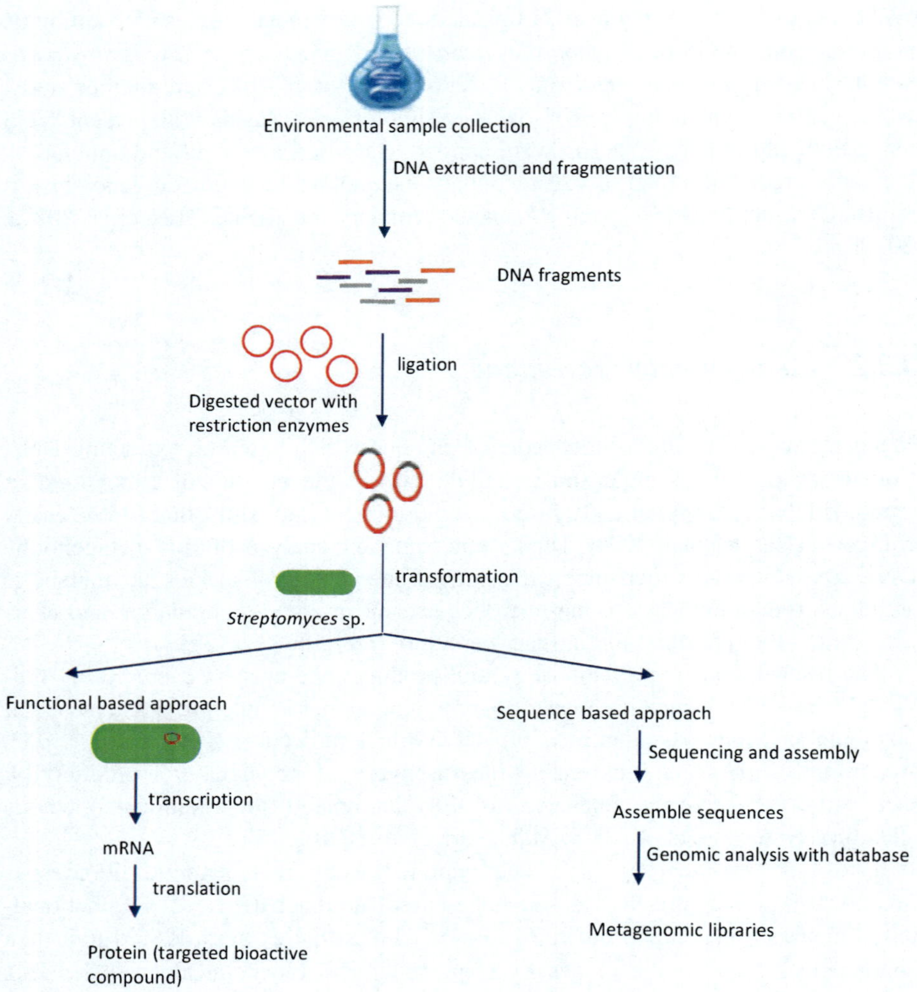

Fig. 3.1 Flowchart of functional and sequence-based metagenomic approaches

Non-ribosomal peptide synthetase (NRPS), antibiotics, and biocatalyst, many obstacles present as poor resolution, misclassification bias and false functional annotation. Others may occur in forms of lack of proper taxonomic context, sequencing errors, lack of efficient database, DNA extract quality, and recovery competence. It may further be complicated when DNA is extracted from extremophiles due to difficulties associated with their cell lyses. Cloning and heterologous expression of metagenomic genes in hosts' cells like *E. coli*, *Pseudomonas putida*, *Bacillus subtilis*, *Streptomyces* sp., and other well-described model vectors may sometimes deviate from its actual design product (Mahapatra et al. 2020; Amin et al. 2019).

3.3.3 CRISPR/Cas 9 for Discovery New Natural Products

In the last decade, editing of genomes has been occurred using traditional methods of genetic engineering which consumed time and need to be validated, such as homologous recombination. Recently, editing of target genome has been demonstrated by the discovery of clustered regularly interspaced short palindromic repeats (CRISPR)/CRISPR-associated (Cas) systems, which form the basis for the CRISPR/Cas9-based genome editing (Alberti and Corre 2019).

The modules of Clustered Regularly Interspaced Short Palindromic Repeats (CRISPR)/CRISPR-associated (Cas) proteins are present in most archaea and many bacteria as adaptive immune systems for defense against foreign and specific DNA or RNA. Briefly, CRISPR-mediated immunization occurs through the uptake of DNA from invasive genetic elements such as plasmids and viruses which called "spacer," followed by its integration into CRISPR loci. These loci are subsequently transcribed and processed into small interfering RNAs that guide nucleases for specific cleavage of complementary sequences (Abudayyeh et al. 2016; Barrangou and Marraffini 2014; Deveau et al. 2010; Koonin and Makarova 2013). Researchers observed high specificity and accuracy of CRISPR/Cas system which make CRISPR/Cas system is an excellent tool for genome editing (Jinek et al. 2012).

CRISPR-Cas systems mainly composed of a CRISPR RNA (crRNA) and Cas proteins. The role of crRNA is to target complementary sequence and thus guides the Cas proteins for the sequence-specific recognition and cleavage. Many CRISPR-Cas technologies have been applied which improve the genome engineering researches. In addition, engineering the Cas proteins to nuclease-deficient Cas (dCas) helps CRISPR-Cas-based systems to be fast, efficient, and multi-target transcriptional repression and activation which enable the control of expression of any targeted genes without genomic sequence manipulation. Fusion of deaminases to dCas, CRISPR-Cas systems can be adapted to allow base editing on DNA and RNA, without requirement of DNA cleavage or any donor templates. Additionally, based on the collateral effect of Cas proteins, CRISPR-Cas systems have been used to detect specific nucleic acids till 10^{-18} of mole (Li et al. 2018).

This technique allows fast and easy genetic manipulation which used by researchers in many applications, including point mutations, deleting of single gene, or two genes or gene clusters in *Streptomyces* spp. (Cobb et al. 2015; Huang et al. 2015; Tong et al. 2015; Zeng et al. 2015). Three different groups successively used CRISPR/Cas9 system for editing strategy in the genome of *Streptomyces coelicolor* M145. The first group succeeded to delete single gene/BGC in *S. coelicolor* while other group used CRISPR/Cas9 system to delete multiplex genes/BGCs. Moreover, the third group cleaved chromosomal DNA at specific site to change the rpsL nucleotides sequence (262–264) from AAG to GAA using CRISPR/Cas9 system. Consequently, glutamic acid (Glu) was produced instead of lysine amino acid (Lys 88) in rpsL (Tao et al. 2018). Other study chose the genes zwf2 and devB encoded oxytetracycline antibiotic produced by *Streptomyces rimosus* and were edited separately single-site mutations, double-site mutations and gene fragment disruptions. The study revealed that a mutant (zwf2-devB-) increase the productivity of oxytetracycline by 36.8% compared to the original strain. These results confirm that CRISPR/Cas9 can successfully serve as a useful targeted genome editing system in *S. rimosus* (Jia et al. 2017).

3.4 Conclusion

Actinobacteria are the most prolific source for novel bioactive compounds. In the past, research studies focus on studying actinobacteria habitats, taxonomy, identification, secondary metabolites, and their applications. The incidence of multidrug-resistant bacteria, resistant fungi, resistant parasites, and severe illness which is accompanied by a drawback in the discovery of novel bioactive compounds represents a serious threat. Consequently, more studies which aim to innovate different, easy, cheap, and efficient techniques to discover novel bioactive compounds are done. Many approaches, traditional or genome based, have the same target which is screening for novel natural products with promising bioactivities and known stable pure chemical structure have been developed. Each approach has its own advantages and limitations. To obtain natural products, it is necessary to use both traditional and genome-guided approaches. Despite the high impact of genomics, all sequence-based developments need to be combined with microbiology and analytical chemistry for identification, isolation, purification, and large-scale production of novel compound.

References

Abdel-Razek AS, El-Naggar ME, Allam A, Morsy OM, Othman SI (2020) Microbial natural products in drug discovery. PRO 8(4):1–19. https://doi.org/10.3390/PR8040470

Abudayyeh OO, Gootenberg JS, Konermann S, Joung J, Slaymaker IM, Cox DBT, Shmakov S, Makarova KS, Semenova E, Minakhin L, Severinov K, Regev A, Lander ES, Koonin EV, Zhang F (2016) C2c2 is a single-component programmable RNA-guided RNA-targeting CRISPR effector. Science 353(6299). https://doi.org/10.1126/science.aaf5573

Albarano L, Esposito R, Ruocco N, Costantini M (2020) Genome mining as new challenge in natural products discovery. Mar Drugs 18(4):1–17. https://doi.org/10.3390/md18040199

Alberti F, Corre C (2019) Editing streptomycete genomes in the CRISPR/Cas9 age. Nat Prod Rep 36(9):1237–1248. https://doi.org/10.1039/c8np00081f

Alberti F, Leng DJ, Wilkening I, Song L, Tosin M, Corre C (2019) Triggering the expression of a silent gene cluster from genetically intractable bacteria results in Scleric acid discovery. Chem Sci 10(2):453–463. https://doi.org/10.1039/c8sc03814g

Amin DH, Abolmaaty A, Borsetto C, Tolba S, Abdallah NA, Wellington EMH (2019) In silico genomic mining reveals unexplored bioactive potential of rare actinobacteria isolated from Egyptian soil. Bull Natl Res Cent 43(1):78. https://doi.org/10.1186/s42269-019-0121-y

Andrade MA, Ponting CP, Gibson TJ, Bork P (2000) Homology-based method for identification of protein repeats using statistical significance estimates. J Mol Biol 298(3):521–537. https://doi.org/10.1006/jmbi.2000.3684

Angelov A, Mientus M, Liebl S, Liebl W (2009) A two-host fosmid system for functional screening of (meta)genomic libraries from extreme thermophiles. Syst Appl Microbiol 32(3):177–185. https://doi.org/10.1016/j.syapm.2008.01.003

Arango C, Acosta-Gonzalez A, Parra-Giraldo CM, Sánchez-Quitian ZA, Kerr R, Díaz LE (2018) Characterization of actinobacterial communities from Arauca River sediments (Colombia) reveals antimicrobial potential presented in low abundant isolates. The Open Microbiology Journal 12:181–194. https://doi.org/10.2174/1874285801812010181

Baltz R (2005) Antibiotic discovery from actinomycetes: will a renaissance follow the decline and fall? SIM News 55:186–196

Baltz RH (2006) Marcel Faber roundtable: is our antibiotic pipeline unproductive because of starvation, constipation or lack of inspiration? J Ind Microbiol Biotechnol 33(7):507–513. https://doi.org/10.1007/s10295-005-0077-9

Barrangou R, Marraffini LA (2014) CRISPR-Cas systems: prokaryotes upgrade to adaptive immunity. Mol Cell 54(2):234–244. https://doi.org/10.1016/j.molcel.2014.03.011

Beld J, Sonnenschein EC, Vickery CR, Noel JP, Burkart MD (2014) The phosphopantetheinyl transferases: catalysis of a post-translational modification crucial for life. Nat Prod Rep 31(1):61–108. https://doi.org/10.1039/c3np70054b

Bentley SD, Chater KF, Cerdeño-Tárraga AM, Challis GL, Thomson NR, James KD, Harris DE, Quail MA, Kieser H, Harper D, Bateman A, Brown S, Chandra G, Chen CW, Collins M, Cronin A, Fraser A, Goble A, Hidalgo J, Hopwood DA (2002) Complete genome sequence of the model actinomycete *Streptomyces coelicolor* A3(2). Nature 417(6885):141–147. https://doi.org/10.1038/417141a

Bérdy J (2005) Bioactive microbial metabolites. J Antibiot 58(1):1–26. https://doi.org/10.1038/ja.2005.1

Blin K, Shaw S, Steinke K, Villebro R, Ziemert N, Lee SY, Medema MH, Weber T (2019) antiSMASH 5.0: updates to the secondary metabolite genome mining pipeline. Nucleic Acids Res 47(W1):W81–W87. https://doi.org/10.1093/nar/gkz310

Butler MS, Blaskovich MA, Cooper MA (2017) Antibiotics in the clinical pipeline at the end of 2015. J Antibiot 70(1):3–24. https://doi.org/10.1038/ja.2016.72

Carlson S, Marler L, Nam S-J, Santarsiero BD, Pezzuto JM, Murphy BT (2013) Potential chemopreventive activity of a new macrolide antibiotic from a marine-derived Micromonospora sp. Mar Drugs 11(4):1152–1161. https://doi.org/10.3390/md11041152

Chao R, Yuan Y, Zhao H (2015) Recent advances in DNA assembly technologies. FEMS Yeast Res 15(1):1–9. https://doi.org/10.1111/1567-1364.12171

Chitte RR, Dey S (2002) Production of a fibrinolytic enzyme by thermophilic *Streptomyces* species. World J Microbiol Biotechnol 18(4):289. https://doi.org/10.1023/A:1015252607118

Cobb RE, Wang Y, Zhao H (2015) High-efficiency multiplex genome editing of *Streptomyces* species using an engineered CRISPR/Cas system. ACS Synth Biol 4(6):723–728. https://doi.org/10.1021/sb500351f

Craney A, Ahmed S, Nodwell J (2013) Towards a new science of secondary metabolism. J Antibiot 66(7):387–400. https://doi.org/10.1038/ja.2013.25

Dashti Y, Grkovic T, Abdelmohsen UR, Hentschel U, Quinn RJ (2014) Production of induced secondary metabolites by a co-culture of sponge-associated actinomycetes, *Actinokineospora* sp. EG49 and *Nocardiopsis* sp. RV163. Mar Drugs 12(5):3046–3059. https://doi.org/10.3390/md12053046

de Jong A, van Hijum SAFT, Bijlsma JJE, Kok J, Kuipers OP (2006) BAGEL: a web-based bacteriocin genome mining tool. Nucleic Acids Res 34:W273–W279. https://doi.org/10.1093/nar/gkl237

Deveau H, Garneau JE, Moineau S (2010) CRISPR/Cas system and its role in phage-bacteria interactions. Annu Rev Microbiol 64:475–493. https://doi.org/10.1146/annurev.micro.112408.134123

Donkor ES (2013) Sequencing of bacterial genomes: principles and insights into pathogenesis and development of antibiotics. Genes 4(4):556–572. https://doi.org/10.3390/genes4040556

Edelvio de BG, Leo RLD, & Rita de CM de M (2018) Actinomycetes bioactive compounds: Biological control of fungi and phytopathogenic insect. Afr J Biotechnol 17(17): 552–559. https://doi.org/10.5897/ajb2017.16323

Esquenazi E, Yang YL, Watrous J, Gerwick WH, Dorrestein PC (2009) Imaging mass spectrometry of natural products. Nat Prod Rep 26(12):1521–1534. https://doi.org/10.1039/b915674g

Feng Z, Kallifidas D, Brady SF (2011) Functional analysis of environmental DNA-derived type II polyketide synthases reveals structurally diverse secondary metabolites. PNAS 108(31):12629–12634. https://doi.org/10.1073/pnas.1103921108

Gan M, Liu B, Tan Y, Wang Q, Zhou H, He H, Ping Y, Yang Z, Wang Y, Xiao C (2015) Saccharothrixones A-D, tetracenomycin-type polyketides from the marine-derived actinomycete Saccharothrix sp. 10-10. J Nat Prod 78(9):2260–2265. https://doi.org/10.1021/acs.jnatprod.5b00577

Gomez-Escribano JP, Bibb MJ (2014) Heterologous expression of natural product biosynthetic gene clusters in *Streptomyces coelicolor*: from genome mining to manipulation of biosynthetic pathways. J Ind Microbiol Biotechnol 41(2):425–431. https://doi.org/10.1007/s10295-013-1348-5

Gomez-Escribano JP, Alt S, Bibb MJ (2016) Next generation sequencing of Actinobacteria for the discovery of novel natural products. Mar Drugs 14(4):78. https://doi.org/10.3390/md14040078

Guo H, Benndorf R, Leichnitz D, Klassen JL, Vollmers J, Görls H, Steinacker M, Weigel C, Dahse H-M, Kaster A-K, de Beer ZW, Poulsen M, Beemelmanns C (2017) Isolation, biosynthesis and chemical modifications of Rubterolones A-F: rare Tropolone alkaloids from *Actinomadura* sp. 5-2. Chemistry 23(39):9338–9345. https://doi.org/10.1002/chem.201701005

Hodkinson BP, Grice EA (2015) Next-generation sequencing: a review of technologies and tools for wound microbiome research. Adv Wound Care (New Rochelle) 4(1):50–58. https://doi.org/10.1089/wound.2014.0542.PMID:25566414;PMCID:PMC4281878

Handelsman J (2004) Metagenomics : application of genomics to uncultured microorganisms. Microbiol Mol Biol Rev 68(4):669–685. https://doi.org/10.1128/MBR.68.4.669

Henke MT, Kelleher NL (2016) Modern mass spectrometry for synthetic biology and structure-based discovery of natural products. Nat Prod Rep 33(8):942–950. https://doi.org/10.1039/c6np00024j.Modern

Ho S-T, Ho Y-N, Lin C, Hsu W-C, Lee H-J, Peng C-C, Cheng H-T, Yang Y-L (2021) Integrated omics strategy reveals cyclic Lipopeptides Empedopeptins from *Massilia* sp. YMA4 and their biosynthetic pathway. Mar Drugs 19(4). https://doi.org/10.3390/md19040209

Huang H, Zheng G, Jiang W, Hu H, Lu Y (2015) One-step high-efficiency CRISPR/Cas9-mediated genome editing in *Streptomyces*. Acta Biochimica et Biophysica Sinica 47(4):231–243. https://doi.org/10.1093/abbs/gmv007

Jahanshah G, Yan Q, Gerhardt H, Pataj Z, Lämmerhofer M, Pianet I, Josten M, Sahl H-G, Silby MW, Loper JE, Gross H (2019) Discovery of the cyclic lipopeptide gacamide a by genome mining and repair of the defective GacA regulator in *Pseudomonas fluorescens* Pf0–1. J Nat Prod 82(2):301–308. https://doi.org/10.1021/acs.jnatprod.8b00747

Jakubiec-Krzesniak K, Ranisz-Mateustiak A, Guspiel A, Ziemska J, Solecka J (2018) Secondary metabolites of actinomycetes and their antibacterial, antifungal and antiviral properties. Pol J Microbiol 67(2544–4646):259–272. https://doi.org/10.21307/pjm-2018-048

Jia H, Zhang L, Wang T, Han J, Tang H, Zhang L (2017) Development of a CRISPR/Cas9-mediated gene-editing tool in *Streptomyces rimosus*. Microbiol 163(8):1148–1155. https://doi.org/10.1099/mic.0.000501

Jinek M, Chylinski K, Fonfara I, Hauer M, Doudna JA, Charpentier E (2012) A programmable dual-RNA-guided DNA endonuclease in adaptive bacterial immunity. Science 337(6096):816–821. https://doi.org/10.1126/science.1225829

Kakirde KS, Wild J, Godiska R, Mead DA, Wiggins AG, Goodman RM, Szybalski W, Liles MR (2011) Gram negative shuttle BAC vector for heterologous expression of metagenomic libraries. Gene 475(2):57–62. https://doi.org/10.1016/j.gene.2010.11.004

Kämpfer P (2006) The family Streptomycetaceae, Part I: taxonomy. In: The prokaryotes. Springer, New York, NY, pp 538–604

Koonin EV, Makarova KS (2013) CRISPR-Cas: evolution of an RNA-based adaptive immunity system in prokaryotes. RNA Biol 10(5):679–686. https://doi.org/10.4161/rna.24022

Kurapova AI, Zenova GM, Sudnitsyn II, Kizilova AK, Manucharova NA, Norovsuren Z, Zvyagintsev DG (2012) Thermotolerant and thermophilic actinomycetes from soils of Mongolia desert steppe zone. Microbiology 81(1):98–108. https://doi.org/10.1134/S0026261712010092

Langille MGI, Zaneveld J, Caporaso JG, McDonald D, Knights D, Reyes JA, Clemente JC, Burkepile DE, Vega Thurber RL, Knight R, Beiko RG, Huttenhower C (2013) Predictive functional profiling of microbial communities using 16S rRNA marker gene sequences. Nat Biotechnol 31(9):814–821. https://doi.org/10.1038/nbt.2676

Lazzarini A, Cavaletti L, Toppo G, Marinelli F (2001) Rare genera of actinomycetes as potential producers of new antibiotics. Antonie Van Leeuwenhoek 79(3–4):399–405

Lee JA, Uhlik MT, Moxham CM, Tomandl D, Sall DJ (2012) Modern phenotypic drug discovery is a viable, neoclassic pharma strategy. J Med Chem 55(10):4527–4538. https://doi.org/10.1021/jm201649s

Lee N, Hwang S, Kim J, Cho S, Palsson B, Cho B-K (2020a) Mini review: genome mining approaches for the identification of secondary metabolite biosynthetic gene clusters in *Streptomyces*. Comput Struct Biotech 18:1548–1556. https://doi.org/10.1016/j.csbj.2020.06.024

Lee N, Kim W, Hwang S, Lee Y, Cho S, Palsson B, Cho BK (2020b) Thirty complete Streptomyces genome sequences for mining novel secondary metabolite biosynthetic gene clusters. Sci Data 7(1):1–9. https://doi.org/10.1038/s41597-020-0395-9

Li JW, Vederas JC (2009) Drug discovery and natural products. Science 10 325(5935):161–165. https://doi.org/10.1126/science.1168243

Li MHT, Ung PMU, Zajkowski J, Garneau-Tsodikova S, Sherman DH (2009) Automated genome mining for natural products. BMC Bioinformatics 10:185. https://doi.org/10.1186/1471-2105-10-185

Li SY, Cheng QX, Liu JK, Nie XQ, Zhao GP, Wang J (2018) CRISPR-Cas12a has both cis- and trans-cleavage activities on single-stranded DNA. Cell Res 28(4):491–493. https://doi.org/10.1038/s41422-018-0022-x

Liu R, Li X, Lam KS (2017) Combinatorial chemistry in drug discovery. Curr Opin Chem Biol 38: 117–126. https://doi.org/10.1016/j.cbpa.2017.03.017

Mahapatra GP, Raman S, Nayak S, Gouda S, Das G, Patra JK (2020) Metagenomics approaches in discovery and development of new bioactive compounds from marine actinomycetes. Curr Microbiol 77(4):645–656. https://doi.org/10.1007/s00284-019-01698-5

Majer HM, Ehrlich RL, Ahmed A, Earl JP, Ehrlich GD, Beld J (2021) Whole genome sequencing of *Streptomyces actuosus* ISP-5337, *Streptomyces sioyaensis* B-5408, and *Actinospica acidiphila* B-2296 reveals secondary metabolomes with antibiotic potential. Biotech Rep (AMS) 29: e00596. https://doi.org/10.1016/j.btre.2021.e00596

Medema MH, Kottmann R, Yilmaz P, Cummings M, Biggins JB, Blin K, De Bruijn I, Chooi YH, Claesen J, Coates RC, Cruz-Morales P, Duddela S, Düsterhus S, Edwards DJ, Fewer DP, Garg N, Geiger C, Gomez-Escribano JP, Greule A, Glöckner FO (2015) Minimum information about a biosynthetic gene cluster. Nat Chem Biol 11(9):625–631. https://doi.org/10.1038/nchembio.1890

Metzker ML (2010) Sequencing technologies the next generation. Nat Rev Genet 11(1):31–46. https://doi.org/10.1038/nrg2626

Moffat JG, Vincent F, Lee JA, Eder J, Prunotto M (2017) Opportunities and challenges in phenotypic drug discovery: an industry perspective. Nat Rev Drug Discov 16(8):531–543. https://doi.org/10.1038/nrd.2017.111

Mulani MS, Kamble EE, Kumkar SN, Tawre MS, Pardesi KR (2019) Emerging strategies to combat ESKAPE pathogens in the era of antimicrobial resistance: a review. Front Microbiol 10:539. https://doi.org/10.3389/fmicb.2019.00539

Nett M, Ikeda H, Moore BS (2009) Genomic basis for natural product biosynthetic diversity in the actinomycetes. Nat Prod Rep 26(11):1362–1384. https://doi.org/10.1039/b817069j

Ochi K (2017) Insights into microbial cryptic gene activation and strain improvement: principle, application and technical aspects. J Antibiot 70(1):25–40. https://doi.org/10.1038/ja.2016.82

Onaka H (2017) Novel antibiotic screening methods to awaken silent or cryptic secondary metabolic pathways in actinomycetes. J Antibiot 70(8):865–870. https://doi.org/10.1038/ja.2017.51

Padma PN, Rao AB, Yadav JS, Reddy G (2002) Optimization of fermentation conditions for production of glycopeptide antibiotic vancomycin by amycolatopsis orientalis. Appl Biochem Biotechnol 102–103(1):395–405. https://doi.org/10.1385/ABAB:102-103:1-6:395

Paulus C, Rebets Y, Tokovenko B, Nadmid S, Terekhova LP, Myronovskyi M, Zotchev SB, Rückert C, Braig S, Zahler S, Kalinowski J, Luzhetskyy A (2017) New natural products identified by combined genomics-metabolomics profiling of marine *Streptomyces* sp. MP131-18. Sci Rep 7:42382. https://doi.org/10.1038/srep42382

Paulus C, Rebets Y, Zapp J, Rückert C, Kalinowski J, Luzhetskyy A (2018) New Alpiniamides from *Streptomycetes* sp. 11-12 assembled by an unusual hybrid-nonribosomal peptide synthetase-transAT-polyketidesynthase enzyme. Front Microbiol 9:1959. https://doi.org/10.3389/fmicb.2018.01959. eCollection 2018

Pham JV, Yilma MA, Feliz A, Majid MT, Maffetone N, Walker JR, Kim E, Cho HJ, Reynolds JM, Song MC, Park SR, Yoon YJ (2019) A review of the microbial production of bioactive natural products and biologics. Front Microbiol 10:1–27. https://doi.org/10.3389/fmicb.2019.01404

Qin S, Li W-J, Dastager SG, Hozzein WN (2016) Editorial: actinobacteria in special and extreme habitats: diversity, function roles, and environmental adaptations. Front Microbiol 7:1415. https://doi.org/10.3389/fmicb.2016.01415

Qin Z, Munnoch JT, Devine R, Holmes NA, Seipke RF, Wilkinson KA, Wilkinson B, Hutchings MI (2017) Formicamycins, antibacterial polyketides produced by *Streptomyces formicae* isolated from African Tetraponera plant-ants. Chem Sci 8(4):3218–3227. https://doi.org/10.1039/c6sc04265a

Rhoads A, Au KF (2015) PacBio sequencing and its applications. Genom Proteom Bioinfo13(5): 278–289. https://doi.org/10.1016/j.gpb.2015.08.002

Salwan R, Sharma V (2020) Molecular and biotechnological aspects of secondary metabolites in actinobacteria. Microbiol Res 231:126374. https://doi.org/10.1016/j.micres.2019.126374

Sarmiento-Vizcaíno A, Braña AF, Pérez-Victoria I, Martín J, de Pedro N, Cruz M, de la Díaz C, Vicente F, Acuña JL, Reyes F, García LA, Blanco G (2017) Paulomycin G, a new natural product with cytotoxic activity against tumor cell lines produced by Deep-Sea sediment derived micromonospora matsumotoense M-412 from the Avilés canyon in the Cantabrian Sea. Mar Drugs 15(9):271. https://doi.org/10.3390/md15090271

Scherlach K, Hertweck C (2009) Triggering cryptic natural product biosynthesis in microorganisms. Org Biomol Chem 7(9):1753–1760. https://doi.org/10.1039/b821578b

Schloss PD, Handelsman J (2003) Biotechnological prospects from metagenomics. Curr Opin Biotechnol 14(3):303–310. https://doi.org/10.1016/s0958-1669(03)00067-3

Schultz AW, Oh D-C, Carney JR, Williamson RT, Udwary DW, Jensen PR, Gould SJ, Fenical W, Moore BS (2008) Biosynthesis and structures of cyclomarins and cyclomarazines, prenylated cyclic peptides of marine actinobacterial origin. J Am Chem Soc 130(13):4507–4516. https://doi.org/10.1021/ja711188x

Schulze CJ, Navarro G, Ebert D, DeRisi J, Linington RG (2015) Salinipostins A-K, long-chain bicyclic phosphotriesters as a potent and selective antimalarial chemotype. J Org Chem 80(3):1312–1320. https://doi.org/10.1021/jo5024409

Sebak M, Saafan AE, AbdelGhani S, Bakeer W, El-Gendy AO, Espriu LC, Duncan K, Edrada-Ebel R (2019) Bioassay- and metabolomics-guided screening of bioactive soil actinomycetes from the ancient city of Ihnasia, Egypt. PLoS One 14(12):1–29. https://doi.org/10.1371/journal.pone.0226959

Shivlata L, Tulasi S (2015) Thermophilic and alkaliphilic actinobacteria: biology and potential applications. Front Microbiol 6:1014. https://doi.org/10.3389/fmicb.2015.01014

Sivalingam P, Hong K, Pote J, Prabakar K (2019) Extreme environment *Streptomyces*: potential sources for new antibacterial and anticancer drug leads? Int J Microbiol 2019. https://doi.org/10.1155/2019/5283948

Solanki R, Kapur MK, Lal R (2008) Bioactive compounds from marine actinomycetes. Indian J Microbiol 48:410–431. https://doi.org/10.1007/s12088-008-0052-z

Starcevic A, Zucko J, Simunkovic J, Long PF, Cullum J, Hranueli D (2008) ClustScan: an integrated program package for the semi-automatic annotation of modular biosynthetic gene clusters and in silico prediction of novel chemical structures. Nucleic Acids Res 36(21):6882–6892. https://doi.org/10.1093/nar/gkn685

Subramani R, Aalbersberg W (2013) Culturable rare actinomycetes: diversity, isolation and marine natural product discovery. App Microbiol Biotechnol 97(21):9291–9321. https://doi.org/10.1007/s00253-013-5229-7

Subramani R, Sipkema D (2019) Marine rare actinomycetes: a promising source of structurally diverse and unique novel natural products. Mar Drugs 17(5):249. https://doi.org/10.3390/md17050249

Tan ZQ, Leow HY, Lee DCW, Karisnan K, Song AAL, Mai CW, Yap WS, Lim SHE, Lai KS (2019) Co-culture systems for the production of secondary metabolites: current and future prospects. Open Biotechnol J 13(1):18–26. https://doi.org/10.2174/1874070701913010018

Tao W, Yang A, Deng Z, Sun Y (2018) CRISPR/Cas9-based editing of *Streptomyces* for discovery, characterization, and production of natural products. Front Microbiol 9:1660. https://doi.org/10.3389/fmicb.2018.01660

Tishkov S (2001) Bioactive products from actinomycetes–antibiotics, enzyme inhibitors, immunomodulators. In: Moncheva P, Tishkov S, Chipeva V (eds) Innovative aspects in biotechnology of prokaryotes, vol 111. National Bank for Industrial Microorganisms and Cell Cultures, Sofia, pp 111–138

Tiwari K, Gupta RK (2012) Rare actinomycetes: a potential storehouse for novel antibiotics. Crit Rev Biotechnol 32(2):108–132. https://doi.org/10.3109/07388551.2011.562482

Tiwari K, Gupta RK (2014) Bioactive metabolites from rare actinomycetes. In: Studies in natural products chemistry, vol 41, 1st edn. Elsevier B.V. https://doi.org/10.1016/B978-0-444-63294-4.00014-0

Tong Y, Charusanti P, Zhang L, Weber T, Lee SY (2015) CRISPR-Cas9 based engineering of actinomycetal genomes. ACS Synth Biol 4(9):1020–1029. https://doi.org/10.1021/acssynbio.5b00038

Uchiyama T, Miyazaki K (2009) Functional metagenomics for enzyme discovery: challenges to efficient screening. Curr Opin Biotechnol 20(6):616–622. https://doi.org/10.1016/j.copbio.2009.09.010

van Elsas JD, Speksnijder AJ, van Overbeek LS (2008) A procedure for the metagenomics exploration of disease-suppressive soils. J Microbiol Methods 75(3):515–522. https://doi.org/10.1016/j.mimet.2008.08.004

Walsh CT (2004) Polyketide and nonribosomal peptide antibiotics: modularity and versatility. Science 303(5665):1805–1810. https://doi.org/10.1126/science.1094318

Ward AC, Bora N (2006) Diversity and biogeography of marine actinobacteria. Curr Opin Microbiol 9(3):279–286. https://doi.org/10.1016/j.mib.2006.04.004

Weber T, Rausch C, Lopez P, Hoof I, Gaykova V, Huson DH, Wohlleben W (2009) CLUSEAN: a computer-based framework for the automated analysis of bacterial secondary metabolite biosynthetic gene clusters. J Biotechnol 140(1–2):13–17. https://doi.org/10.1016/j.jbiotec.2009.01.007

Weber T, Blin K, Duddela S, Krug D, Kim HU, Bruccoleri R, Lee SY, Fischbach MA, Müller R, Wohlleben W, Breitling R (2015) AntiSMASH 3.0—a comprehensive resource for the genome mining of biosynthetic gene clusters. Nucl Acids Res 43(W1):W237–W243. https://doi.org/10.1093/nar/gkv437

Williams GJ (2013) Engineering polyketide synthases and nonribosomal peptide synthetases. Curr Opin Struct Biol 23(4):603–612. https://doi.org/10.1016/j.sbi.2013.06.012

Wohlleben W, Mast Y, Stegmann E, Ziemert N (2016) Antibiotic drug discovery. Microb Biotechnol 9(5):541–548. https://doi.org/10.1111/1751-7915.12388

Wright GD (2013) Q&A: antibiotic resistance: what more do we know and what more can we do? Is the problem of antibiotic resistance getting worse? BMC Biol 8:123. (2010). https://doi.org/10.1186/1741-7007-8-123

Xuan J, Yu Y, Qing T, Guo L, Shi L (2013) Next-generation sequencing in the clinic: promises and challenges. Cancer Lett 340(2):284–295. https://doi.org/10.1016/j.canlet.2012.11.025

Yang C-L, Wang Y-S, Liu C-L, Zeng Y-J, Cheng P, Jiao R-H, Bao S-X, Huang H-Q, Tan R-X, Ge H-M (2017) Strepchazolins A and B: two new alkaloids from a marine *Streptomyces chartreusis* NA02069. Mar Drugs 15(8). https://doi.org/10.3390/md15080244

Yu M, Li Y, Banakar SP, Liu L, Shao C, Li Z, Wang C (2019) New metabolites from the co-culture of marine-derived Actinomycete *Streptomyces rochei* MB037 and fungus *Rhinocladiella similis* 35. Front Microbiol 10:915. https://doi.org/10.3389/fmicb.2019.00915

Zarins-Tutt JS, Barberi TT, Gao H, Mearns-Spragg A, Zhang L, Newman DJ, RJM G (2016) Prospecting for new bacterial metabolites: a glossary of approaches for inducing, activating and upregulating the biosynthesis of bacterial cryptic or silent natural products. Nat Prod Rep 33(1):54–72. https://doi.org/10.1039/c5np00111k

Zeng H, Wen S, Xu W, He Z, Zhai G, Liu Y, Deng Z, Sun Y (2015) Highly efficient editing of the actinorhodin polyketide chain length factor gene in *Streptomyces coelicolor* M145 using CRISPR/Cas9-CodA(sm) combined system. Appl Microbiol Biotechnol 99(24):10575–10585. https://doi.org/10.1007/s00253-015-6931-4

Zenova GM, Manucharova NA, Zvyagintsev DG (2011) Extremophilic and extremotolerant actinomycetes in different soil types. Eurasian Soil Sci 44(4):417–436. https://doi.org/10.1134/S1064229311040132

Zerikly M, Challis GL (2009) Strategies for the discovery of new natural products by genome mining. Chembiochem Eur J Chem Biol 10(4):625–633. https://doi.org/10.1002/cbic.200800389

Zhang W, Liu Z, Li S, Lu Y, Chen Y, Zhang H, Zhang G, Zhu Y, Zhang G, Zhang W, Liu J, Zhang C (2012) Fluostatins I-K from the South China Sea-derived *Micromonospora rosaria* SCSIO N160. J Nat Prod 75(11):1937–1943. https://doi.org/10.1021/np300505y

Zhen X, Gong T, Liu F, Zhang P-C, Zhou W-Q, Li Y, Zhu P (2015) A new analogue of echinomycin and a new cyclic dipeptide from a marine-derived *Streptomyces* sp. LS298. Mar Drugs 13(11):6947–6961. https://doi.org/10.3390/md13116947

Zhou T, Komaki H, Ichikawa N, Hosoyama A, Sato S, Igarashi Y (2015) Biosynthesis of akaeolide and lorneic acids and annotation of type I polyketide synthase gene clusters in the genome of *Streptomyces* sp. NPS554. Mar Drugs 13(1):581–596. https://doi.org/10.3390/md13010581

Chapter 4
The Relationship between Actinobacteria and Rice

Fengli Zhang and Dabing Zhang

Abstract Rice (*Oryza sativa*) is one of the most important cereal crops, but study on its symbiotic microbiome, particularly actinobacteria functional properties is very limited. Actinobacteria, one of the largest bacterial phyla, which has an extensive secondary metabolism, that can produce about two-thirds of natural clinical medicine, as well as many anticancer, anthelmintic, and antifungal compounds. In this chapter, rice actinobacterial biodiversity, rice endophyte actinobacteria, beneficiations of actinobacteria to rice growth, and rice actinobacterial activities have been summarized. In the future research, the following questions should be considered and explored: (1) How does interaction between actinomycetes and host rice? (2) What are the secondary metabolites produced by actinomycetes and how are secondary metabolite biosynthesized? (3) Is there gene horizontal transfer between actinomycetes and rice?

Keywords Rice endophyte · Actinobacteria · Actinomycetes · Streptomyces · Rice (*Oryza sativa* L.)

4.1 Introduction

As rice (*Oryza sativa* L.), a staple food crop for about a half of the global population, is cultivated under a wide range of ecosystems, it is a very urgent task to improve rice productivity and modern agricultural techniques (Elert 2014). Genetic variation in the 3000 rice genomes (3 K-RG) that focuses on important aspects of *O. sativa* genome sequence diversity, including single nucleotide polymorphisms (SNPs) and structural variation with deletions, duplications, inversions, and translocations has been presented (Wang et al. 2018). Moreover, because rice genome sequence is small and can be used as gene representative for other gramineous species, rice has

F. Zhang (✉) · D. Zhang
School of Life Sciences and Biotechnology, Shanghai Jiao Tong University, Shanghai, China
e-mail: zhangfengli@sjtu.edu.cn

L. Karthik (ed.), *Actinobacteria*, https://doi.org/10.1007/978-981-16-5835-8_4

been considered as a model plant to study on monocotyledonous plants (Phillips et al. 2007). Symbiotic microorganisms from rice root play an important role in rice nutrient acquisition, biomass production, and environmental stress tolerance (Long and Yao 2020). Therefore, microbes, especially endophytic microbes, have potential applications in maintaining plants health and environmental tolerance for agricultural sustainability (Rana et al. 2020). Endophytic microbes belong to different phyla such as Actinobacteria, Bacteroidetes, Deinococcus-Thermus, Firmicutes, Proteobacteria, and Verrucomicrobia (Rana et al. 2020). Actinobacteria constitutes a major phylum and is divided into six classes: *Actinobacteria*, *Acidimicrobia*, *Coriobacteriia*, *Nitriliruptoria*, *Rubrobacteria*, and *Thermoleophilia* (Barka et al. 2016; Gao and Gupta 2012). Actinobacteria are Gram-positive mostly aerobic bacteria. Actinobacteria typically have a high G + C content (>50%) in their genomes and are commonly known for their remarkable capacity to produce bioactive compounds. Two-thirds of the bioactive substances from microorganisms have been produced the phylum-Actinobacteria (Barka et al. 2016; Berdy 2005), including antibiotics, antiviral agents, anticancer agents, immunosuppressive agents, antioxidants, enzymes, plant growth hormones, etc. and that are highly important for applications in medicine, industry, and agriculture (Assad et al. 2021; Strobel and Daisy 2003). Based on molecular and chemical composition criteria, actinomycetes have been divided into 14 suborders: Actinomycineae, Actinopolysporineae, Catenulisporineae, Corynebacterineae, Frankineae, Glycomycineae, Jiangellineae, Kineosporineae, Micrococcineae, Micromonosporineae, Propionibacterineae, Pseudonocardineae, Streptomycineae, and Streptosporangineae (Garrity et al. 2004). To achieve rice yield without the mass use of chemical fertilizers and pesticides, it is very crucial for application of microorganisms. This chapter presents an updated research progress on rice actinobacterial biodiversity, rice actinobacterial endophytes, rice rhizosphere actinomycetes, and function of actinomycetes on rice (symbionts and pathogens in rice-associated microbial communities).

4.2 Biodiversity of Actinobacteria from Jasmine Rice (*O. Sativa*)

Complex microbial communities from rice play important roles in rice growth and health. To pursue distribution of actinomycetes in rice species, many studies on rice actinobacteria biodiversity have been investigated. Plant genotype, soil type, nutrient, phytochemicals, and age have affected on composition of microbes' communities (Hardoim et al. 2011; Ikeda et al. 2014; Nakaew and Sungthong 2018; Wagner et al. 2016). Despite the progress in rice genome studies, little is known about the putative interactions between rice and their associated bacteria (Mano and Morisaki 2008; Nakaew and Sungthong 2018). Soil environment, such as nutrient elements application affected both the rice root endospheric and soil rhizosphere microbial community composition, especially species related to nutrient elements cycling—such as P-cycling (Long and Yao 2020).

Biodiversity of endophytic actinobacteria from jasmine rice grown in Thailand showed that there were three different genera: *Streptomyces*, *Microbispora*, and *Kineococcus* (Kampapongsa and Kaewkla 2016). *Streptomyces* have been shown to be enriched in the roots and rhizosphere of rice (Edwards et al. 2015). The preponderant endophytic rice actinomycetes, namely *Streptomyces,* isolated from two different districts in Guangdong province, South China were studied (Tian et al. 2004). Tian et al. (2007) characterized the diversity and composition of actinobacterial communities in rice stems and roots by a dual approach consisting of cultivation and molecular cloning of actinobacterial 16S rRNA gene, and the results have showed *Streptomyces* is the most frequently isolated genus from rice stems and roots. The community structures of cultivable actinobacteria from Thai pigmental rice have been investigated by seed phytochemicals, the results showed that *Microbispora* (76%) and *Streptomyces* (73%) were the predominant endophytic actinobacteria of Luem Pua glutinous rice and Hom Nin rice, respectively (Nakaew and Sungthong 2018).

The results of dynamic changes in the rice microbiome have showed that the seedling compartments were the driving factor for microbial community composition (Wang et al. 2020). Core bacteria genera identified as *Paenibacillus, Pantoea, Rhizobium,* and *Sphingomonas* have been served as plant growth-promoting bacteria; and the results showed that Proteobacteria and Actinobacteria composed the entire rice seeds bacterial community (Wang et al. 2020). Twenty-five actinobacteria isolates were isolated from various rice straw compost sources and some isolates exhibited enzymatic degradation of starch, cellulose, and lignin (Kausar et al. 2011). Using metagenomics analysis, the study has shown that endophytic bacteria in rice *(O. sativa)* shoots were possessed by Alphaprobacteria (51–52%), Actinobacteria (11–15%), Gammaproteobacteria (9–10%), and Betaproteobacteria (4–10%) (Okubo et al. 2014).

In the future, the diversity of rice symbiotic actinomycetes in various tissues and organs during various developmental stages needs to be systematic studied.

4.3 Actinobacterial Endophytes in Rice

The three major phyla Actinobacteria, Proteobacteria, and Firmicutes belong to the most predominant and studied plant endophytes (Coutinho et al. 2015). Endophytic actinobacteria in the inner tissues of living plants have increased attention to agronomists, chemists, ecologists, evolutionary biologists, and taxonomists (Qin et al. 2011) Endophytic actinobacteria have been isolated from healthy crop plants including rice (Qin et al. 2011). From agricultural and economic perspective, rice is one of the most important crops, but the study on the endophytic bacteria of rice has been limited (Okunishi et al. 2005). Inside the seed of cultivated rice (Kusatsu city, Shiga Prefecture, Japan), three endophytes' genera *Bacillus, Sphingomonas,* and

Pantoea have been isolated, but from surface-sterilized rice seed, actinobacteria also has been isolated (Okunishi et al. 2005). Tian et al. (2007) explored diverse communities of endophytic actinobacteria within rice stems and roots. *Actinoallomurus oryzae* sp. nov., an endophytic actinomycete has been isolated from roots of a Thai jasmine rice plant (Indananda et al. 2011). An endophytic *Streptomyces* sp. GMKU 3100 isolated from rice (*Oryza sativa* L. cv. KDML105) roots showed siderophore production (Rungin et al. 2012). An actinomycete strain S16-07T was isolated from surface-sterilized stems of rice (*O. sativa* L.) (Mingma et al. 2015). Endophytic bacteria play a crucial role in rice survival and growth. Compared to the rhizobacteria and free-living bacteria in environment, endophytic bacteria living inside the plants can be better protected from various biotic and abiotic stresses (Rana et al. 2020). Nakaew and Sungthong (2018) examined that abundance, bioactivity of cultivably endophytic actinobacteria isolated from two Thai pigmented rice cultivars. Therefore, in the future these endophytic bacteria as commercial biofertilizers and biocontrol agents for rice and other crops can promote sustainable development of agriculture.

4.4 Actinobacteria as Biological Control agents Against Rice Pathogens

The actinobacteria are widely used as potential biocontrol agents because they are major producers of bioactive compounds (Doumbou et al. 2001). There are many studies on the activity of actinobacteria as biological control agents to colonize plant surface, synthesize antibiotics against plant pathogens, synthesize particular extracellular proteins to degrade phytotoxins (Doumbou et al. 2001).

Rice sheath blight (ShB) caused by *Rhizoctonia solani* is considered as one of the most destructive diseases can reduce both grain yield and quality (Kumar et al. 2009). Streptomyces are well-known species to produce a wide range of secondary metabolites that can inhibit growth of phytopathogens (Newitt et al. 2019). *Streptomyces* sp. (Suryawanshi et al. 2020), *S. philanthi* RM-1-138 (Boukaew and Prasertsan 2014), *S. aurantiogrieus* VSMGT1014b (Harikrishnan et al. 2014), and *Micromonospora* sp. SF-1917 (Shomura et al. 1983) have exhibited inhibitory activity against *R. solani* (Table 4.1). The study (Suryawanshi et al. 2020) has explored six different *Streptomyces* sp. strains that show more than 85% inhibitory activity against *R. solani*. The compounds from *S. philanthi* RM-1-138 have suppressed the growth of *R. solani* causing rice sheath blight disease on the greenhouse experiment (Boukaew and Prasertsan 2014). Fifty-seven actinomycetes strains isolated from different rice rhizosphere soils of Tami Nadu, India, have showed inhibition activity against (Harikrishnan et al. 2014). Micromonospora species have been recognized as an important microbe for biomedicine, biocontrol agents, and potential biofuels (Hirsch and Valdés 2010). Shomura et al. (1983) found

Table 4.1 Studies on different actinobacteria strains as biocontrol agents against rice disease

Actinomycete strains	Types of rice diseases	Inhibition activity against pathogen	References
Streptomyces strain BG2–53	Rice blast disease	*Streptomyces* BG2–53 showed highest fungi *Magnaporthe oryzae* control (98%) than Blasticidin-S (86%) and Tricyclazole (96%)	Lee et al. (2002)
Endophytic *Streptomyces* (*S. griseofuscus*, *S. hygroscopicus*, *S. globisporus*, *S. aureus*, *S. albosporus*)	Rice blast disease	*S. griseofuscus* and *S. hygroscopicus* exhibited strongest inhibition against *M. oryzae*	Tian et al. (2004)
S. vinaceusdrappus	Rice blast disease	Mycelial growth inhibition of *M. oryzae* (53.5%)	Ningthoujam et al. (2009)
S. sindeneusis isolate 263 *S. sindeneusis* isolate 339	Rice blast disease	Antifungal activity against *M. oryzae*	Zarandi et al. (2009)
S. globisporus JK-1	Rice blast disease	Highest fungal control (88.3%) against *M. oryzae* than Tricyclazole (79.4%), as compared to the inoculated control	Li et al. (2011)
S. philanthi RM-1-138	Rice blast disease	Mycelial growth inhibition of *M. oryzae* (88.73%)	Boukaew and Prasertsan (2014)
S. flavotricin	Rice blast disease	Antifungal activity against *M. oryzae* (40 mm inhibition zone)	Khalil et al. (2014)
Streptomyces sp.	Rice sheath blight disease	85% inhibitory activity against *Rhizoctonia solani*	Suryawanshi et al. (2020)
S. philanthi RM-1-138	Rice sheath blight disease	100% inhibitory activity against *R. solani*	Boukaew and Prasertsan (2014)
S. aurantiogriseus VSMGT1014 b	Rice sheath blight disease	Crude metabolites from VSMGT1014 b with inhibitory activity against R. solani at 5 μg/ml	Harikrishnan et al. (2014)
Micromonospora sp. SF-1917	Rice sheath blight disease	Highly effective activity against *R. solani*	Shomura et al. (1983)
Twenty-four percent of endophytic actinobacteria	/	Anti-Gram-positive (*Bacillus subtilis* DMST 5871 and *Staphylococcus aureus* ATCC 25923) and antifungal ((*Pyricularia* sp. F2 (MF946553), *Exserohilum* sp. F3 (MF946554), *Arthrinium* sp. F4 (MF946555), *Colletotrichum* sp. F5 (MF946556), *Rhizopus* sp. F6 (MF946557)) activities	Nakaew and Sungthong (2018)

(continued)

Table 4.1 (continued)

Actinomycete strains	Types of rice diseases	Inhibition activity against pathogen	References
S. shenzhenensis TKSC3 and S. sp. SS8 co-treatment	Rice bacterial leaf streak (BLS) disease	Co-treatment TKSC3 + SS8 showed suppression efficiency (81.02%) against *Xanthomonas oryzae* pv. oryzicola (Xoc)	Hata et al. (2021)

the nucleoside antibiotic-dapiramicin from *Micromonospora* sp. SF-1917 had inhibited the growth of *R. solani* causing rice ShB disease.

Rice blast fungus *Magnaporthe oryzae* (anamorph *Pyricularia oryzae*) is one of the most vital pathogens in rice, and many Streptomyces, such as *S. vinaceusdrappus*, *S. philanthi* RM1–138, *S. griseofuscus*, *S. hygroscopicus*, *S.* sp. 339, and *S. flavotricini* are potential candidates as biocontrol agents against rice blast disease as they possess inhibitory activity against fungus *M. oryzae* (Law et al. 2017) (Table 4.1).

Biological control and plant growth promotion agents from plant beneficial microbes have been considered as an effective alternative to chemical pesticides and fertilizers, and actinobacteria can be used as microbial inoculants in rice (Palaniyandi et al. 2013). Such microbial inoculants will be suitable for the environment without causing much trouble to natural microbial ecology. Actinobacteria that were purified from rhizosphere samples indicate the biocontrol of *Streptomyces* sp. against sheath blight and plant growth promotion potential in rice (Suryawanshi et al. 2020). The experiments on *Streptomyces* as biocontrol against *M. oryzae* have been explored in greenhouse and in vitro (Edwards et al. 2015).

BLS (Bacterial Leaf Streak), one of the major bacterial diseases in rice, can significantly cause the decrease of rice yield. *Xanthomonas oryzae* pv. *Oryzicola* (Xoc) can cause rice BLS disease. The samples from *S. shenzhensis TKSC3* and *S. sp.SS8 consortium* treatment showed suppression efficiency (81.02%) against *X. oryzae pv. oryzicola* (*Xoc*) (Hata et al. 2021).

4.5 Actinobacteria Beneficiate Rice Growth by Synthesis of Protein

Through diverse biochemical mechanisms, plant growth promoting (PGP) and plant health-promoting bacteria (Bulgarelli et al. 2013) are served as a novel trait that has extended plants adapt to the environment. Because of improving soil and plant health, eco-friendliness, low production cost, and consumption of non-renewable resources, PGP bacteria have become one of the attractive strategies for developing sustainable agricultural systems (Sathya et al. 2017).

Six *Streptomyces* strains have enhanced rice growth during producing a series of proteins, such as siderophores, chitinase, lipase, protease, and 1–3-glucanase

(Gopalakrishnan et al. 2014). Single and consortium *Streptomyces* (*S. shenzhenensis* TKSC3 and *S.* sp. SS8) treatments demonstrate enormous promise in promoting plant growth on rice by increasing synthesis of peroxidase (POX), polyphenol oxidase, phenylalanine ammonia-lyase, and β,1–3-glucanase (GLU) (Hata et al. 2021).

4.6 Actinobacteria as Rice Pathogens

Comparison to other bacteria, actinomycetes play a relatively minor role causing plant diseases. However, actinomycetes represent major pathogens of certain crops in particular areas, and under special conditions (Barka et al. 2016). In rice, the study on actinobacteria as rice pathogens has not been reported.

4.7 Prospect of Rice Actinobacteria

In the further research, there are several research points that need to be strengthened and explored. (1) The biosynthesis mechanism of compounds from rice actinomycetes should be explored and analyzed. (2) Interaction mechanism of gene, gene products, and metabolites between rice and actinobacteria, especially horizontal gene transfer and secondary metabolites communication should be investigated.

Acknowledgments This work was supported by the National Natural Sciences Foundation of China (81973230).

References

Assad BM, Savi DC, Biscaia SM, Mayrhofer BF, Iantas J, Mews M, de Oliveira JC, Trindade ES, Glienke C (2021) Endophytic actinobacteria of Hymenachne amplexicaulis from the Brazilian Pantanal wetland produce compounds with antibacterial and antitumor activities. Microbiol Res 248:126768. https://doi.org/10.1016/j.micres.2021.126768

Barka EA, Vatsa P, Sanchez L, Gaveau-Vaillant N, Jacquard C, Klenk H-P, Clément C, Ouhdouch Y, van Wezel GP (2016) Taxonomy, physiology, and natural products of Actinobacteria. Microbiol Mol Biol Rev 80:1–43. https://doi.org/10.1128/MMBR.00019-15

Berdy J (2005) Bioactive microbial metabolites. J Antibiot 58:1–26. https://doi.org/10.1038/ja.2005.1

Boukaew S, Prasertsan P (2014) Suppression of rice sheath blight disease using a heat stable culture filtrate from Streptomyces philanthi RM-1-138. Crop Prot 61:1–10

Bulgarelli D, Schlaeppi K, Spaepen S, Van Themaat EVL, Schulze-Lefert P (2013) Structure and functions of the bacterial microbiota of plants. Annu Rev Plant Biol 64:807–838. https://doi.org/10.1146/annurev-arplant-050312-120106

Coutinho BG, Licastro D, Mendonça-Previato L, Cámara M, Venturi V (2015) Plant-influenced gene expression in the rice endophyte Burkholderia kururiensis M130. Mol Plant Microbe Interact 28:10–21. https://doi.org/10.1094/MPMI-07-14-0225-R

Doumbou CL, Hamby Salove M, Crawford DL, Beaulieu C (2001) Actinomycetes, promising tools to control plant diseases and to promote plant growth. Phytoprotection 82:85–102. https://doi.org/10.7202/706219ar

Edwards J, Johnson C, Santos-Medellín C, Lurie E, Podishetty NK, Bhatnagar S, Eisen JA, Sundaresan V (2015) Structure, variation, and assembly of the root-associated microbiomes of rice. Proc Natl Acad Sci 112:E911–E920. https://doi.org/10.1073/pnas.1414592112

Elert E (2014) Rice by the numbers: a good grain. Nature 514:S50–S51

Gao B, Gupta RS (2012) Phylogenetic framework and molecular signatures for the main clades of the phylum Actinobacteria. Microbiol Mol Biol Rev 76:66–112. https://doi.org/10.1128/MMBR.05011-11

Garrity GM, Bell JA, Lilburn TG (2004) Taxonomic outline of the prokaryotes. Bergey's manual of systematic bacteriology. New York. https://doi.org/10.1007/bergeysoutline200405

Gopalakrishnan S, Vadlamudi S, Bandikinda P, Sathya A, Vijayabharathi R, Rupela O, Kudapa H, Katta K, Varshney RK (2014) Evaluation of Streptomyces strains isolated from herbal vermicompost for their plant growthpromotion traits in rice. Microb Physiol 169:40–48. https://doi.org/10.1016/j.micres.2013.09.008

Hardoim PR, Andreote FD, Reinhold-Hurek B, Sessitsch A, van Overbeek LS, van Elsas JD (2011) Rice root-associated bacteria: insights into community structures across 10 cultivars. FEMS Microbiol Ecol 77:154–164. https://doi.org/10.1111/j.1574-6941.2011.01092.x

Harikrishnan H, Shanmugaiah V, Balasubramanian N, Sharma MP, Kotchoni SO (2014) Antagonistic potential of native strain Streptomyces aurantiogriseus VSMGT1014 against sheath blight of rice disease. World J Microbiol Biotechnol 30:3149–3161. https://doi.org/10.1007/s11274-014-1742-9

Hata EM, Yusof MT, Zulperi D (2021) Induction of systemic resistance against bacterial leaf streak disease and growth promotion in rice plant by Streptomyces shenzhenesis TKSC3 and Streptomyces sp. SS8. Plant Pathol J 37:173. https://doi.org/10.5423/PPJ.OA.05.2020.0083

Hirsch AM, Valdés M (2010) Micromonospora: an important microbe for biomedicine and potentially for biocontrol and biofuels. Soil Biol Biochem 42:536–542. https://doi.org/10.1016/j.soilbio.2009.11.023

Ikeda S, Sasaki K, Okubo T, Yamashita A, Terasawa K, Bao Z, Liu D, Watanabe T, Murase J, Asakawa S (2014) Low nitrogen fertilization adapts rice root microbiome to low nutrient environment by changing biogeochemical functions. Microbes Environ 29:50–59. https://doi.org/10.1264/jsme2.ME13110

Indananda C, Thamchaipenet A, Matsumoto A, Inahashi Y, Duangmal K, Takahashi Y (2011) Actinoallomurus oryzae sp. nov., an endophytic actinomycete isolated from roots of a Thai jasmine rice plant. Int J Syst Evol Microbiol 61:737–741. https://doi.org/10.1099/ijs.0.022509-0

Kampapongsa D, Kaewkla O (2016) Biodiversity of endophytic actinobacteria from jasmine rice (Oryza sativa L. KDML 105) grown in Roi-Et Province, Thailand and their antimicrobial activity against rice pathogens. Ann Microbiol 66:587–595. https://doi.org/10.1007/s13213-015-1140-z

Kausar H, Sariah M, Saud HM, Alam MZ, Ismail MR (2011) Isolation and screening of potential actinobacteria for rapid composting of rice straw. Biodegradation 22:367–375. https://doi.org/10.1007/s10532-010-9407-3

Khalil MS, Moubasher H, Hasan FF (2014) Biological control of rice blast disease by Streptomyces flavotricini. Res J Pharm, Biol Chem Sci 5:1453–1461

Kumar KVK, Reddy M, Kloepper J, Lawrence K, Groth D, Miller M (2009) Sheath blight disease of rice (Oryza sativa L.)–an overview. Biosci Biotechnol Res Asia 6:465–480

Law JW-F, Ser H-L, Khan TM, Chuah L-H, Pusparajah P, Chan K-G, Goh B-H, Lee L-H (2017) The potential of Streptomyces as biocontrol agents against the rice blast fungus, Magnaporthe oryzae (Pyricularia oryzae). Front Microbiol 8:3. https://doi.org/10.3389/fmicb.2017.00003

Lee C-H, Kim B-J, Choi G-J, Cho K-Y, Yang H-j, Shin C-S, Min S-Y, Lim Y-G (2002) Streptomyces with antifungal activity against rice blast causing fungus, Magnaporthe grisea. J Microbiol Biotechnol 12:1026–1028

Li Q, Jiang Y, Ning P, Zheng L, Huang J, Li G, Jiang D, Hsiang T (2011) Suppression of Magnaporthe oryzae by culture filtrates of Streptomyces globisporus JK-1. Biol Control 58:139–148. https://doi.org/10.1016/j.biocontrol.2011.04.013

Long X-E, Yao H (2020) Phosphorus input alters the assembly of rice (Oryza sativa L.) root-associated communities. Microb Ecol 79:357–366. https://doi.org/10.1007/s00248-019-01407-6

Mano H, Morisaki H (2008) Endophytic bacteria in the rice plant. Microbes Environ 23:109–117. https://doi.org/10.1264/jsme2.23.109

Mingma R, Duangmal K, Thamchaipenet A, Trakulnaleamsai S, Matsumoto A, Takahashi Y (2015) Streptomyces oryzae sp. nov., an endophytic actinomycete isolated from stems of rice plant. J Antibiot 68:368–372. https://doi.org/10.1038/ja.2014.166

Nakaew N, Sungthong R (2018) Seed phytochemicals shape the community structures of cultivable actinobacteria-inhabiting plant interiors of Thai pigmented rice. Microbiology 7:e00591. https://doi.org/10.1002/mbo3.591

Newitt JT, Prudence SM, Hutchings MI, Worsley SF (2019) Biocontrol of cereal crop diseases using streptomycetes. Pathogens 8:78. https://doi.org/10.3390/pathogens8020078

Ningthoujam S, Sanasam S, Tamreihao K, Nimaich S (2009) Antagonistic activities of local actinomycete isolates against rice fungal pathogens. Afr J Microbiol Res 3:737–742. https://doi.org/10.5897/AJMR.9000038

Okubo T, Ikeda S, Sasaki K, Ohshima K, Hattori M, Sato T, Minamisawa K (2014) Phylogeny and functions of bacterial communities associated with field-grown rice shoots. Microbes Environ 29:329–332. https://doi.org/10.1264/jsme2.ME14077

Okunishi S, Sako K, Mano H, Imamura A, Morisaki H (2005) Bacterial flora of endophytes in the maturing seed of cultivated rice (Oryza sativa). Microbes Environ 20:168–177. https://doi.org/10.1264/jsme2.20.168

Palaniyandi SA, Yang SH, Zhang L, Suh J-W (2013) Effects of actinobacteria on plant disease suppression and growth promotion. Appl Microbiol Biotechnol 97:9621–9636. https://doi.org/10.1007/s00253-013-5206-1

Phillips R, Odland W, Kahler A (2007) Rice as a reference genome and more. In: Rice genetics V. World Scientific, pp 3–15. https://doi.org/10.1142/9789812708816_0001

Qin S, Xing K, Jiang J-H, Xu L-H, Li W-J (2011) Biodiversity, bioactive natural products and biotechnological potential of plant-associated endophytic actinobacteria. Appl Microbiol Biotechnol 89:457–473. https://doi.org/10.1007/s00253-010-2923-6

Rana KL, Kour D, Kaur T, Devi R, Yadav AN, Yadav N, Dhaliwal HS, Saxena AK (2020) Endophytic microbes: biodiversity, plant growth-promoting mechanisms and potential applications for agricultural sustainability. Antonie Van Leeuwenhoek 113:1075–1107. https://doi.org/10.1007/s10482-020-01429-y

Rungin S, Indananda C, Suttiviriya P, Kruasuwan W, Jaemsaeng R, Thamchaipenet A (2012) Plant growth enhancing effects by a siderophore-producing endophytic streptomycete isolated from a Thai jasmine rice plant (Oryza sativa L. cv. KDML105). Antonie Van Leeuwenhoek 102:463–472. https://doi.org/10.1007/s10482-012-9778-z

Sathya A, Vijayabharathi R, Gopalakrishnan S (2017) Plant growth-promoting actinobacteria: a new strategy for enhancing sustainable production and protection of grain legumes. *3*. Biotech 7:1–10. https://doi.org/10.1007/s13205-017-0736-3

Shomura T, Nishizawa N, Iwata M, Yoshida J, Ito M, Amano S, Koyama M, Kojima M, Inouye S (1983) Studies on a new nucleoside antibiotic, dapiramicin I. producing organism, assay method and fermentation. J Antibiot 36:1300–1304. https://doi.org/10.7164/antibiotics.36.1300

Strobel G, Daisy B (2003) Bioprospecting for microbial endophytes and their natural products. Microbiol Mol Biol Rev 67:491–502. https://doi.org/10.1128/MMBR.67.4.491-502.2003

Suryawanshi PP, Krishnaraj PU, Suryawanshi MP (2020) Evaluation of actinobacteria for biocontrol of sheath blight in rice. J Pharmacogn Phytochem 9:371–376

Tian X, Cao L, Tan H, Zeng Q, Jia Y, Han W, Zhou S (2004) Study on the communities of endophytic fungi and endophytic actinomycetes from rice and their antipathogenic activities in vitro. World J Microbiol Biotechnol 20:303–309. https://doi.org/10.1023/B:WIBI.0000023843.83692.3f

Tian X, Cao L, Tan H, Han W, Chen M, Liu Y, Zhou S (2007) Diversity of cultivated and uncultivated actinobacterial endophytes in the stems and roots of rice. Microb Ecol 53:700–707. https://doi.org/10.1007/s00248-006-9163-4

Wagner MR, Lundberg DS, del Rio TG, Tringe SG, Dangl JL (2016) Host genotype and age shape the leaf and root microbiomes of a wild perennial plant. Nat Commun 7:12151. https://doi.org/10.1038/ncomms12151

Wang W, Mauleon R, Hu Z, Chebotarov D, Tai S, Wu Z, Li M, Zheng T, Fuentes RR, Zhang F (2018) Genomic variation in 3,010 diverse accessions of Asian cultivated rice. Nature 557:43–49. https://doi.org/10.1038/s41586-018-0063-9

Wang M, Eyre AW, Thon MR, Oh Y, Dean RA (2020) Dynamic changes in the microbiome of rice during shoot and root growth derived from seeds. Front Microbiol 11:2183. https://doi.org/10.3389/fmicb.2020.559728

Zarandi ME, Bonjar GS, Dehkaei FP, Moosavi SA, Farokhi PR, Aghighi S (2009) Biological control of rice blast (*Magnaporthe oryzae*) by use of Streptomyces sindeneusis isolate 263 in greenhouse. Am J Appl Sci 6:194–199. https://doi.org/10.3844/ajassp.2009.194.199

Chapter 5
Nonribosomally and Ribosomally Synthesized Bioactive Peptides (NRPS and RiPPs) from Actinobacteria

Kamarudheen Neethu, Loganathan Karthik, and Zhiyong Li

Abstract From time immemorial, natural products (NPs) are a major source of drugs in the pharmaceutical industry ranging from antibiotics, anti-depressants, anti-inflammatory, anti-tumor, and anti-cancer agents to name a few. Actinobacteria and their biosynthetic pathways contribute substantially to these drugs present in the laboratory level, clinical phase trials, and in the market. The majority of the secondary metabolites secreted are produced from the nonribosomal peptide synthetase (NRPS) and polyketide synthetase (PKS) enzymes. Apart from the nonribosomal peptides produced, ribosomally produced peptides that undergo drastic post-translational modifications also play a part in contributing to the essential drugs produced. The chapter deals with both nonribosomal peptides (NRPs) and ribosomally synthesized and post-translationally modified peptides (RiPPs).

Keywords NRPs · RiPPs · Actinobacteria · Secondary metabolites · Antibiotics

5.1 Introduction

The idea of a microbe fighting and eliminating another microbe was revolutionary and ever since the shift towards natural drugs has gained significant momentum. Natural products play a pivotal role in the drug market accounting to around nonribosomal peptide synthetases (NRPS) are factories, huge multi-modular biological catalysts, that are responsible for the production of a variety of remarkable peptides. Considering the type of functions these peptides perform, they are ideally

K. Neethu
Department of Pharmaceutical Sciences, Oregon State University, Corvallis, OR, USA

L. Karthik (✉)
Synthetic Biology, R&D Center, Salem Microbes Private Limited, Salem, India

Z. Li
State Key Laboratory of Microbial Metabolism, School of Life Sciences and Biotechnology, Shanghai Jiao Tong University, Shanghai, P.R. China

L. Karthik (ed.), *Actinobacteria*, https://doi.org/10.1007/978-981-16-5835-8_5

not limited to the 20 amino acids, but alpha hydroxy acids, other non-proteogenic amino acids and fatty acids (Caboche et al. 2007). These functional molecules not only add to the structural variety but also enhance their functionality. The nonribosomal peptide synthetases consist of three major domains for their functioning: the A domain for adenylation; P domain, a peptidyl-carrier domain for the thioester binding of the adenylated amino acid; a C domain generating new peptide bonds between two amino acids on two P domains.

The uncovering of NRPS was rather lately, the past 50 years and are found associated to fungi and bacteria. The history of NRPS discovery was in the interesting observation by Tatum and colleagues when they deduced that the biological synthesis of tyrocidine, a cyclic peptide was unaffected by aureomycin, a ribosome inhibitor. The subsequent contributions of teams of Fritz Lipmann, and of Søren Laland, led to the unraveling of NRPS (Mach et al. 1963; Berg et al. 1965). Although some eukaryotes were suspected to have NRPS, the assumptions were later proved wrong except for nemamide synthetase from *Caenorhabditis elegans* and NRPS Ebony from *Drosophila melanogaster*. The bacterial contribution is significantly accounted to Actinobacteria followed by alpha, beta, and gamma Proteobacteria along with Firmicutes. Fungal contribution is assigned mostly to Ascomycetes, study of which is often hindered due to the large genome size (Shou et al. 2016; Richardt et al. 2003). However, NRPS is relatively well explored in actinomycetes, and the distribution was seen higher in the proficient bioactive metabolite producer, *Streptomyces* sp.

5.2 Nonribosomal Peptides in Actinobacteria

Actinomycetes are substantial producers of bioactive secondary metabolites which exhibit astounding variety. The contribution of actinobacteria into the pool of microbial bioactive metabolites is essentially above 50%. Figures indicate that out of >20,000 microbial natural drugs over 10,100 is sourced from actinobacteria (Solecka et al. 2012). The production of these secondary metabolites, including anti-microbials, anti-cancer, anti-tumor, anti-parasitic, immunosuppressive agents, are all essentially attributed to the genes encoding them located on specific gene clusters. Identifying these Biosynthetic Gene Clusters (BCGs) were revolutionary in creating the idea that even the estimate that 1,50,000 metabolites produced by *Streptomyces* sp. was an underestimation (Watve et al. 2001). The identification of genes of NRPS and PKS revealed a gripping evidence stating that a large range of bioactive metabolites are produced as a result of NRPS and PKS gene clusters (Zhou et al. 2011).

5.2.1 NRPS Gene Clusters Attributed to Bioactivity

As reported, each of *Streptomyces* strain has chromosomes that encode 15–20 biosynthetic gene clusters which ideally can produce a large number of secondary metabolites. Thus, in the real scenario the number of metabolites produced is far fewer than capabilities. From genomic analysis it is evident that NRPS constitutively large among the BGCs of actinobacteria followed by polyketide synthases, and therefore most of the secondary metabolites can be attributed to these groups (Nett et al. 2009). Genome mining was the approach used to analyze these BCGs to effectively isolate novel secondary metabolites (Challis 2008) (Table 5.1).

5.3 Methods for Discovering Nonribosomal Peptides: Proteomic Approach

Secondary metabolite drug discovery by the traditional methods of identification and characterization of one compound/peptide per time is very time consuming. These results could also result in redundancy of the secondary metabolites increasing the cost and wasting of time (Li et al. 2009). Thus, genome mining became popular as tools to detect the biosynthetic gene cluster that potentially produce novel bioactive NRPs are also available. Another famous technique is peptidogenomics that allows the high-throughput mass spectrometry to detect the presence of novel NRPs (Bouslimani et al. 2014; Li and Vederas 2009).

5.3.1 NRPquest

Although nonribosomal peptides (NRPs) are extremely important, the techniques involved in the sequencing of these peptides are in its early stages. NRPquest is one such technique for the identification which involves multi-tolerant (mutation and modification) data search against a putative NRPs database. This tool is a potent discovery tool which can uncover 100 NRPs in a study thus making MS-based NRPs identification a new and potential technique. NRPquest tool (www.cyclo.ucsd.edu) performs its first step of putative NRPs database generation which was obtained from the genome sequence by the help of nonribosomal code. NRPquest is a joint approach of mass spectrometry and genome mining which radically transforms the earlier denovo sequencing/dereplication methodology to MS/MS database search method. The tool uses genome sequence data and mass spectral data as input and performs NRPS identification by NRP prediction tool which is NRPSpredictor2 and construct a database thereafter (Mohimani et al. 2011). In the second step, the mass spectral data is matched against the NRPs database content in a blind fashion. Thirdly, it statistically computes the significance of the resultant peptide-spectrum

Table 5.1 Under characterized bioactive NRPs across various actinobacterial isolates

Actinobacteria species	Location	Source	Bioactivity	Primer used	Reference
Streptomyces sp BDUSMP 02	Pitchavaram mangrove, India	Soil sediments	Antibacterial	F -CCAACSGGCNNNCCSAAGGGCGT R -ACCCTCSGTSGGSCCGTA	Sivalingam et al. (2019)
Aeromicrobium kwangyangens	Beibu Gulf, South China Sea	Sponge	Antibacterial	A3F GCSTACSYSATSTACACSTCSGG A7R SASGTCVCCSGTSCGGTAS	Liu et al. (2019)
Brachybacterium tyrofermentans	Beibu Gulf, South China Sea	Sponge	Antibacterial	A3F GCSTACSYSATSTACACSTCSGG A7R SASGTCVCCSGTSCGGTAS	Liu et al. (2019)
Pseudonocardia carboxydivorans	Beibu Gulf, South China Sea	Sponge	Antibacterial	A3F GCSTACSYSATSTACACSTCSGG A7R SASGTCVCCSGTSCGGTAS	Liu et al. (2019)
Streptomyces olivaceus	Beibu Gulf, South China Sea	Sponge	Antibacterial	A3F GCSTACSYSATSTACACSTCSGG A7R SASGTCVCCSGTSCGGTAS	Liu et al. (2019)
Streptomyces sp. TP-A0598	Toyama Bay, Japan,	Sea water	Antibacterial	–	Komaki et al. (2020)
Streptomyces parvisporogenes S2A-04	Florida	Soil	Fungicidal	A3F-GCSTACSYSATSTACACSTCSGG-3 A7R 50-SAS GTC VCC SGT SGC GTA S-30	Escalante-Réndiz et al. (2019)
Micromonospora sp. ACM2–092	Indian Ocean, Mozambique	Soft coral	Anti-tumor	MTF2-GCNGGYGGYGCNTAYGTNCC PS4-SAGSAGGSWGTGGCCGCCSAGCTCGAAGAA	Lombó et al. (2006)
Micromonospora sp. ML1	Indian Ocean, Mozambique	Mollusc	Anti-tumor	MTF2-GCNGGYGGYGCNTAYGTNCC PS4-SAGSAGGSWGTGGCCGCCSAGCTCGAAGAA	Lombó et al. (2006)
Catenulispora strains	Forest Gerenzano, Italy	Soil	Antibacterial	ADEdom5-ACSGGCNNNCCSAAGGGCGT ADEdom3 - CTCSGTSGGSCCGTA	Busti et al. (2006)

matches (PSM) and as a final step it constructs a network of the identified PSMs and shows the families of related NRPs via spectral network dereplication (Bandeira et al. 2007; Mohimani et al. 2014).

5.3.2 Pep2Path

Pep2Path acts as a gap to accelerate the peptidogenomic approach by automatically identifying the biosynthetic gene clusters for peptides analyzed by tandem MS approaches. It matches the mass shift sequences from mass spectrum to the gene clusters that would encode the corresponding peptide. Pep2Path comes with a program called Nrp2Path which does the probability matching between the mass shifts to the NRPS gene cluster. Nrp2Path by default comes with a database for the same purpose. The input data for Nrp2Path can be a list of amino acids or the order of mass shifts, the latter gets converted into amino acid sequence tags using Kersten et al. table (Kersten et al. 2011). In the functioning of the program, Nrp2Path searches for BCGs within the selected NRP database on how close is the probability of the BCG encoding the peptide corresponding to the sequence tag. The user can filter data by selecting the taxonomy with the available knowledge of the peptide (Medema et al. 2014).

5.4 Ribosomally Synthesized and Post-Translationally Modified Peptides-Lantipeptides

Well known is the concept that actinomycetes are prolific producers of specialized secondary metabolites which includes polyketides; aromatic and macrocyclic, nonribosomally synthesized peptides, and ribosomally synthesized ones. Lantipeptides are one among these ribosomally synthesized peptides are well studied among ribosomally synthesized and post-translationally modified peptides (RiPPs). Synthesized using conventional amino acids, they undergo interesting post-translational modifications (PTMs) enriching the structural and bioactivity diversity. The extensive PTMs include but are not limited to epimerization, dehydration, acylation, glycosylation, and halogenation (McIntosh et al. 2009; Nolan and Walsh 2009). The feature of two thioether cross-linked amino acid bridge, namely meso-lathionine (Lan) and (2S,3S,6R)-3-methyllanthionine (Meso) mark to distinguish lantipeptides. Among the lantipeptides, a huge number are antibacterial in nature which gave it the term, lantibiotic, but was later no longer in use to bring all the lantipeptides performing various under the same umbrella. The distribution of lantipeptides was believed to be in a group of Gram+ve bacteria, including *Bacillus*, *Streptomyces*, *Lactococcus*, *Staphylococcus*, and *Streptococcus*. However, with extended genomic analysis this concept was disapproved as many other members of the phylum Actinobacteria showed the production of lantipeptides. With further advent of using bioinformatic tools, it was identified that even bacteroidetes, chlamydiae, proteobacteria, and cyanobacteria possessed the genes for lantipeptide production (Marsh et al. 2010).

5.4.1 Biosynthetic Machinery

The basic biosynthetic mechanism in production of lantipeptides is made possible by the action of a gene lanA which encodes for a precursor protein. Any lantipeptide is made into a final natural bioactive compound by processing and modifying the precursor peptide which is produced by the ribosome which was genetically coded for production. The precursor protein which has an N-terminal leader peptide undergoes PTMs and results in carrying the lan-MeLan bridge, formed by the dehydration reaction of the serine and threonine residues. This leader portion at the N-terminus plays a role in enzyme recognition and activity, and the core peptide gets processed to the natural product upon proteolysis of the N-terminal leader peptide (Fig. 5.1) (Hegemann and Süssmuth 2020; Arnison et al. 2013; Bierbaum and Sahl 2009).

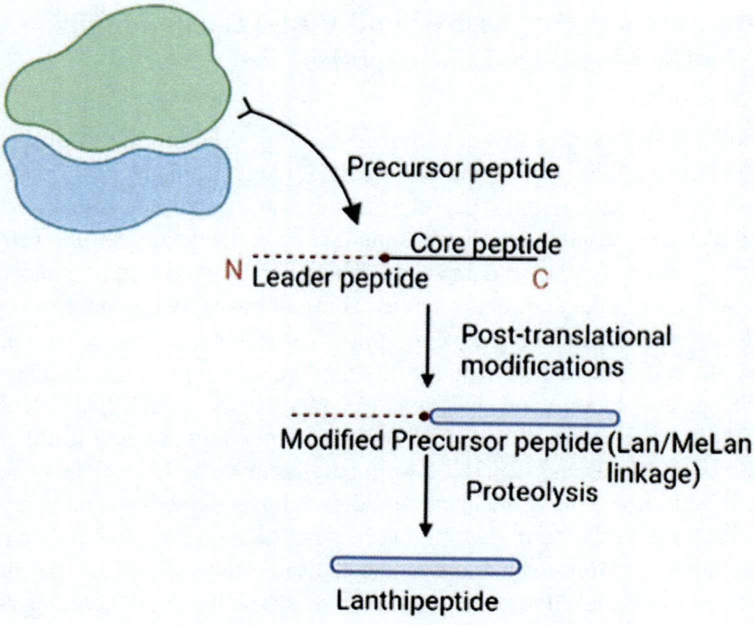

Fig. 5.1 General Lanthipeptide pathway

5.4.2 Classes of Lantipeptides

Based on the biosynthetic machinery, lantipeptides are divided into four classes.

5.4.2.1 Class I

This class of lantipeptides possesses separate enzymes for performing the basic mechanism of thioether cross-links; LanB enzyme aminoacyl-tRNA-dependent dehydratases dehydrate threonine and serine residues, and LanC helps in cyclization. In the next phase, the export of the modified peptides is made possible by a less specific, an ATP-binding cassette transmembrane transporter LanT. Thereafter, the leader peptide is removed by a subtilisin-like serine protease, LanP, but this is not present uniformly in all species (Li et al. 2006; Ortega et al. 2015). An important example of this class from the actinobacteria *Microbispora corallina* is microbisporicin which possesses 1 MeLan and 3 Lan residues. It is an efficient antibacterial against drug-resistant Gram-positive bacteria (Castiglione et al. 2008). Yet another prominent example is planosporicin which is also from a rare actinomycetes *Planomonospora* sp. (Castiglione et al. 2007).

5.4.2.2 Class II

Lantipeptides belonging to this class carry a bifunctional LanM enzyme which performs the functions of dehydratase and cyclase enzymes. The N-terminal ATP-dependent dehydratase is not similar to LanB structurally, but the C-terminal cyclase is identical by around 25% to LanC. The functions of transporting the peptide and the proteolytic leader peptide processing are both performed by LanT which has a conserved N-terminal cysteine protease domain (Chatterjee et al. 2005; Pag and Sahl 2002). An example of the class is Variacin which is isolated from *Kocuria varians*. It contains 1 Lan and 1 MeLan ring and exhibits activity against food pathogens, namely *Listeria monocytogenes*, *Enterococcus* sp. *Clostridium* spp. and *Bacillus* spp. (Pridmore et al. 1996).

5.4.2.3 Class III

Class III lantipeptides were studied in detail with the discovery of labyrinthopeptins isolated from *Actinomadura namibiensis* and produce LabA1, LabA2, and LabA3. The lantipeptides have an NTP-dependent 3 domain N-terminal lyase which has resemblances to serine/threonine protein kinase, a kinase and C-terminal cyclase which lack zinc binding motifs unlike Class I, II, and IV cyclases (Hegemann et al. 2019; Wang and Van Der Donk 2012). They lack a transporter protein dedicated to serve the function. The single trifunctional enzyme is LabKC and makes use of GTP but not ATP for kinase activity (Müller et al. 2010). LabA1 is capable of exhibiting anti-viral activity against HSV and HIV.

5.4.2.4 Class IV

Lantipeptides share similar features as to class III. The enzyme VenL carries an N-terminal lyase and a central serine and threonine kinase domains. The presence of a LanT-like transporter was observed but lacked any specific gene for protease. VenL used ATP and magnesium chloride for dehydration. *Streptomyces venezuelae* produced venezuelin (Chen et al. 2019) (Table 5.2).

5.5 Generic Lantipeptides of Actinobacteria and Applications

Lantipeptides are essentially applied as antimicrobial agents with wide spectrum of action. Nevertheless, they also perform other activities like antitumor, anti-inflammatory, anti-allergic, etc. The diversity of lantipeptides from actinobacteria has seen a widening ever since genome mining came into play expanding the knowledge of its presence to various rare actinomycetes. Among the lantipeptides, microbisporicin or NAI-107 is the strongest with a wide range of activities. Another version of the same designated as NAI-108 was found to be produced by *Microbispora* spp. and *Actinoallomurus* spp. upon growth in potassium bromide supplemented medium and is the first bromine containing lantibiotic. This rendered

Table 5.2 Classes of lantipeptides

	Biosynthetic machinery			Function	Examples
Classes	Modifying enzymes	Transporter enzyme	Protease/ protease domain		
Class I	LanB dehydratase LanC cyclase	LanT	LanP	Antibacterial	Microbisporicin, planosporicin
Class II	LanM bifunctional	LanT	LanT	Antibacterial	Variacin B
Class III	LabKC	LanT-like transporters	None	No antibiotic activity, but neuropathic in mice anti-viral properties against HSV, HIV, dengue virus, and Zika virus	Labyrinthopeptin A2
Class IV	VenL	LanT-liketransporter	None	–	Venezuelin

the lantipeptide a higher activity than observed in its parent peptide. It was found to destroy the pathogen *S. aureus* leading to the 5-$\log_{10}$ reduction in the viable cell count within 30 min at a concentration of 0.25 μg/mL (Cruz et al. 2015). *Planomonospora* sp. DSM 14920 produces a cell wall inhibitor, namely planosporicin which is active against *Staphylococci* sp., *Streptococci* sp. and *Enterococci* sp. Slight differences in the structural characteristics of planosporicin make it lesser potent than microbisporicin (Vasile et al. 2012).

One of the class II lantibiotics is actagardine which was first discovered as gardimycin is another cell wall inhibitor which acts by blocking transglycosylation. It is produced by *Actinoplanes garbadinensis* and *Actinoplanes liguriae*. It is found effective against *Streptococcus* spp. and *Clostridium* spp. and was found to be as effective as ampicillin and cephaloridine when tested in vivo (Arioli et al. 1976). *Actinoplanes liguriae* ATCC 31048 produces Ala(0)-actagardine which is active against *Staphylococcus* spp. Another similar lantipeptide is produced by *A. liguriae* NCIMB41362 which is deoxyactagardine B and is antimicrobial in nature and under clinical studies (Boakes et al. 2010). *A. liguriae* produced NVB302—aminoheptylamido deoxyactagardine is clearing its phase I clinical trial for treatment against *C. difficile*. Yet another actagardine type lantipeptide is NAI-802, produced by *Actinoplanes* sp. ID104802 and ID104771 and has an improved activity against *Streptococcus* spp. with MIC 0.5 to 32 μg/mL in comparison to 2 to 64 μg/mL for actagardine. In addition, it also showed the activity against *C difficile*, *C butyricum*, *C perfringens*, and *Peptostreptococcus asaccharolyticus* (MIC 0.25–2 μg/mL) (Simone et al. 2013). *Clavibacter michiganensis* subsp. *Michiganensis* produces a distinctive plant disease control lantibiotic. This heat stable lantipeptide known as Michiganin A is active against *Clavibacter michiganensis* subsp. *Sepedonicus,* a causative of potato ring rot disease by inhibiting cell wall synthesis (Holtsmark et al. 2006).

Other examples of Class II lantipeptides from actinobacteria include cinnamycin and duramycin. Cinnamycin an exclusive product of *Streptomyces* spp. is a potent inhibitor of *Bacillus subtilis*, fungi, yeasts, and HSV. Additionally, it is also an inhibitor of angiotensin-converting enzyme, thus regulating blood pressure (Kido et al. 1983). It also controls inflammation, treats cystic fibrosis, atherosclerosis, diabetes, and cancer (Mouchlis and Dennis 2016). Duramycin isolated from *Streptomyces cinnamoneus* is a potent plant pathogen controlling agent which controls *Rhodococcus fascians* which causes leafy gall (Goethals et al. 2001). Duramycin has also proved to be useful in cystic fibrosis clearing phase II clinical trials as a nebulization treatment. Duramycin has shown its wide range of efficacy as an antiviral and anti-tumor agent by inhibiting the viral entry of West Nile virus, dengue virus, and Ebola virus and by inhibiting the cell proliferation of pancreatic cells (Richard et al. 2015; Yates et al. 2012). A class III example of lantipeptide is NAI-112 from *Actinoplanes* sp. DSM 24059. It is a cell wall inhibitor which exhibits inhibitory actions against *Staphylococcus* spp. and *Streptococcus* spp. It also is found to relieve pain in vivo in animal models without exerting toxicity (Iorio et al. 2014).

5.6 Tools for Identification of Lantipeptides

5.6.1 BAGEL3

The peptides that lack any previous homology are hard to be identified and thus get easily overlooked. Here, BAGEL3 is devised to perform a dual approach which uses direct mining of genes and indirect mining through context genes. DNA sequences in the form of FASTA are fed to BAGEL. These sequences are analyzed simultaneously by two different ways, by finding the genes commonly seen producing RiPPs and searching the gene itself. BAGEL is accompanied by three databases—modified peptides, unmodified, and peptides which underwent PTMs which could be bactericidal or non-bactericidal (van Heel et al. 2013).

References

Arioli V, Berti M, Silvestri L (1976) Gardimycin, a new antibiotic from actinoplanes III. biological properties. J Antibiot 29:511–515. https://doi.org/10.7164/antibiotics.29.511

Arnison PG, Bibb MJ, Bierbaum G, Bowers AA, Bugni TS, Bulaj G, Camarero JA, Campopiano DJ, Challis GL, Clardy J (2013) Erratum: ribosomally synthesized and post-translationally modified peptide natural products: overview and recommendations for a universal nomenclature. Nat Prod Rep 30:1568. https://doi.org/10.1039/C2NP20085F

Bandeira N, Tsur D, Frank A, Pevzner PA (2007) Protein identification by spectral networks analysis. Proc Natl Acad Sci U S A 104:6140–6145. https://doi.org/10.1073/pnas.0701130104

Berg TL, Froholm LO, Laland SG (1965) The biosynthesis of gramicidin S in a cell-free system. Biochem J 96:43–52. https://doi.org/10.1042/bj0960043

Bierbaum G, Sahl H-G (2009) Lantibiotics: mode of action, biosynthesis and bioengineering. Curr Pharm Biotechnol 10:2–18. https://doi.org/10.2174/138920109787048616

Boakes S, Appleyard AN, Cortés J, Dawson MJ (2010) Organization of the biosynthetic genes encoding deoxyactagardine B (DAB), a new lantibiotic produced by Actinoplanes liguriae NCIMB41362. J Antibiot 63:351–358. https://doi.org/10.1038/ja.2010.48

Bouslimani A, Sanchez LM, Garg N, Dorrestein PC (2014) Mass spectrometry of natural products: current, emerging and future technologies. Nat Prod Rep 31:718–729. https://doi.org/10.1039/c4np00044g

Busti E, Monciardini P, Cavaletti L, Bamonte R, Lazzarini A, Sosio M, Donadio S (2006) Antibiotic-producing ability by representatives of a newly discovered lineage of actinomycetes. Microbiology 152:675–683. https://doi.org/10.1099/mic.0.28335-0

Caboche S, Pupin M, Leclere V, Fontaine A, Jacques P, Kucherov G (2007) NORINE: a database of nonribosomal peptides. Nucleic Acids Res 36:D326–D331. https://doi.org/10.1093/nar/gkm792

Castiglione F, Cavaletti L, Losi D, Lazzarini A, Carrano L, Feroggio M, Ciciliato I, Corti E, Candiani G, Marinelli F (2007) A novel lantibiotic acting on bacterial cell wall synthesis produced by the uncommon actinomycete Planomonospora sp. Biochem J 46:5884–5895. https://doi.org/10.1021/bi700131x

Castiglione F, Lazzarini A, Carrano L, Corti E, Ciciliato I, Gastaldo L, Candiani P, Losi D, Marinelli F, Selva E (2008) Determining the structure and mode of action of microbisporicin, a potent lantibiotic active against multiresistant pathogens. Chem Biol 15:22–31. https://doi.org/10.1016/j.chembiol.2007.11.009

Challis GL (2008) Mining microbial genomes for new natural products and biosynthetic pathways. Microbiology 154:1555–1569. https://doi.org/10.1099/mic.0.2008/018523-0

Chatterjee C, Paul M, Xie L, Van Der Donk WA (2005) Biosynthesis and mode of action of lantibiotics. Chem Rev 105:633–684. https://doi.org/10.1021/cr030105v

Chen S, Xu B, Chen E, Wang J, Lu J, Donadio S, Ge H, Wang H (2019) Zn-dependent bifunctional proteases are responsible for leader peptide processing of class III lanthipeptides. Proc Natl Acad Sci U S A 116:2533–2538. https://doi.org/10.1073/pnas.1815594116

Cruz JC, Iorio M, Monciardini P, Simone M, Brunati C, Gaspari E, Maffioli SI, Wellington E, Sosio M, Donadio S (2015) Brominated variant of the lantibiotic NAI-107 with enhanced antibacterial potency. J Nat Prod 78:2642–2647. https://doi.org/10.1021/acs.jnatprod.5b00576

Escalante-Réndiz D, de-la-Rosa-García S, Tapia-Tussell R, Martín J, Reyes F, Vicente F, Gamboa-Angulo M (2019) Molecular identification of selected streptomyces strains isolated from mexican tropical soils and their anti-candida activity. Int J Environ Res Public Health 16:1913. https://doi.org/10.3390/ijerph16111913

Goethals K, Vereecke D, Jaziri M, Van Montagu M, Holsters M (2001) Leafy gall formation by Rhodococcus fascians. Annu Rev Phytopathol 39:27–52. https://doi.org/10.1146/annurev.phyto.39.1.27

Hegemann JD, Süssmuth RD (2020) Matters of class: coming of age of class III and IV lanthipeptides. RSC Chem Biol 1:110–127. https://doi.org/10.1039/D0CB00073F

Hegemann JD, Shi L, Gross ML, Van Der Donk WA (2019) Mechanistic studies of the kinase domains of Class IV lanthipeptide synthetases. ACS Chem Biol 14:1583–1592. https://doi.org/10.1021/acschembio.9b00323

Holtsmark I, Mantzilas D, Eijsink V, Brurberg M (2006) Purification, characterization, and gene sequence of michiganin A, an actagardine-like lantibiotic produced by the tomato pathogen Clavibacter michiganensis subsp. michiganensis. Appl Environ Microbiol 72:5814–5821. https://doi.org/10.1128/AEM.00639-06

Iorio M, Sasso O, Maffioli SI, Bertorelli R, Monciardini P, Sosio M, Bonezzi F, Summa M, Brunati C, Bordoni R (2014) A glycosylated, labionin-containing lanthipeptide with marked antinociceptive activity. ACS Chem Biol 9:398–404. https://doi.org/10.1021/cb400692w

Kersten RD, Yang Y-L, Xu Y, Cimermancic P, Nam S-J, Fenical W, Fischbach MA, Moore BS, Dorrestein PC (2011) A mass spectrometry–guided genome mining approach for natural product peptidogenomics. Nat Chem Biol 7:794–802. https://doi.org/10.1038/nchembio.684

Kido Y, Hamakado T, Yoshida T, Anno M, Motoki Y (1983) Isolation and characterization of ancovenin, a new inhibitor of angiotensin I converting enzyme, produced by actinomycetes. J Antibiot 36:1295–1299. https://doi.org/10.7164/antibiotics.36.1295

Komaki H, Hosoyama A, Igarashi Y, Tamura T (2020) Streptomyces lydicamycinicus sp. nov. and its secondary metabolite biosynthetic gene clusters for polyketide and nonribosomal peptide compounds. Microorganisms 8:370. https://doi.org/10.3390/microorganisms8030370

Li JW-H, Vederas JC (2009) Drug discovery and natural products: end of an era or an endless frontier? Science 325:161–165. https://doi.org/10.1126/science.1168243

Li B, Yu JPJ, Brunzelle JS, Moll GN, Van der Donk WA, Nair SK (2006) Structure and mechanism of the lantibiotic cyclase involved in nisin biosynthesis. Science 311:1464–1467. https://doi.org/10.1126/science.1121422

Li MH, Ung PM, Zajkowski J, Garneau-Tsodikova S, Sherman DH (2009) Automated genome mining for natural products. BMC Bioinform 10:1–10. https://doi.org/10.1186/1471-2105-10-185

Liu T, Wu S, Zhang R, Wang D, Chen J, Zhao J (2019) Diversity and antimicrobial potential of Actinobacteria isolated from diverse marine sponges along the Beibu Gulf of the South China Sea. FEMS Microbiol Ecol 95(7):fiz089. https://doi.org/10.1093/femsec/fiz089

Lombó F, Velasco A, Castro A, De la Calle F, Braña AF, Sánchez-Puelles JM, Méndez C, Salas JA (2006) Deciphering the biosynthesis pathway of the antitumor thiocoraline from a marine actinomycete and its expression in two Streptomyces species. Chembiochem 7:366–376. https://doi.org/10.1002/cbic.200500325

Mach B, Reich E, Tatum E (1963) Separation of the biosynthesis of the antibiotic polypeptide tyrocidine from protein biosynthesis. Proc Natl Acad Sci U S A 50:175. https://doi.org/10.1073/pnas.50.1.175

Marsh AJ, O'Sullivan O, Ross RP, Cotter PD, Hill C (2010) In silico analysis highlights the frequency and diversity of type 1 lantibiotic gene clusters in genome sequenced bacteria. BMC Genomics 11:1–21. https://doi.org/10.1186/1471-2164-11-679

McIntosh JA, Donia MS, Schmidt EW (2009) Ribosomal peptide natural products: bridging the ribosomal and nonribosomal worlds. Nat Prod Rep 26:537–559. https://doi.org/10.1039/B714132G

Medema MH, Paalvast Y, Nguyen DD, Melnik A, Dorrestein PC, Takano E, Breitling R (2014) Pep2Path: automated mass spectrometry-guided genome mining of peptidic natural products. PLoS Comput Biol 10:e1003822. https://doi.org/10.1371/journal.pcbi.1003822

Mohimani H, Liu W-T, Yang Y-L, Gaudêncio SP, Fenical W, Dorrestein PC, Pevzner PA (2011) Multiplex de novo sequencing of peptide antibiotics. J Comput Biol 18:1371–1381. https://doi.org/10.1089/cmb.2011.0158

Mohimani H, Liu W-T, Kersten RD, Moore BS, Dorrestein PC, Pevzner PA (2014) NRPquest: coupling mass spectrometry and genome mining for nonribosomal peptide discovery. J Nat Prod 77:1902–1909. https://doi.org/10.1021/np500370c

Mouchlis VD, Dennis EA (2016) Membrane and inhibitor interactions of intracellular phospholipases A2. Adv Biol Regul 61:17–24. https://doi.org/10.1016/j.jbior.2015.11.011

Müller WM, Schmiederer T, Ensle P, Süssmuth RD (2010) In vitro biosynthesis of the prepeptide of Type-III lantibiotic labyrinthopeptin A2 including formation of a C-C bond as a post-translational modification. Angew Chem Int Ed 49:2436–2440. https://doi.org/10.1002/anie.200905909

Nett M, Ikeda H, Moore BS (2009) Genomic basis for natural product biosynthetic diversity in the actinomycetes. Nat Prod Rep 26:1362–1384. https://doi.org/10.1039/b817069j

Nolan EM, Walsh CT (2009) How nature morphs peptide scaffolds into antibiotics. Chembiochem 10:34–53. https://doi.org/10.1002/cbic.200800438

Ortega MA, Hao Y, Zhang Q, Walker MC, van der Donk WA, Nair SK (2015) Structure and mechanism of the tRNA-dependent lantibiotic dehydratase NisB. Nature 517:509–512. https://doi.org/10.1038/nature13888

Pag U, Sahl H-G (2002) Multiple activities in lantibiotics-models for the design of novel antibiotics? Curr Pharm Des 8:815–833. https://doi.org/10.2174/1381612023395439

Pridmore D, Rekhif N, Pittet A, Suri B, Mollet B (1996) Variacin, a new lanthionine-containing bacteriocin produced by Micrococcus varians: comparison to lacticin 481 of Lactococcus lactis. Appl Environ Microbiol 62:1799–1802. https://doi.org/10.1128/aem.62.5.1799-1802.1996

Richard AS, Zhang A, Park S-J, Farzan M, Zong M, Choe H (2015) Virion-associated phosphatidylethanolamine promotes TIM1-mediated infection by Ebola, dengue, and West Nile viruses. Proc Natl Acad Sci U S A 112:14682–14687. https://doi.org/10.1073/pnas.1508095112

Richardt A, Kemme T, Wagner S, Schwarzer D, Marahiel MA, Hovemann BT (2003) Ebony, a novel nonribosomal peptide synthetase for β-alanine conjugation with biogenic amines in Drosophila. J J Biol Chem 278:41160–41166. https://doi.org/10.1074/jbc.M304303200

Shou Q, Feng L, Long Y, Han J, Nunnery JK, Powell DH, Butcher RA (2016) A hybrid polyketide-nonribosomal peptide in nematodes that promotes larval survival. Nat Chem Biol 12:770–772. https://doi.org/10.1038/nchembio.2144

Simone M, Monciardini P, Gaspari E, Donadio S, Maffioli SI (2013) Isolation and characterization of NAI-802, a new lantibiotic produced by two different actinoplanes strains. J Antibiot 66:73–78. https://doi.org/10.1038/ja.2012.92

Sivalingam P, Muthuselvam M, Pote J, Prabakar K (2019) Phylogenetic insight of nonribosomal peptide synthetases (NRPS) Adenylate domain in antibacterial potential Streptomyces BDUSMP 02 isolated from Pitchavaram Mangrove. Bioinformation 15:412–418. https://doi.org/10.6026/97320630015412

Solecka J, Zajko J, Postek M, Rajnisz A (2012) Biologically active secondary metabolites from Actinomycetes. Open Life Sci 7:373–390. https://doi.org/10.2478/s11535-012-0036-1

van Heel AJ, de Jong A, Montalbán-López M, Kok J, Kuipers OP (2013) BAGEL3: Automated identification of genes encoding bacteriocins and (non-)bactericidal posttranslationally modified peptides. Nucleic Acids Res 41:W448–W453. https://doi.org/10.1093/nar/gkt391

Vasile F, Potenza D, Marsiglia B, Maffioli S, Donadio S (2012) Solution structure by nuclear magnetic resonance of the two lantibiotics 97518 and NAI-107. J Pept Sci 18:129–134. https://doi.org/10.1002/psc.1425

Wang H, Van Der Donk WA (2012) Biosynthesis of the class III lantipeptide catenulipeptin. ACS Chem Biol 7:1529–1535. https://doi.org/10.1021/cb3002446

Watve MG, Tickoo R, Jog MM, Bhole BD (2001) How many antibiotics are produced by the genus Streptomyces? Arch Microbiol 176:386–390. https://doi.org/10.1007/s002030100345

Yates KR, Welsh J, Udegbunam NO, Greenman J, Maraveyas A, Madden LA (2012) Duramycin exhibits antiproliferative properties and induces apoptosis in tumour cells. Blood Coagul Fibrinolysis 23:396–401. https://doi.org/10.1097/mbc.0b013e3283538875

Zhou K, Zhang X, Zhang F, Li Z (2011) Phylogenetically diverse cultivable fungal community and polyketide synthase (PKS), non-ribosomal peptide synthase (NRPS) genes associated with the south china sea sponges. Microb Ecol 62:644–654. https://doi.org/10.1007/s00248-011-9859-y

Chapter 6
Genome Data Mining, Chemistry and Bioactivity of Sesquiterpenes from Actinobacteria

Gangarapu Kiran, Loganathan Karthik, Gouthami Thumma, and Arivarasan Vishnukirthi

Abstract Actinobacteria are a potent source of secondary metabolites. These metabolites have unique metabolic activities which make them an important component in the development of a drug. Sesquiterpene belongs to the class of terpenes and are large structurally diverse group of marine natural products with C-15 carbon skeleton. Whole genome sequence information provides challenges and opportunities for the functional assignment of numerous enzymes that could construct the carbon skeleton of sesquiterpene. They possess versatile chemical scaffolds, suitable biological activities, and unique modes of action, making them potent leads for cancer. Sesquiterpenes possess numerous biological activities such as anti-inflammatory, CNS, cytotoxic activities. There are more than 100 such metabolites of sesquiterpenes mainly derived from marine sources. In this review, we summarized different sesquiterpenes isolated from various sources of actinobacteria and whole genome phylogenomic analysis of actinobacteria. In addition, the review also illustrated the different routes of biosynthetic pathways and various techniques employed for structural characterization and pharmacological effects of sesquiterpene alkaloids.

Keywords Actinobacteria · Sesquiterpenes · Genome data mining · Phylogenomic analysis

---

Gangarapu Kiran and Loganathan Karthik contributed equally with all other contributors.

G. Kiran
School of Pharmacy, Anurag Group of Institutions, Hyderabad, Telangana, India

L. Karthik (✉)
Synthetic Biology, R&D Center, Salem Microbes Private Limited, Salem, India

G. Thumma
Department of Pharmaceutics, University College of Pharmaceutical Sciences, Kakatiya University, Warangal, Telangana, India

A. Vishnukirthi
Department of Microbiology, Lovely Professional University, Punjab, India

L. Karthik (ed.), *Actinobacteria*, https://doi.org/10.1007/978-981-16-5835-8_6

6.1 Introduction

Nowadays, the drug discovery from the marine sources have rapidly expanding due to the huge biodiversity of marine organisms and production many bioactive constituents (Cragg and Newman 2013; Bakiu 2020). Among the class of marine natural products, Sesquiterpene offers an important class of marine natural products and have wide structural variability and identified from all kingdoms of life (Lorigooini et al. 2020; Chakraborty et al. 2018; Rahman 2016). They have a wide range of pharmacological activities such as anticancer (Naine et al. 2016), antiviral (Naine et al. 2016), neuroprotective (Fu et al. 2019), anti-HIV, antibiotic (Banks et al. 2020), immunosuppressant (Isaka et al. 2020), insecticidal and antifungal (Tripathi et al. 2020) activities. Many therapeutically active agents were produced by marine organisms like sponges, molluscs, corals, and tunicates as secondary metabolites (Hassan et al. 2017; Lindequist 2016). These secondary metabolites have diverse structures and could be useful in the development of lead molecules.

About 50,000 terpenoids have identified in all forms of life as secondary metabolites. Among the largest class of terpenoids in natural secondary metabolites, sesquiterpenoids have structural diversity with more than 100 compounds and have potent biological activities such as anticancer, antiviral, antioxidant, antiviral, and CNS activities (Ludwiczuk et al. 2017). Among the different classes of terpenoids, sesquiterpenoids are biosynthesized from FPP (farnesyl diphosphate) leads to synthesis of cyclic terpene cyclases into different hydrocarbons, ethers, or alcohols (Huynh et al. 2018; Reddy et al. 2020).

6.2 Chemistry of Sesquiterpenes

Sesquiterpenes are the largest class of terpenes that contain a core of three isoprene units with $C_{15}H_{24}$ empirical formula and may be cyclic or acyclic in nature (Perveen and Al-Taweel 2018; Böttger et al. 2018). On oxidation or molecular rearrangement, reactions by biochemical modification produce sesquiterpenoids. Any compounds which obey the isoprene rule (C_5H_8) will belong to the class of terpenoids (Ludwiczuk et al. 2017); monoterpenes consist of two isoprene units and have the molecular formula $C_{10}H_{16}$; sesquiterpenes has three isoprene units and have the molecular formula $C_{15}H_{24}$; and similarly diterpenes, triterpenes, tetraterpenes, and polyterpenes will have four, six, and "*n*" number of isoprene units, respectively. The isoprene units are combined by "head to tail" in which C1 unit is joined by C4 of another isoprene unit.

Sesquiterpenes are classified according to chemical structure as acyclic (linear), mono-, bi-, tri-tetracyclic systems and majority are mono- and bicyclic products (Zeng et al. 2019). The farnesyl pyrophosphate (FPP) is biosynthesized with the reaction between GPP (geranyl pyrophosphate) and isopentenyl pyrophosphate (IPP); FPP is the substrate for biosynthesis of different sesquiterpenes (Abdallah

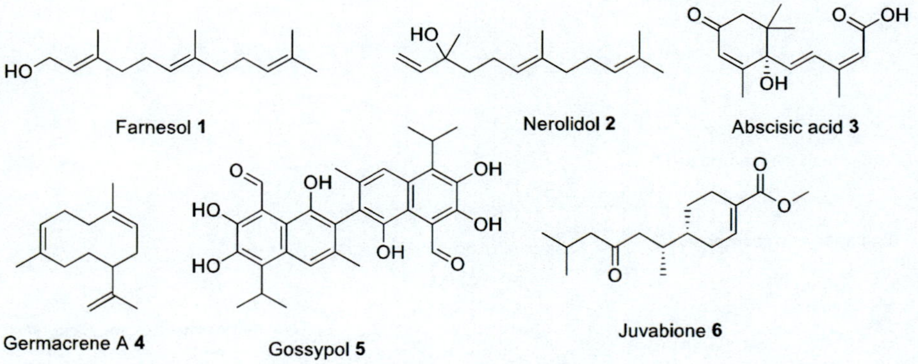

Fig. 6.1 Chemical structures of acyclic and cyclic sesquiterpenes

and Quax 2017). The biosynthesis of cyclic sesquiterpenoids is based on the cyclization of FPP, and these reactions are catalyzed by terpene synthase and produce the secondary metabolites with stereospecificity (Zografos and Anagnostaki 2016). Sesquiterpene have numerous biological properties like insect antifeedant (Inocente et al. 2019), insect pheromones (Lancaster et al. 2019), and phytoalaxins (Tian et al. 2016) which plant produce antimicrobial compounds and protect against the attack of bacteria, fungi, or virus organisms. The examples of acyclic sesquiterpenes are farnesol (**1**) (Jung et al. 2018) and nerolidol (**2**) (Chan et al. 2016) and cyclic sesquiterpenes are abscisic acid (**3**) (Ashraf et al. 2019), germacrene (**4**) (Nguyen et al. 2016), gossypol (**5**) (Tian et al. 2018), juvabione (**6**) (Zheng et al. 2018), etc. (Fig. 6.1).

6.3 Biosynthesis of Sesquiterpenes from Actinobacteria

Actinobacteria have the potential to biosynthesize large number of potent sesquiterpenes as secondary metabolites (Harir et al. 2018). Sesquiterpenes are biosynthesized from mevalonic acid pathway which involves condensation of DMAPP (C_5) with IPP (C_5) to form GPP (C_{10}) and subsequent reaction of IPP with GPP to yield FPP (C_{15}) with catalytic action of prenyltransferase (Ashour et al. 2018). Terpene cyclase releases phosphate groups from GPP and FPP and generates reactive carbocation intermediate and further undergoes intra-molecular cyclization to yield sesquiterpenes (Fig. 6.2). The catalytic action of terpene cyclase consists of aspartic acid (D)-rich motifs at active site with Mg^{+2} ions and water molecules which interact with diphosphate group.

The bacteria terpene from cultural extracts by classical isolation produce the following chemical constituents, viz., pentalenene (**6**), 2-methylisoborneol (**7**), geosmin (**8**), epicubenol (**9**), with the advancement of recent technology in the genome mining have resulted in the discovery of several bacterial terpene synthase

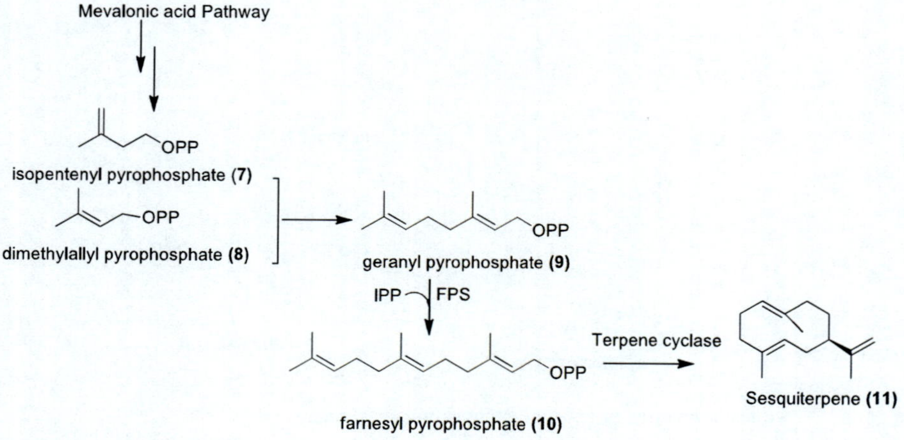

Fig. 6.2 Biosynthesis of Sesquiterpene

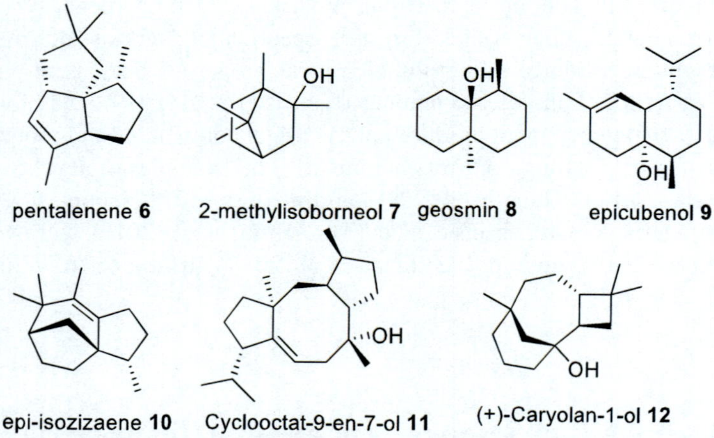

Fig. 6.3 Classical terpenes isolated form actinobacteria

producing epi-isozizaene (**10**), cyclooctat-9-en-7-ol (**11**), caryolan-1-ol (**12**) (Fig. 6.3).

Lukas Lauterbach and Jeroen S. Dickschat have characterized sesquiterpene synthase called bungoene synthase from *S bungoensis* and produce the compound bungoene (**14**). The bungoene synthase composed of two domains and homologs to geosmin synthases from *S. coelicolor* (82% identity) and *S. flavochromogenes* (64% identity), which is obtained in a headspace extract from agar plate cultures of *S. bungoensis*. The mechanism of biosynthesis of bungoene takes place by the isomerization of FPP to nerolidyl diphosphate (NPP) followed by 1,6-ring closer to bisabolyl cation and by1,2-hydride shift, and cyclization forms bungoene (Fig. 6.4)

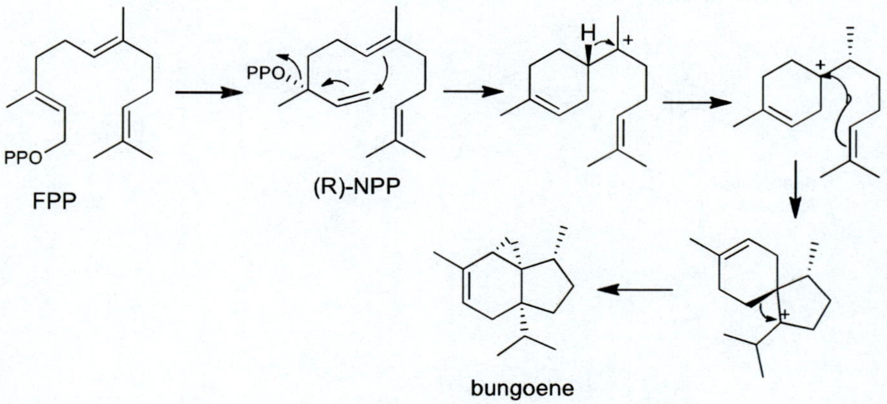

Fig. 6.4 Biosynthesis of bungoene

Tu Cam Le group has isolated Saccharoquinoline from the marine bacterium *Saccharomonospora* sp. CNQ-490 (Le et al. 2019). It is an alkaloidal monoterpene that consists of drimane sesquiterpene with moiety 6,7,8-trihydroxyquinoline-2-carboxylic acid. Saccharoquinoline is a structure similarity to thallusin which has isolated from *Flavobacterium* sp. Thallusin is an open ring structure between C-4′ and C-5′ that could be formed by oxidation of Saccharoquinoline (**15**). The two alkaloids of the biosynthetic pathway were shown in Fig. 6.1. Saccharoquinoline has the cytotoxic activity on colon cancer cell line HCT-116 and shows arrest at G1 phase and causes cell growth inhibition.

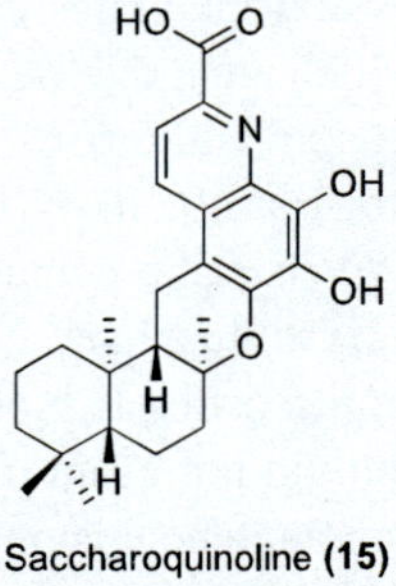

Saccharoquinoline (**15**)

Linear triquinanes composed of hydrocarbon skeleton with three fused 5-membered ring sesquiterpene natural products. Patrick N. Blank et al. have done genome mining with *Streptomyces clavuligerus* and discovered a new terpenoid cyclase with Uniprot Id: B5GLM7 called cucumene synthase (Blank et al. 2018). The enzyme sesquiterpene cyclase catalyzes the stereospecific cyclization of FPP (farnesyl diphosphate) to form a linear triquinane hydrocarbon, (5S,7S,10R,11S)-cucumene (**16**). Cucumene synthase is a class I terpenoid cyclase and consists of four monomers and are identical with less RMS deviation between the monomers with α-helix in their structure. The Mg^{+} ions present in the active site will form

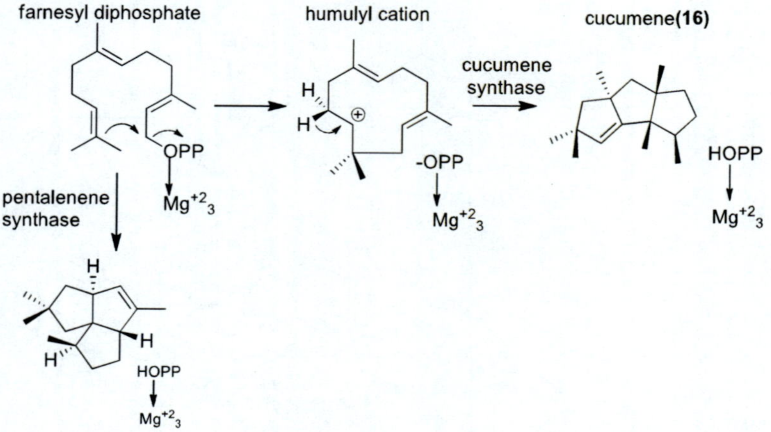

Fig. 6.5 Mechanism of cucumene synthase

interactions with diphosphate which is a good leaving group and simultaneously a bond formation occurs between the C1–C11 and the energy for the bond formation is obtained by bond breaking of diphosphate and thus the process aids in cyclization. This results in formation of cation at C10 and forms an intermediate humulyl cation A and by 1,2-hydride shift humulyl cation B is formed and subsequently carbon–carbon bond results an intermediate protoilludyl cation with 5–6-4 cyclic ring structure. Further alkyl migration yields 5–5-5-cyclic sesquiterpene cucumene is obtained through a sequence of cucumyl cations intermediates. The crystal structure of wild type cucumene synthase at 3.05 A° and T181N variant at 1.96 A° were reposited without ligand bound in protein data bank with PDB ID 6EGK. The active site consists of aromatic residues and facilitates cation-II interactions and forms a stable carbocation intermediate which are a rate-determining step in the biosynthesis of all tricyclic sesquiterpene (Figs. 6.5 and 6.6)

The enzymes that undergo anti-Markovnikov C1–C11 cyclization reaction are cucumene synthase from *S. clavuligerus* and pentalenene synthase from *Streptomyces* UC5319 and these two were crystallized and characterized (Matos et al. 2020).

The synthesis of tricyclic sesquiterpenes is biosynthesized by Markonikov and anti-Markonikov rules with carbenium ion as an intermediate and the reactions are catalyzed by terpene synthase. In pentalenene synthase, the reaction takes place by anti-Markonikov cyclization with FPP as substrate and mechanism of 1,2-hydride shift leads to formation of an intermediate humulyl carbenium cation.

Matos et al. have crystallized apo-pentalenene synthase with 12,13-difluoro farensyl diphosphate as co-crystal ligand and resolved the protein structure with 2.2 A° resolution (Matos et al. 2020). The active site of sesquiterpene cyclase synthase undergoes anti-Markonikov reaction with Mg^+ ion as cofactor. Pentalene synthase is catalyzed by anti-Markonikov cyclization which attacks 10,11 double bonds on C1 of FPP by displacement of diphosphate ions and formation of carbenium ion as an intermediate. The most important amino acid residue Phe76 forms interaction with C9 of FPP and forms cation on C10. The C9 of prenyl chain is

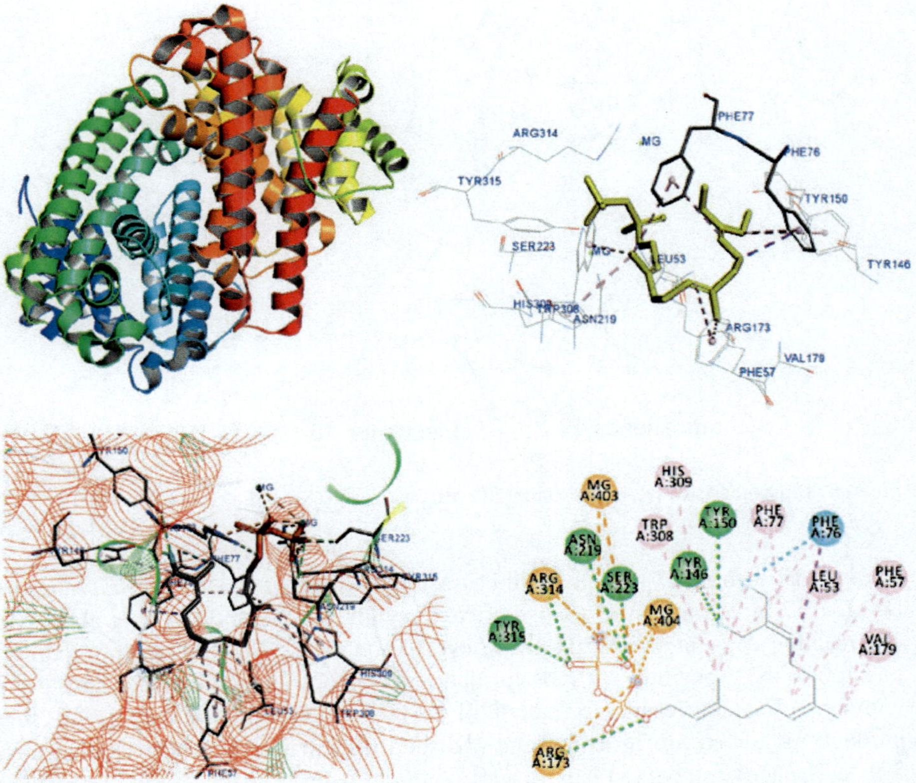

Fig. 6.6 Interactions of cucumene synthase

positioned above the benzene ring residue of Phe76, and this pi-alkyl hydrophobic interaction will stabilize the cation at C10 carbon by hyperconjugation. The other residues which also involved in catalytic action are Phe77 and I177. The crystal structure of pentalenene synthase complexes with 12,13-difluorofarnesyl diphosphate is reposited in protein data bank with PDB ID: 6WKD (https://www.rcsb.org/structure/6WKD) (Fig 6.7).

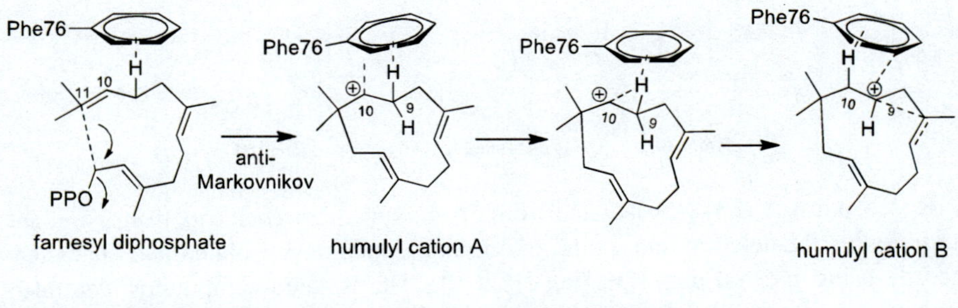

Fig. 6.7 Scheme showing cyclization with concerted 1,2-hydride shift

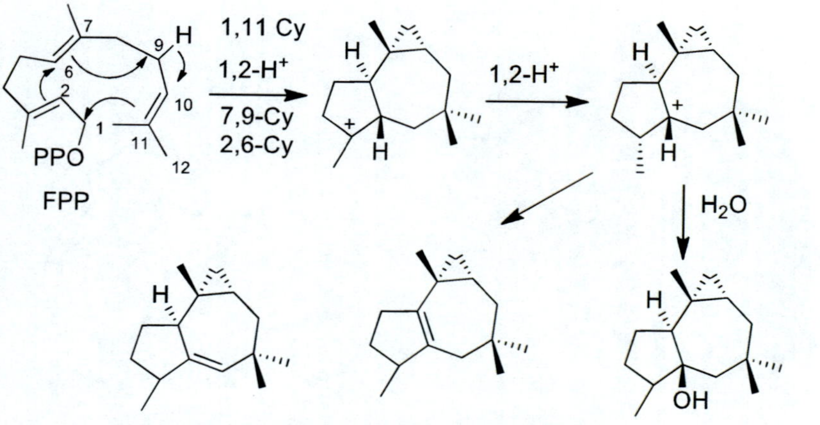

Fig. 6.8 Biosynthesis of (+)-isoafricanol (**17**), african-1-ene (**18**), and african-2(6)-ene (**19**)

(+)-Isoafricanol (**17**) isolated and characterized from *Streptomyces malaysiensis* DSM 4137 by the action of (+)-isoafricanol synthases enzyme (Rabe et al. 2017). The reaction steps involved are C1-C11 cyclization with 1,2-hydride shift from ninth position to tenth position with the formation of cations as intermediates and in the presence of water results formation of main product (+)-isoafricanol and side products produced are african-1-ene (**18**) and african-2(6)-ene (**19**) (Rabe et al. 2017). The active site of (+)-isoafricanol synthases have aromatic amino acid residue and catalysis takes place by cation-II interactions (Fig. 6.8).

Three novel sesquiterpene synthase with catalytic action by α-eudesmol synthase were isolated from *Streptomyces chartreusis* (Kracht et al. 2019). The crystal structure of enzyme α-eudesmol synthase (SCNRRL3882_07544) was crystallized and produced α-eudesmol (**20**), 10-epi-γ-eudesmol (**21**) and elemol (**22**). All these bioactive constituents show repellent activity against mosquitoes and ticks. The crystal structure of α-eudesmol synthase has reposited with 1.83 A° resolution in apo form with PDB entry 1PS1. The structure is homologous to germacradien-4-ol synthase (PDB: 5DW7) from *S. citricolor* shows 33% identity and 25% identity with 4 MC8 with putative sesquiterpene cyclase from *Kitasatospora setae*.

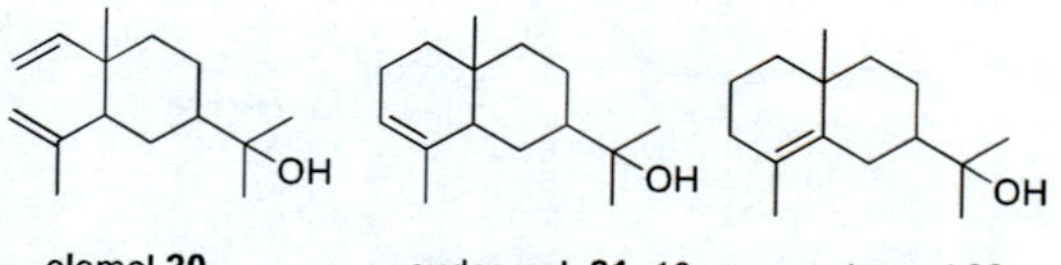

elemol **20** α-eudesmol **21** 10-ep-γ–eudesmol **22**

Roseosporol A (**23**) was isolated from *Streptomyces roseosporus* from a mutant strain with *Lsr2*-deletion and a total of 13 compounds have isolated and characterized by using spectral data (Deng et al. 2019). The isolated compounds were also evaluated for cytotoxic activity against human cancer cell lines, viz., gastric, breast, lung, and hepatic carcinoma. The structural elucidation of Roseosporol A shows 15 resolved peaks in ^{13}C NMR spectrum and identified 12 carbons with 23 protons and three quaternary carbons.

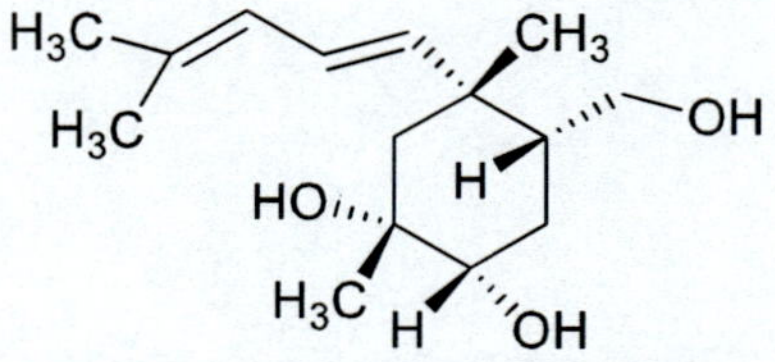

Roseosporol A **23**

Two sesquiterpene ethers (Corvol ethers A and B) **24** and **25** were isolated from *Kitasatospora setae* from the precursor FPP by catalyzing action with enzyme sesquiterpene cyclase (Rabe et al. 2016). Both sesquiterpenes, corvol ether A and B biosynthetic mechanism takes place from FPP to NPP by the removal of diphosphate and cation formation. Then 1,10 cyclization by 1,3-hydride shift and attack of water forms germacrene D-4-ol and neutral intermediate. The product **2** is formed by Wagner–Meerwein rearrangement reaction and product 1 is formed by 1,2-hydride shift and intra-molecular attack of hydroxyl group (Rinkel et al. 2016) (Fig. 6.9).

Neopentalenoketolactone (**26**) the biosynthetic gene cluster was isolated from *S. avermitilis* and is highly similar to pentalenolactone **27** (Gao et al. 2018). They

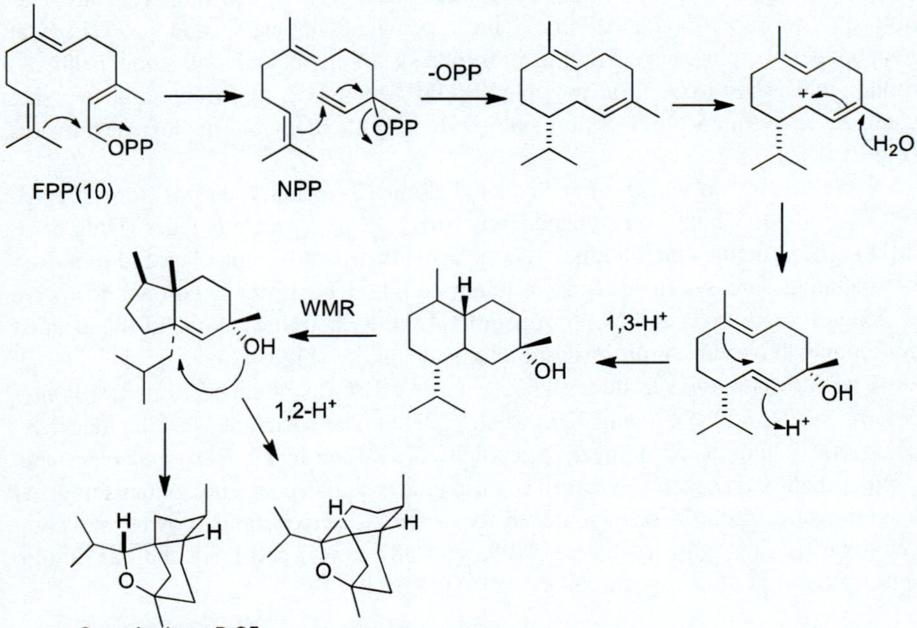

Fig. 6.9 Biosynthesis of corvol ethers A and B

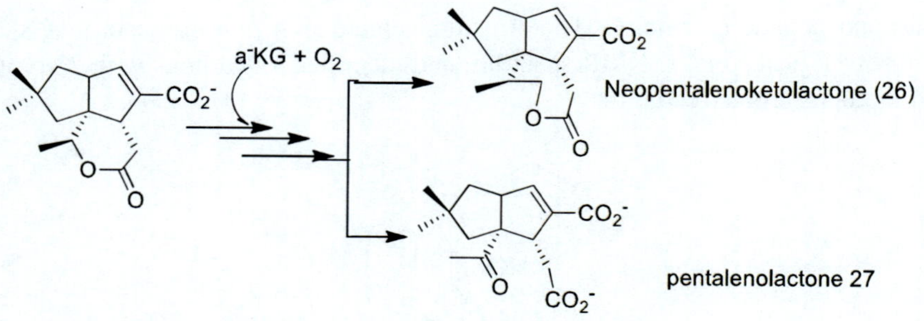

Fig. 6.10 Pentalenolactone (**27**) and neopentalenoketolactone (**26**) biosynthetic pathways

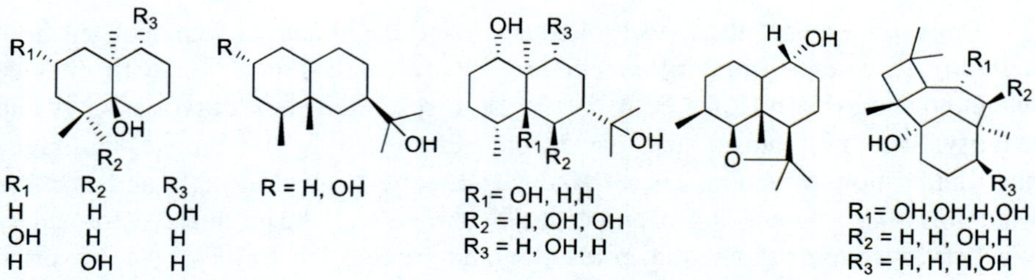

Fig. 6.11 Sesquiterpenoids from *Streptomyces anulatus*

biosynthetic pathway differ at flavin-dependent Baeyer-Villiger monoxygenase catalyzed reaction (PenE/PntE in pentalenolactone and Pt1E in neopentalenoketolactone). For the biosynthesis of neopentalenolactone pathway, mononuclear non-heme iron enzyme (Pt1D) which is a multifunctional enzyme catalyzes the sequence of reaction and involves in the biosynthesis of two pathways (Fig. 6.10).

Nan Ding and co-workers have isolated about 13 sesquiterpenoids from *Streptomyces anulatus* which was isolated from *Giraffa camelopardalis* feces (Ding et al. 2018). The structure elucidation was done using the spectroscopic data and found six new analogs and seven known sesquiterpenoids. The isolated compounds were screened for cytotoxic and NO production inhibitory activities. None of the isolated compounds have shown promising biological activity (Fig 6.11).

A new zizaane-type sesquiterpene, Antartin (B1) was obtained from *Streptomyces sp.* SCO736 by Dayoung Kim et al. (2018a) The structure was elucidated by using NMR and stereochemistry specificity was done using NOE and electronic circular dichroism spectra. Antartin is a tricyclic sesquiterpene and contains unusual phenyl group. Antartin were screened for cytotoxic activity and showed moderate cytotoxic activity against A549, H1299, and U87 cancer cell lines and has mitotic activity arrest at G1 phase of cell cycle (Table 6.1).

Table 6.1 Source, chemical name, and biological properties of sesquiterpenes

Name of the organisms	Name of the compounds	Properties	References
Streptomyces sp. SCO736	1. Antartin	Cytotoxic activity against cancer cell lines	Kim et al. (2018b)
Streptomyces sp. (YIM 56130)	2. Caryolane-1,7 α-diol 3. 1,6,11-eudesmanetriol; (1α,6β)-form 4. 11-eudemane-1,6-diol;(1α,6β)-form		Yang et al. (2011)
Streptomyces violascens ISP 5183 T	5. Albaflavenone 6. Albaflavenol	Cytotoxic activity against tumor cells	Moody et al. (2012)
Streptomyces sp. M491	7. Amorpha-4,11-diene		Wu et al. (2007)
Verrucosispora gifhornensis	8. Cyperusol C		Shirai et al. (2010)
Streptomyces sp. Hd7–21	9. Pyrrolosesquiterpene	Cytotoxic activity	Liu and Liang (2014)
Streptomyces sp. RM-14-6	10. Isopterocarolone		Shaaban et al. (2014)
Streptomyces sp SCSIO 10355	11. Strepsesqutriol	Inhibitory activity against TNF α	Yang et al. (2013)
Streptomyces sp. CNT-372	12. Farneside A 13. Farneside B	Antimalarial activity	Zafrir Ilan et al. (2013)
Streptomyces sp. HKI0595	14. Xiamycin B 15. Indosespene 16. Sespenine 17. Xiamycin A 18. Kandenol A 19. Kandenol B 20. Kandenol C 21. Kandenol D 22. Kandenol E	Antimicrobial activity	Ding et al. (2011)
Streptomyces sp.	23. Pentalenic acid		Takamatsu et al. (2011)
Streptomyces Sp. KS84	24. Oridamycin A and B	Anti S. parasitica activity against phytopathogenic fungus	Takada et al. (2010)
Streptomyces sp. M491	25. 15-hydroxy-T-muurolol 26. 11,15-dihydroxy-T-muurolol	Cytotoxic activity against tumor cells	Ding et al. (2009)
Streptomyces sp. QD518	27. Selina-4(14),7(11)-diene-8,9-diol		Wu et al. (2006)
Streptomyces sp. NPS008187	28. Glaciapyrroles A, B, and C		Macherla et al. (2005)
Actinomyces isolate CNH-099	29. Marinones	Antibiotic	Pathirana et al. (1992)

(continued)

Table 6.1 (continued)

Name of the organisms	Name of the compounds	Properties	References
Streptomyces fradiae IMRU 3535	30. Geosmin	Antibiotic	Gerber and Lechevalier (1977)
Streptomyces albidoflavus	31. (+)-epi-isozizaene	Antibiotic	Moody et al. (2012)
Streptomyces niveus SCSIO 3406	32. Marfuraquinocin 33. Marfuraquinocin D 34. Phenaziterpene B 35. Marfuraquinocin C 36. CHEMBL3092697 37. Marinophenazine B	Cytotoxicity	Song et al. (2013)
Streptomyces arenae	38. Pentalenolactone F	Antibiotic	Wang et al. (2018)

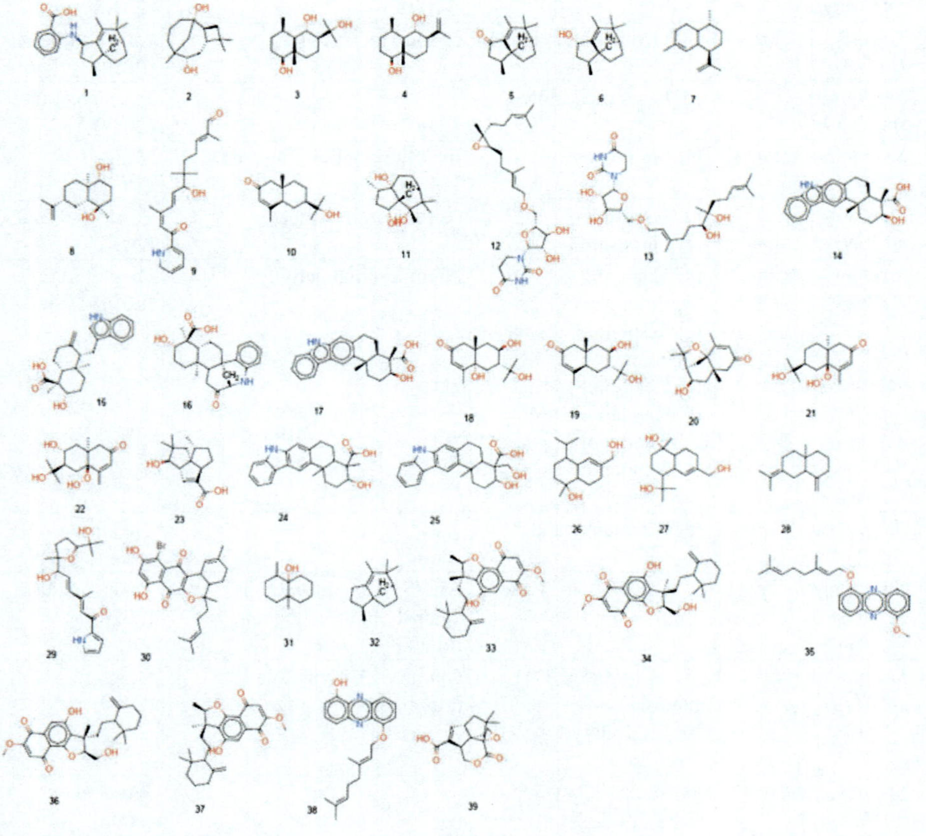

Structures of sesquiterpenes

6.4 Genome Mining of Sesquiterpene from Actinobacteria

Verified and confirmed full-length sequence of sesquiterpene synthase gene of *Streptomyces hygroscopicus* was retrieved from NCBI gene database. The putative sesquiterpene synthase genes were mined from genomes of all microbes available at IMG/M database using BLASTP. The gene equivalents including the hypothetical and putative protein entries were retrieved as an amino acid FASTA file. The amino acid sequences were clustered with 100% similarity using **uclust** algorithm in **USEARCH** command line tool (V-11.0.667). The multiple sequence alignments of the individual clusters were done using CLUSTALW algorithm in MEGA X. The phylogenetic tree data was supplied to Interactive Tree of Life webserver to visualize the phylogenetic tree. The phylogenetic tree clearly indicates that lot of novel sesquiterpenes gene clusters present in the actinobacteria. In future, the novel gene should be expressed using synthetic biology approaches, and its compounds will be used against pathogenic microorganisms.

The gene mining for putative sesquiterpene synthase from all of the microbial genomes returned 497 hits.

Serial no.	Putative gene/protein	Count
1	Hypothetical protein	107
2	Avermitilol synthase	34
3	Germacradienol/geosmin synthase	159
4	Pentalenene synthase	166
5	Terpene cyclase (incl. putative entries)	7
6	Entries belonging broadly under terpene synthase family	22
7	pntA	1
8	penA	1
	Total	**497**

The phylogenetic tree of the complete sequences before clustering is displayed below as Fig. 6.1.

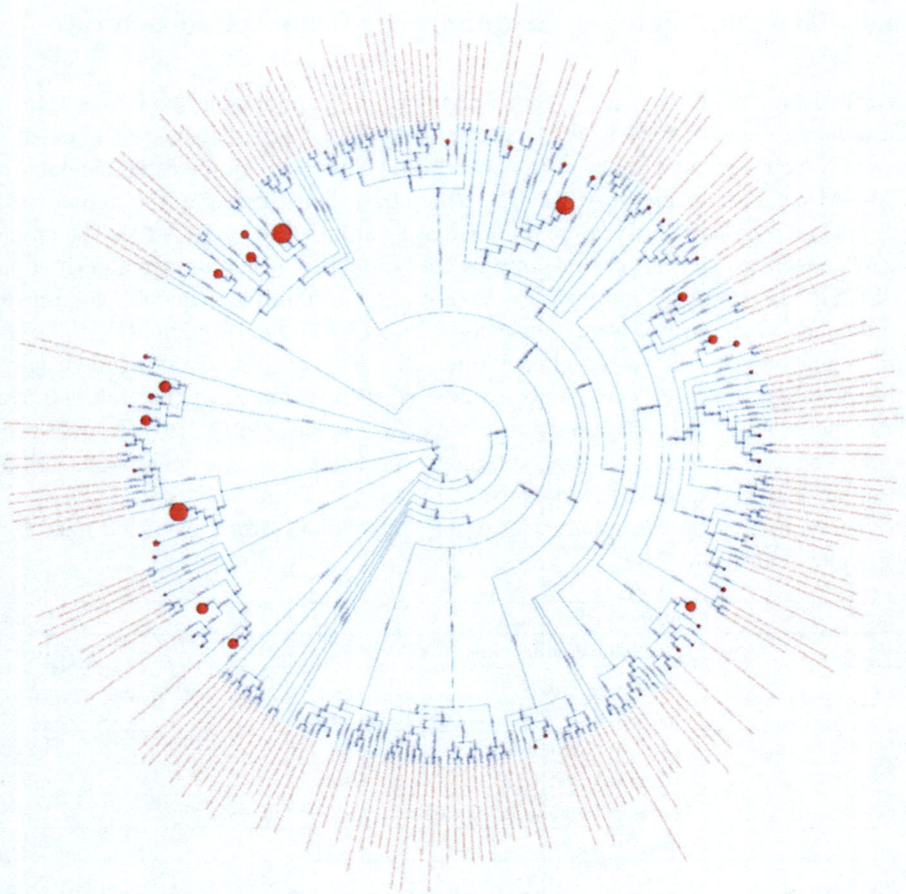

A minimal illustrative phylogenetic tree was created by clubbing the branches with branch length less than 0.05 as shown in Fig. 6.2.

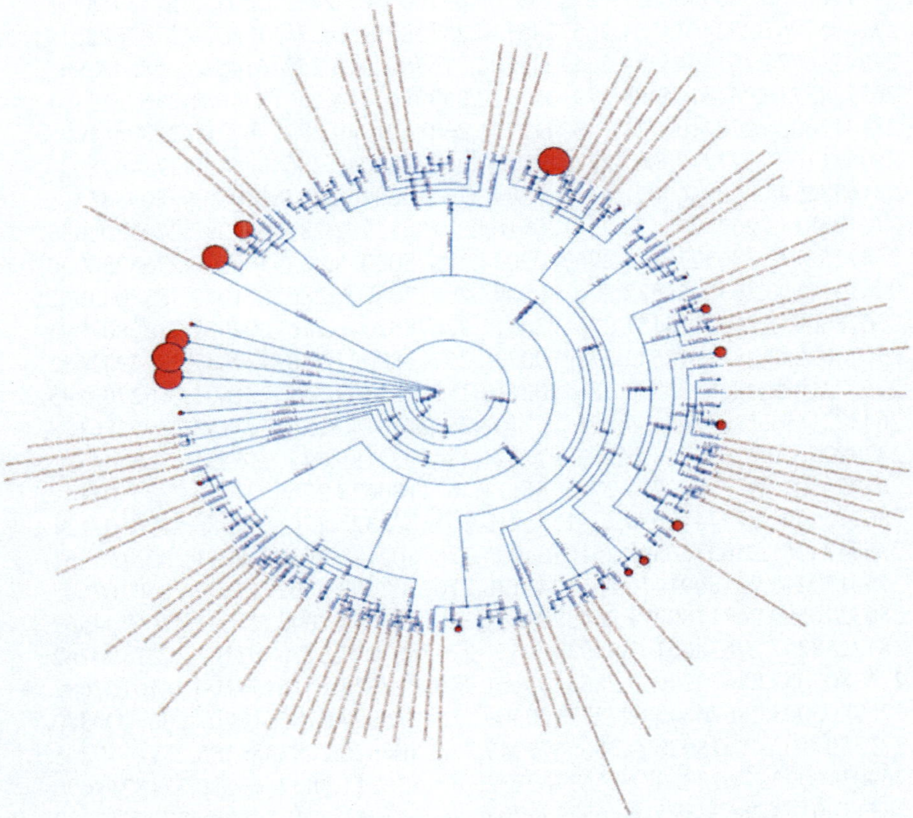

Genome IDs to retrieve metadata—Works only with GOLD database

2872980847,2667527450,2602042089,2775507095,2872807650,2767802136,
2863298641,2731957799,2693429886,2521172697,2528311104,2693429886,28
97521255,2528311104,2528311104,2693429886,2718217699,2863144022,2775
507097,2863155039,2863132943,2744054422,2863121742,2739368071,277385
7764,2775507103,2508501039,2862666418,2687453738,2775507079,27678020
29,2873012545,2645728072,2582581771,2582581028,2648501849,2872922060,
2675903055,2865833183,2616644847,2868405050,2547132121,2862489564,
2767802003,2808606982,2582581764,2856635278,2867592373,2802429656,
2765236465,2791355407,2675902999,2773857921,2551306127,2785510763,
2524614559,2758568662,2767802090,2862705112,2775507138,646862346,
2791355431,2791355412,2875420249,2862407089,2502894011,2775507090,
2877719830,2872793772,647533233,651324105,2693429751,2873077840,
2867260348,2897770829,2867039408,2896961825,2554235020,2524614564,
2873063797,2773857985,2872800169,2873266570,2834008668,2863263905,
2690315863,2867695843,2820999004,2785510740,2866691068,2513237107,
2811995172,2547132111,2636416184,2765236003,2617271283,2718217652,
2765235981,2818991273,2724679104,2775507140,2816332348,2862500152,

2711768182,2609460021,2875405368,637000304,2765236023,2875398477,
2773857967,2524614581,2657245714,2513237100,2751185739,2767802014,
2784746770,2518645584,2657245708,2767802016,2524614552,2585427945,
2657245716,2873144488,2791355422,2639762784,2861534081,2862553101,
2814123086,2772190771,2863412751,2648501849,2872967250,2582581028,
2806311005,2772190667,2772190660,2872922060,2863349733,2724678988,
2816332348,2609460021,637000304,2767802096,2767802008,645058855,
2711768182,2675903139,2551306164,2773857982,2873250297,2767802038,
2547132122,2765236021,2767802034,2767802058,2875367630,2765236030,
2561511190,2630968572,2767802095,2867703208,2865671049,2868033862,
2767802037,2667528177,2765236015,2767802046,2767802051,2767802005,
2767802093,2873012545,2772190735,2671180027,2802429347,2872942632,
2765236022,2866725952,2767802112,2554235367,2643221601,2862793195,
2643221631,2863249894,2513237107,2862739142,2862793195,2795386108,
2873225630,2791355405,2896977621,2862272186,642791623,2856873457,
2639763012,2773857715,2795386111,2671180628,2866571616,2775507142,
2867727567,2737472028,2791355419,2657245328,2767802086,2775507121,
2865494261,2563366508,2515154095,2767802123,2739368071,2873151551,
2756170278,2818991476,2657245138,2767802003,2862215599,2860167608,
2863240559,2671180271,2643221601,2765236017,2643221631,2706794562,
2873258297,2767802102,2693429651,2775507117,2775507131,2772190762,
2767802052,2862224959,2856693151,2865217878,2515154100,2870767688,
2767802023,2867064541,2772190669,2765235994,2785510707,2867009320,
2775507102,2767802069,2865671049,2862507626,2795386105,2574179795,
2690316023,2767802072,2663762722,2681813211,2867663276,2718217699,
2671180173,2863389319,2657245707,2862626331,2711768086,2718217689,
2865968507,2862634903,2675903140,2862799924,2775507138,2657245709,
2765235994,2772190669,2522572192,2795386109,2866412304,2791355431,
2765235983,2767802090,2648501282,2772190736,2773857764,2788500557,
2785510702,2862207470,2728369488,2744054956,2862247409,2862523906,
2857435884,2865841310,2873019354,2866231997,2772190682,2675903055,
2687453810,2645727759,2791355422,2765236039,2765236028,2767802067,
648861016,2865327074,2767802020,2862799924,2654587546,2873040929,
2811995077,2639762732,2767802111,2767802135,2767802018,2765236013,
2767802043,2862458749,2773857986,2765236032,2751185749,2519103181,
2775507126,2791355429,2775507085,2706794555,2785510707,2767802018,
2767802114,2863231158,2767802063,2863276171,2860720294,2684622553,
2547132120,2765235991,2675903135,2844147185,2862523906,2767802011,
2863231158,2574179761,2574179794,2767802124,2863389319,2663762755,
2739367638,2818991472,2862207470,2856722873,2875420249,647533233,
2731957744,2862537573,651324105,2770939519,2671180946,2502894011,
2788500557,2545824776,2866435396,2765235990,2775507108,2767802050,
2671180344,2767802033,2867325999,2863185108,2706794556,2814123243,
2765236027,2561511090,2767802036,2767802054,2895795890,2767802039,
2615840648,2802429331,2765236012,2684622644,2862489564,2875420249,

2767802065,2724679019,2523533533,651324105,2877719830,647533233,
2767802117,2872793772,2867346516,2867353192,2827722778,2565956830,
2502894011,2862603961,2767802045,2872891511,2863083887,2524614673,
2675903050,2690315621,2872898917,2772190740,2654587765,2772190661,
651324106,2643221647,2802429271,2606217189,2867332405,2775507141,
2657245729,2870730386,2795386112,2739368067,2865833183,2868405050,
2867134129,2667527448,2547132121,2823942327,2852152698,2643221548,
2767802035,2788499816,2643221682,2862359521,2765235987,2872793772,
2502894011,2868068591,2856758937,2865272216,2561511106,2875420249,
2877719830,2684622646,2740891882,647533233,651324105,2772190773,
2524614567,2791355431,2568526584,2513237109,2597490183,2524614543,
2639763142,2877684891,2791355412,2718217663,2524614542,2657245720,
2657245702,2873005499,2515154165,2872935496,2657245715,2775507137,
2657245706,2523533511,2896115756,2781125622,646862346,2524614575,
2791355424,2875427445,645058822,645058827,2657245714,2862413135,
2765236030,2862207470,2862731443,2657245707,2687453805,2524614673,
2773857656,2868451029,2775507138,2657245722,2897095264,2518645584,
2767802090,2773857982,2857776322,2648501119,2767802093,2513237100,
2597490081,2524614581,2744054895,2862650393,2765235998,2579778653,
2706794554,2671180027,2865825894,2765236023,2585427945,2802429347,
2574179785,2561511190,2667528177,2773857967,2630968572,2767802095,
2775507099

6.5 Future Perspectives

The genome mining reveals that a lot of novel sesquiterpenes gene cluster located in the actinobacteria genome. Using modern technologies (promoter engineering, synthetic biology, etc.), these silent genes should be expressed to reveal its real biological potentials.

Acknowledgments Thanks to the management of Anurag Group of Institutions and Salem Microbes Private Limited, India.

References

Abdallah II, Quax WJ (2017) A glimpse into the biosynthesis of Terpenoids. KnE Life Sciences:81–98

Ashour M, Wink M, Gershenzon J (2018) Biochemistry of terpenoids: monoterpenes, sesquiterpenes and diterpenes. Ann Plant Rev:258–303

Ashraf SM, Sebastian J, Rathinasamy K (2019) Zerumbone, a cyclic sesquiterpene, exerts antimitotic activity in HeLa cells through tubulin binding and exhibits synergistic activity with vinblastine and paclitaxel. Cell Prolif 52(2):e12558

Bakiu R (2020) Biomedicine developments based on marine biodiversity: present and future. In: Biodiversity and biomedicine. Elsevier, pp 63–79

Banks AM, Song L, Challis GL, Bailey AM, Foster GD (2020) Bovistol B, bovistol D and strossmayerin: Sesquiterpene metabolites from the culture filtrate of the basidiomycete Coprinopsis strossmayeri. PLoS One 15(4):e0229925

Blank PN, Pemberton TA, Chow J-Y, Poulter CD, Christianson DW (2018) Crystal structure of cucumene synthase, a terpenoid cyclase that generates a linear triquinane sesquiterpene. Biochemistry 57(44):6326–6335

Böttger A, Vothknecht U, Bolle C, Wolf A (2018) Terpenes and terpenoids. Lessons on Caffeine, Cannabis & Co. Springer, New Yor, pp 153–170

Chakraborty, K.; Salas, S.; Joy, M.; Francis, P.; Dhara, S.; Antony, T., Classification of marine natural products-chemistry and bioactivity. 2018

Chan W-K, Tan LT-H, Chan K-G, Lee L-H, Goh B-H (2016) Nerolidol: a sesquiterpene alcohol with multi-faceted pharmacological and biological activities. Molecules 21(5):529

Cragg GM, Newman DJ (2013) Natural products: a continuing source of novel drug leads. Biochim Biophys Acta Gen Subj 1830(6):3670–3695

Deng L, Wang R, Wang G, Liu M, Liao Z, Liao G, Chen M (2019) Roseosporol A, the first isolation of a novel sesquiterpenoid from Streptomyces roseosporus. Nat Prod Res 33(14):2038–2043

Ding L, Pfoh R, Rühl S, Qin S, Laatsch H (2009) T-muurolol sesquiterpenes from the marine *Streptomyces sp. M491* and revision of the configuration of previously reported amorphanes. J Nat Prod 72(1):99–101

Ding L, Maier A, Fiebig HH, Lin WH, Hertweck C (2011) A family of multicyclic indolosesquiterpenes from a bacterial endophyte. Org Biomol Chem 9(11):4029–4031

Ding N, Han L, Jiang Y, Li G, Liu J, Mu Y, Huang X (2018) Sesquiterpenoids from Streptomyces anulatus isolated from Giraffa camelopardalis feces. Magn Reson Chem 56(5):352–359

Fu Y, Wang W, Gong Q, Zhang H, Zhao W (2019) Neuroprotective dihydro-β-agarofuran-type sesquiterpenes from the seeds of euonymus maackii. J Nat Prod 82(11):3096–3103

Gao S-S, Naowarojna N, Cheng R, Liu X, Liu P (2018) Recent examples of α-ketoglutarate-dependent mononuclear non-haem iron enzymes in natural product biosyntheses. Nat Prod Rep 35(8):792–837

Gerber NN, Lechevalier HA (1977) Production of geosmin in fermentors and extraction with an ion-exchange resin. Appl Environ Microbiol 34(6):857–858. https://doi.org/10.1128/AEM.34.6.857-858.1977

Harir M, Bendif H, Bellahcene M, Fortas Z, Pogni R (2018) Streptomyces secondary metabolites. Basic Biol Applic Actinobacteria 6:99–122

ul Hassan SS, Anjum K, Abbas SQ, Akhter N, Shagufta BI, Shah SAA, Tasneem U (2017) Emerging biopharmaceuticals from marine actinobacteria. Environ Toxicol Pharmacol 49:34–47

Huynh F, Miller DJ, Allemann RK (2018) Sesquiterpene synthase-catalyzed conversion of a farnesyl diphosphate analogue to a nonnatural Terpenoid ether. In: Methods in enzymology, vol 608. Elsevier, pp 83–95

Inocente EA, Nguyen B, Manwill PK, Benatrehina A, Kweka E, Wu S, Cheng X, Rakotondraibe LH, Piermarini PM (2019) Insecticidal and antifeedant activities of Malagasy medicinal plant (Cinnamosma sp.) extracts and drimane-type sesquiterpenes against Aedes aegypti mosquitoes. Insects 10(11):373

Isaka M, Yangchum A, Choeyklin R, Anaphon S (2020) Acetylenic sesquiterpenoids from cultures of the basidiomycete Stereum cf. hirsutum BCC 26597. Nat Prod Res:1–7

Jung YY, Hwang ST, Sethi G, Fan L, Arfuso F, Ahn KS (2018) Potential anti-inflammatory and anti-cancer properties of farnesol. Molecules 23(11):2827

Kim D, Lee EJ, Lee J, Leutou AS, Shin Y-H, Choi B, Hwang JS, Hahn D, Choi H, Chin J (2018a) Antartin, a cytotoxic zizaane-type Sesquiterpenoid from a Streptomyces sp. Isolated from an antarctic marine sediment. Mar Drugs 16(4):130

Kim D, Lee EJ, Lee J, Leutou AS, Shin YH, Choi B, Hwang JS, Hahn D, Choi H, Chin J, Cho SJ, Hong YD, Ko J, Seong CN, Maloney KN, Oh DC, Yang I, Hwang H, Nam SJ (2018b) Antartin, a cytotoxic zizaane-type sesquiterpenoid from a Streptomyces sp. Isolated from an antarctic marine sediment. Mar Drugs 16(4)

Kracht ON, Cordeiro RSC, Håkansson M, Stockmann J, Sander D, Bandow J, Senges CH, Logan DT, Kourist R (2019) Discovery of three novel sesquiterpene synthases from Streptomyces chartreusis NRRL 3882 and crystal structure of an α-eudesmol synthase. J Biotechnol 297:71–77

Lancaster J, Lehner B, Khrimian A, Muchlinski A, Luck K, Köllner TG, Weber DC, Gundersen-Rindal DE, Tholl D (2019) An IDS-type sesquiterpene synthase produces the pheromone precursor (Z)-α-bisabolene in Nezara viridula. J Chem Ecol 45(2):187–197

Le TC, Lee EJ, Lee J, Hong A, Yim C-Y, Yang I, Choi H, Chin J, Cho SJ, Ko J (2019) Saccharoquinoline, a cytotoxic alkaloidal meroterpenoid from marine-derived bacterium Saccharomonospora sp. Mar Drugs 17(2):98

Lindequist U (2016) Marine-derived pharmaceuticals–challenges and opportunities. Biomol Ther 24(6):561

Liu DZ, Liang BW (2014) A new pyrrolosesquiterpene from the terrestrial *Streptomyces sp. Hd7–21*. Nat Prod Commun 9(4):451–452

Lorigooini Z, Jamshidi-kia F, Dodman S (2020) Analysis of sesquiterpenes and sesquiterpenoids. In: Recent advances in natural products analysis. Elsevier, pp 289–312

Ludwiczuk A, Skalicka-Woźniak K, Georgiev M (2017) Terpenoids. In: *Pharmacognosy*. Elsevier, pp 233–266

Macherla VR, Liu J, Bellows C, Teisan S, Nicholson B, Lam KS (2005) Glaciapyrroles A, B, and C, pyrrolosesquiterpenes from a *Streptomyces sp*. isolated from an Alaskan marine sediment. J Nat Prod 68(5):780–783

Matos JO, Kumar RP, Ma AC, Patterson M, Krauss IJ, Oprian DD (2020) Mechanism underlying anti-markovnikov addition in the reaction of pentalenene synthase. Biochemistry 59(35): 3271–3283

Moody SC, Zhao B, Lei L, Nelson DR, Mullins JG, Waterman MR et al (2012) Investigating conservation of the albaflavenone biosynthetic pathway and CYP170 bifunctionality in streptomycetes. FEBS J 279(9):1640–1649

Naine SJ, Devi CS, Mohanasrinivasan V, Doss CGP, Kumar DT (2016) Binding and molecular dynamic studies of sesquiterpenes (2R-acetoxymethyl-1, 3, 3-trimethyl-4t-(3-methyl-2-buten-1-yl)-1t-cyclohexanol) derived from marine Streptomyces sp. VITJS8 as potential anticancer agent. Appl Microbiol Biotechnol 100(6):2869–2882

Nguyen T-D, Faraldos JA, Vardakou M, Salmon M, O'Maille PE, Ro D-K (2016) Discovery of germacrene a synthases in Barnadesia spinosa: the first committed step in sesquiterpene lactone biosynthesis in the basal member of the Asteraceae. Biochem Biophys Res Commun 479(4): 622–627

Pathirana C, Jensen JC, Fenical W (1992) Marinone and debromomarinone: antibiotic sesquiterpenoid naphthoquinones of a new structure class from a marine bacterium. Tetrahedron Lett 33(50):7663–7666

Perveen, S.; Al-Taweel, A., Terpenes and terpenoids. BoD–Books on Demand: 2018

Rabe P, Janusko A, Goldfuss B, Dickschat JS (2016) Experimental and theoretical studies on Corvol ether biosynthesis. Chembiochem 17(2):146–149

Rabe P, Samborskyy M, Leadlay PF, Dickschat JS (2017) Isoafricanol synthase from Streptomyces malaysiensis. Org Biomol Chem 15(11):2353–2358

Rahman A-U (2016) Studies in natural products chemistry: bioactive natural products (part XII). Elsevier

Reddy GK, Leferink NG, Umemura M, Ahmed ST, Breitling R, Scrutton NS, Takano E (2020) Exploring novel bacterial terpene synthases. PLoS One 15(4):e0232220

Rinkel J, Rabe P, Garbeva P, Dickschat JS (2016) Lessons from 1, 3-hydride shifts in sesquiterpene cyclizations. Angew Chem Int Ed 55(43):13593–13596

Shaaban KA, Singh S, Elshahawi SI, Wang X, Ponomareva LV, Sunkara M et al (2014) The native production of the sesquiterpene isopterocarpolone by Streptomyces sp. RM-14-6. Nat Prod Res 28(5):337–339

Shirai M, Okuda M, Motohashi K, Imoto M, Furihata K, Matsuo Y, Katsuta A, Shizuri Y, Seto H (2010) Terpenoids produced by actinomycetes: isolation, structural elucidation and biosynthesis of new diterpenes, gifhornenolones A and B from *Verrucosispora gifhornensis YM28–088*. J Antibiot (Tokyo) 63(5):245–250

Song Y, Huang H, Chen Y, Ding J, Zhang Y, Sun A, Zhang W, Ju J (2013) Cytotoxic and antibacterial marfuraquinocins from the deep South China Sea-derived Streptomyces niveus SCSIO 3406. J Nat Prod 76(12):2263–2268. https://doi.org/10.1021/np4006025

Takada K, Kajiwara H, Imamura N (2010) Oridamycins A and B, anti-Saprolegnia parasitica indolosesquiterpenes isolated from *Streptomyces sp. KS84*. J Nat Prod 73(4):698–701

Takamatsu S, Xu LH, Fushinobu S, Shoun H, Komatsu M, Cane DE, Ikeda H (2011) Pentalenic acid is a shunt metabolite in the biosynthesis of the pentalenolactone family of metabolites: hydroxylation of 1-deoxypentalenic acid mediated by CYP105D7 (SAV_7469) of *Streptomyces avermitilis*. J Antibiot (Tokyo) 64(1):65–71

Tian X, Ruan J, Huang J, Fang X, Mao Y, Wang L, Chen X, Yang C (2016) Gossypol: phytoalexin of cotton. Sci China Life Sci 59(2):122–129

Tian X, Ruan J-X, Huang J-Q, Yang C-Q, Fang X, Chen Z-W, Hong H, Wang L-J, Mao Y-B, Lu S (2018) Characterization of gossypol biosynthetic pathway. Proc Natl Acad Sci 115(23):E5410–E5418

Tripathi SK, Feng Q, Liu L, Levin DE, Roy KK, Doerksen RJ, Baerson SR, Shi X, Pan X, Xu W-H (2020) Puupehenone, a marine-sponge-derived sesquiterpene quinone, potentiates the antifungal drug caspofungin by disrupting Hsp90 activity and the cell wall integrity pathway. Msphere 5(1)

Wang X, Shi J, Liu Y (2018) Oxidative rearrangement mechanism of Pentalenolactone F catalyzed by cytochrome P450 CYP161C2 (PntM). Inorg Chem 57(15):8933–8941. https://doi.org/10.1021/acs.inorgchem.8b00860

Wu SJ, Fotso S, Li F, Qin S, Kelter G, Fiebig HH, Laatsch H (2006) N-carboxamido-staurosporine and selina-4(14),7(11)-diene-8,9-diol, new metabolites from a marine *Streptomyces sp*. J Antibiot (Tokyo) 59(6):331–337

Wu SJ, Fotso S, Li F, Qin S, Laatsch H (2007) Amorphane sesquiterpenes from a marine Streptomyces sp.,1. J Nat Prod 70(2):304–306

Yang Z, Yang Y, Yang X, Zhang Y, Zhao L, Xu L, Ding Z (2011) Sesquiterpenes from the secondary metabolites of Streptomyces sp.(YIM 56130). Chem Pharm Bull 59(11):1430–1433

Yang X, Peng K, Liu Z, Zhang G, Li J, Wang N, Steinmetz A, Liu Y (2013) Strepsesquitriol, a rearranged zizaane-type sesquiterpenoid from the Deep-Sea-derived actinomycete *Streptomyces sp SCSIO 10355*. J Nat Prod 76(12):2360–2363

Zafrir Ilan E, Torres MR, Prudhomme J, Le Roch K, Jensen PR, Fenical W (2013) Farnesides A and B, sesquiterpenoid nucleoside ethers from a marine-derived *Streptomyces sp., strain CNT-372*, Fiji. J Nat Prod 76(9):1815–1818

Zeng T, Liu Z, Liu H, He W, Tang X, Xie L, Wu R (2019) Exploring chemical and biological space of terpenoids. J Chem Inf Model 59(9):3667–3678

Zheng J, Margarita C, Krajangsri S, Andersson PG (2018) Asymmetric total synthesis of (−)-Juvabione via sequential Ir-catalyzed hydrogenations. Org Lett 20(18):5676–5679

Zografos AL, Anagnostaki EE (2016). From acetate to mevalonate and deoxyxylulose phosphate biosynthetic pathways: an introduction to terpenoids. From Biosynthesis to Total Synthesis: Strategies and Tactics for Natural Products. 189

Chapter 7
Mining for Biosynthetic Gene Clusters in *Actinobacteria* Genomes Via Bioinformatics Tools

Omnia Mohamed Abdallah, Amal Mohamed Shawky, Dina Hatem Amin, and Alaa Fayez Elsayed

Abstract Genome mining using next-generation sequencing (NGS) technologies and bioinformatics tools opens the way to understand secondary metabolite biosynthesis potentials of *Actinobacteria*. Reads of different lengths are produced using technologies like Illumina, Ion Torrent, PacBio, and ONT. These reads are used to reconstruct the original genome using Metagenomic assembly approaches. Genome annotation then detects the coding loci and identifies their protein products. Many in silico genome mining softwares allow the identification of secondary metabolites biosynthetic gene clusters (smBGCs) as well as detection of unique genes. These novel technologies will allow the production of new bioactive compounds.

Keywords Next-generation sequencing · Genome mining · Metagenomic assembly · Gene prediction · Annotation

7.1 Introduction

Actinobacteria are Gram-positive bacteria that possess increased G+C DNA content and form one of the biggest bacterial phyla, and they are distributed widely in aquatic and terrestrial environments. They are producing about two-thirds of all naturally produced antibiotics in present clinical use (Barka et al. 2016). Streptomycetes family of *Actinobacteria* produces about 75% of antibiotics like streptomycin, ivermectin, tetracycline, and nystatin (Basnet et al. 2006). They also produce several

O. M. Abdallah · A. F. Elsayed
Microbiology Department, Faculty of Science, Ain Shams University, Cairo, Egypt

A. M. Shawky · D. H. Amin (✉)
Microbiology Department, Egyptian Drug Authority (EDA), Giza, Egypt
e-mail: dina.hatem@sci.asu.edu.eg

L. Karthik (ed.), *Actinobacteria*, https://doi.org/10.1007/978-981-16-5835-8_7

anticancer, anthelmintic, and antifungal compounds. Therefore, *Actinobacteria* are of great value for medicine, biotechnology, and agriculture (Barka et al. 2016).

Biosynthetic gene clusters (BGCs) encode biologically active compounds. The understanding of secondary metabolite biosynthesis potentials of Actinobacteria has increased by the advancement of next-generation sequencing (NGS) technologies (Nouioui et al. 2018). The genomic analysis revealed that Actinobacteria are able to produce many more compounds than those observed in in vitro culture, which indicates that under standard laboratory conditions many of these BGCs are silent or weakly expressed.

Two main strategies in the application of bioinformatic tools are present. First, the rule-based approaches used to determine gene clusters encoding known biosynthetic pathways accurately. Initially, these tools determine genes encoding conserved enzymes/protein domains that possess a function in secondary metabolism. After that, predefined rules are applied to link the appearance of such hits with defined classes of natural products (Blin et al. 2013; Medema et al. 2011; Weber et al. 2015). Depending on the precondition of having defined rules, these algorithms are not able to discover novel pathways which use a variant biochemistry and enzymes. To bypass this constraint, second rule-independent strategies have been developed that apply machine learning-based methods or automated phylogenomics analyses to create their predictions (Cimermancic et al. 2014; Cruz-Morales et al. 2015).

7.2 Next-Generation Sequencing

Next-generation sequencing (NGS) includes second- and third-generation sequencing technologies that provided much accurate insights of microbiomes. Technologies like Illumina and Ion Torrent, which generate several million short reads (150–400 bp) are needed in the second-generation sequencing; while third-generation PacBio and ONT are included in sequencing and generates considerably longer reads (6–20 kb), but much fewer per run.

Illumina technology utilizes the sequence-by-synthesis approach where small DNA fragments are connected to a glass slide in microwave. For clusters production, intensified nucleotides labeled by fluorescence are washed and incorporated across the flow cell, supplementary to the clustered fragment's DNA sequence. Fluorescence from the embedded nucleotides is detected uncovering the sequence of DNA. Illumin, almost definitely, Illumina offers the best throughput, generating relatively short duration reads with the lowest cost per-base, up to 300 bp (van Dijk et al. 2014). Several recent studies have used illumina for genome sequencing (Fernandes et al. 2020; Lin et al. 2020; Nimnoi and Pongsilp 2020).

Ion Torrent technology, DNA fragments are loaded on beads, then single beads are located into micro-wells, each one of the four nucleotides flow and get merged into a complementary strand, an H^+ ion is released that is detected as a change in voltage. The process is repeated. A run can be finished by Ion Torrent technology much faster than other platforms generating up to 400 bp duration reads. Still,

Illumina technology is used more widely, presumably because of the increased rate of homopolymer errors (van Dijk et al. 2014).

Another technology is **Oxford nanopore technology**. In this technology, one single strand of DNA passes across a nanopore protein, and the electrical variations are calculated. A DNA polymer complex used in this technology includes a double-stranded DNA and an enzyme that unwinds the double strand and drives the single strand of DNA across a nanopore. A detectable interruption in the electrical current is generated that determines the bases order at the DNA stand (Lee et al. 2019).

7.3 Metagenomic Assembly

Metagenomic assembly indicates the reconstruction of the original genome from sequenced fragments. Assemblers use a reference-free de novo strategy to build contiguous sequences (contigs). Two major models are used by de novo assembly software tools: overlap-layout-consensus (OLC) and the approach of the de Bruijn graph (Pevzner et al. 2001). For the long reads assembly, the (OLC) approach has been used. Whereas for the assembly of short reads, the de Bruijn graph approach has been used through their conversion to k-mers (sequences of long k bases) (Sohn and Nam 2018) as shown in Fig. 7.1. The resultant graph can be used to create

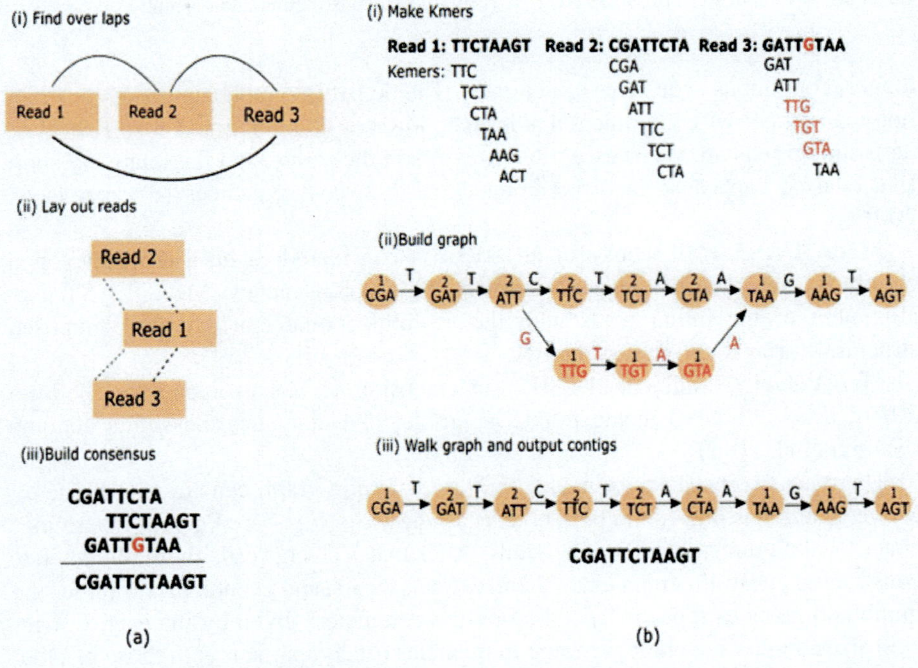

Fig. 7.1 The two different approaches to genome assembly: (a) Overlap, Layout, (b) de Bruijn graph assembly

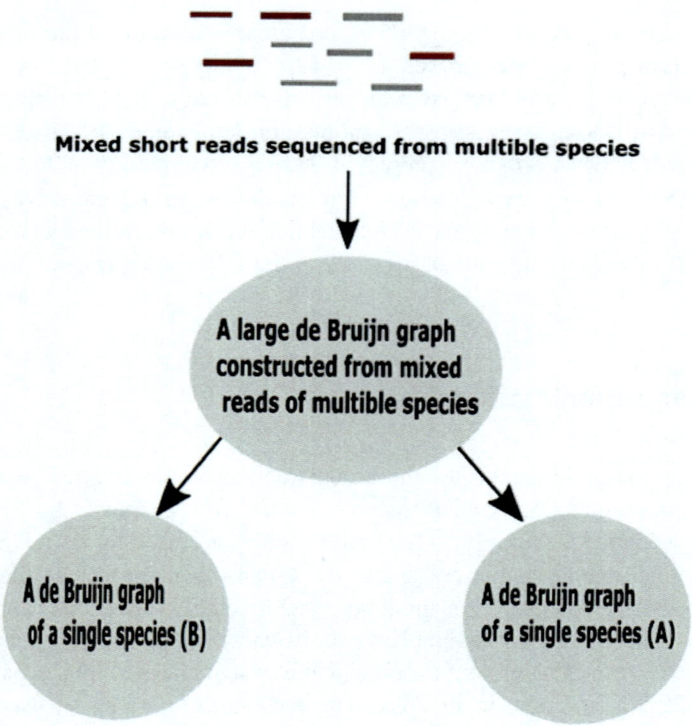

Fig. 7.2 Two main steps in MetaVelvet: de Bruijn graph constructing and hashing

longer, contiguous sequences of genomes. The de Bruijn graph can make construction without pairwise alignment that leads to lowered computational cost. However, de Bruijn graphs are very susceptible to errors in the series, and the relatively short k-mers used can cause incorrect connections between sequences (Pevzner et al. 2001).

MetaQUAST (Mikheenko et al. 2016) is considered to be a tool developed mainly to assess the metagenomics assemblies for consistency. MetaQUAST uses alignment of the initial reads with the assembled data that can find supposed structural variants and misassemblies.

MetaVelvet (Namiki et al. 2012) an extension of Velvet assembler (Zerbino 2010), it is a de novo metagenomic assembler that elongates the single genome (Namiki et al. 2012).

There are two basic steps in MetaVelvet, de Bruijn graph constructing (done by Velvetg) and hashing (done by Velveth) as shown in Fig. 7.2. Velveth determines exact local alignments among the reads. After that Velvetg reads these alignments, constructs a de Bruijn graph using them, extracts errors, and eventually simplifies the graph and resolves repeats depending on the parameters given by the user. Longer N50 sizes, higher coverage, genome frequencies (in comparison with those of other metagenomes and single genomes assemblers), and high predicted protein numbers by program for MetaGene gene discovery was recorded by authors of MetaVelvet (Noguchi et al. 2006).

MetaVelvet-SL (Afiahayati and Sakakibara 2015) is a MetaVelvet extension in assembling metagenomics data which focuses on the identification and classification of chimeric nodes, in the assembly network, MetaVelvet is recorded to be more leading in performance than other assemblers like (IDBA-UD) (Peng et al. 2012) and Ray Meta (Boisvert et al. 2012).

In **IDBA-UD,** contigs are created (Peng et al. 2012(using continuous increasing rise across progressive assembly cycles values in k-mer. In the assembly operation, IDBA-UD iterates from a small *k* to a large *k*. In every repetition, short and low-depth contigs are extracted by depth-relative cutoff thresholds, from low to high, to lower the errors in low-depth and high-depth regions. This process counterbalances for the data loss with de Bruijn graphs built using a single k-mer length. The reads are locally collected to produce some lost k-mers in low-depth regions. The local assembly narrows the gaps and resolve repeats in the de Bruijn graph.

Megahit (Li et al. 2015) uses a very analogous approach to IDBA-UD, besides the brief benefit of de Bruijn graphs (Bowe et al. 2012) and memory specialization lowering Graphics Processing Units (GPUs); moreover, it possesses high speed.

Ray Meta is a measurable software tool which utilizes distributed computing to manage large dataset. Structures depending on de Bruijn graphs are built. The mean coverage depth in parallel assembly processes is estimated by local processes of the distributions coverage of k-mers (Boisvert et al. 2012). The assembled data is verified by aligning it with the reference [MUMmm] Genomes software (Kurtz et al. 2004).

MetAMOS is an example of a module system (Treangen et al. 2013) which merges existing instruments into a metagenomic structure pipeline for research. The pipeline is divided into three stages: the first one is to make (meta) genome assembly with an option. Second, scaffolding is produced using paired-end and mate-pair data by Bambus 2 (Koren et al. 2011). Finally, there is a post-assembly point, where the scaffolds are annotated and determined taxonomically. It is possible to choose the best assembler for a certain application. A basic advantage of MetAMOS is that it can test various assembly tools and find the most appropriate one for a given database.

One of the fundamental features for keeping the scaffolds contiguous is the spotting of genetic variation patterns. MetAMOS has the ability to maintain a contiguous genomic backbone and also maintain a contiguous backbone highlighting regions of variables. An HTML report is constructed by summing up the conclusion of the study findings.

7.4 Gene Prediction and Annotation

Genome annotation is the following step in metagenomic analysis approach which detects the whole coding region and their loci, in addition to the identification of the protein products; the whole process is indicated in Fig. 7.3.

MetaGeneAnnotator (Noguchi et al. 2008) is algorithm of metagenomic gene finding that predicts short-sequence genes using uncharacterized metagenomic groups. It predicts. Statistical models of prophage genes, as well as bacterial and archaeal genes are integrated by the MetaGeneAnnotator, in addition it uses a self-training model from input sequences for predictions. Consequently, it possesses high sensitively which detects typical genes as well as atypical genes, like prophage and horizontally transferred genes in a prokaryotic genome. (Noguchi et al. 2008).

Orphelia uses a machine learning two-step approach. In the first stage, monocodon use-based linear discriminant analysis, Dicodon use and initiation sites for translation are used to determine genomic sequence features. The second stage, artificial neural network is built, integrating the features of phase 1 with open reading frame length details and GC-content calculating the chances of an encoding protein in an ORF (Open reading frame). Orphelia has the advantage of higher levels of specificity but decreased sensitivity in comparison to MetaGeneAnnotator and MetaGene in gene prediction (Noguchi et al. 2006) (a MetaGeneAnnotator precursor) on simulated results.

Glimmer-MG (Kelley et al. 2012) is an extension of the common Glimmer program for prediction of bacterial genes (Delcher et al. 2007). Glimmer-MG begins with data clustering of mostly the same organism by using Phymm (Brady and

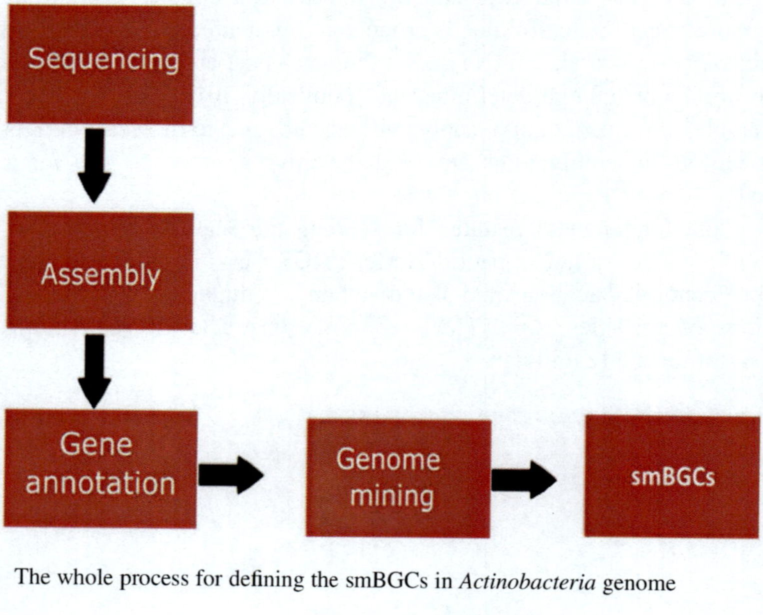

Fig. 7.3 The whole process for defining the smBGCs in *Actinobacteria* genome

Salzberg 2009). Unclassified data are clustered through using Scimm (Kelley and Salzberg 2010).

Gene models on the basis of HMMs (hidden Markov models) through incorporating probabilistic models for genes used to predict genes determine genes length and locate start/stop codons. It was shown that the merging between gene prediction and phylogenetic classification can give more precise predictions. Glimmer-MG detects insertions/deletions in simulated data more precisely and can more accurately predict errors of substitution that affect stop codons.

To discover new secondary metabolites, which can be antibiotics, anticancer, or anthelmintic through genome mining the first step is to make sequencing of the genome using NGS technology, then assembly of the sequences emerged by assembly software, after that comes the genome annotation which define the structural and functional identity of genes; this leads to the outcome of genome mining and location of smBGCs that produces the secondary metabolites.

7.5 Secondary Metabolite Biosynthetic Gene Clusters

Secondary metabolite biosynthetic gene clusters (smBGCs) are biosynthetic gene clusters that encode bacterial secondary metabolites that their production is specific to phylogeny terms called phylotype and ecotype front. Also smBGCs mostly encode genes for enzymes that synthesize specific monomers, transporters, and regulatory elements, besides their host resistance role (Belknap et al. 2020). Mainly there are two pathways for biosynthesizing bacterial secondary metabolites (Lin et al. 1994), the first one is the non-ribosomal peptide synthetase (NRPS) and the second one is the polyketide synthase (PKS) gene clusters.

7.6 In Silico Tools for Genome Mining of smBGCs

DECIPHER was the first microbial natural-product biosynthetic loci database for in silico genome mining of smBGCs, constructed by Ecopia Biosciences Inc. (Zazopoulos et al. 2003). Different free databases and tools are developed for smBGC prediction; they are BAGEL (de Jong et al. 2006), ClustScan (Starcevic et al. 2008), CLUSEAN (Weber et al. 2009), and NP.searcher (Li et al. 2009).

BAGEL enables researchers to identify bacterial (meta-) genomic DNA for bacteriocins that are small antimicrobial peptides have been usually omitted in the annotation process and RiPPs according to based bacteriocin databases and motif databases.

ClustScan program role is to predict chemical structures of metabolites, alongside permitting the user to add specification related to domains. The outputs of analyses are presented in an easy, modifiable graphical interface (Starcevic et al. 2008).

CLUSEAN (CLUster SEquence ANalyzer) is an open-source software pipeline which aids in annotating and analyzing such gene clusters to determine the functional domains and motifs NRPS/type I PKS and the prediction of specificities of NRPS (Weber et al. 2009).

NP.searcher is an open-source software which can be modified through additional programming, it also allows accessing to 2D and 3D molecular structures directly from DNA by tracking of nucleotide sequence input and then use of the output SMILES (Li et al. 2009).

These tools have a manual of their mining software and databases that is available at "Secondary Metabolite Bioinformatics Portal." These tools are restricted to certain classes of secondary metabolites. PRISM and antiSMASH are probes-dependent tools that use Hidden Markov Model (HMM) profile that includes the alignment sequence; they identify smBGCs according to the highly conserved core of biosynthetic enzymes. The data are introduced in FASTA format to provide rapid gene annotation of the bacterial genome with excluding the false positive by applying negative models (e.g., fatty acid synthases are homologous to PKSs). PRISM version 3 provides identification of 22 various kinds of smBGCs, and antiSMASH version 5 provides prediction up to 52 various kinds of smBGCs (Lee et al. 2019).

DeepBGC and Clusterfinder are developed machine learning tools that enable the identification of novel smBGCs using MIBiG database; however, there is a limitation to identify complete novel smBGCs (Cimermancic et al. 2014; Hannigan et al. 2019).

7.7 Characterizing smBGCs Identified by Genome Mining

There are two ways of genome mining of *Streptomyces*: The reverse (metabolites to genes) approach that permits researchers to detect the BGCs of defined secondary metabolites and forward (genes to metabolites) approach that detects the products of unknown smBGCs (Lee et al. 2019).

These tools helped with identification of already discovered antimicrobials like anthracimycin antibiotic that discovered in 1995, through antiSMASH tool where single PKS gene cluster was predicted as the biosynthetic way for the production of anthracimycin (Alt and Wilkinson 2015), so it is useful to use a standard experimentally identified smBGCs for PKSs and NRPSs to help with constructing the basic structure of the compound followed by identification of domains to predict how to be modified and cyclized; these results are aligned to the database to put hands on the compound produced by their novel smBGCs, in addition to the ability of prediction of their chemical structure through antiSMASH and PRISM (Khater et al. 2016).

The precision of chemistry prediction depends on the algorithm and the database mediated to predict the enzyme functional domains and the substrate affinity of the domains (Skinnider et al. 2015). For more accurate predictions, updated versions of antiSMASH and PRISM are developed; antiSMASH version 5 is an improved version to show more developed chemistry prediction, while PRISM can construct a broad range of predicted structures depending on unsureness of tailoring sites. When there are a huge number of unknown genes that may encode bioactive metabolites; the prediction of chemical structure in forward approach has a challenge of unknown genes (Lee et al. 2019).

Some genetic engineering procedures are applied to help in silent smBGCs characterization, in addition to using of synthetic biology tools and using of "Heterologous expression of silent smBGCs in different Streptomyces"((Lee et al. 2019)).

7.8 PK and NRP Biosynthetic Pathways Detection Using Bioinformatics Tools

In silico genetic search in sequenced microbial genomic data to reveal novel NRPS/PKS clusters and consequently identify new (NRPs) have been progressed through the studies on NRPS domains, the understanding of their gene cluster architecture and tailoring enzymes. Adenylation domain is a part of the NRPSs which is selective for the substrate. Sometimes, it requires the MbtH homolog small protein for its ideal activity. The existence of adenylation domain and MbtH homologs in a sequenced genome possibly identifies the new secondary metabolite producers. (Singh et al. 2017).

The biosynthetic pathways PK and NRP are also detected by hidden Markov models (HMMs), HMMs as statistical models produced from multiple sequences. They are preferable than pairwise search methods such as BLAST in revealing far linked homologs. Proteins profile from type I, type II, and type III PK and NRP biosynthetic pathways are detected using advanced applications of HMMs. These tools have been involved into web-based search tools like antiSMASH, NP.searcher (Li et al. 2009), NaPDoS (Ziemert et al. 2012), and PKMiner (Kim and Yi 2012) (that focuses particularly on type II PKs).

AntiSMASH (antibiotics secondary metabolites analysis shell) is a fully automated approach to explore bacterial and fungal genome for smBGCs (Weber and Kim 2016). It defines clusters as sets of signature genes separated by 10 kb between them and elongates the cluster 20 kb on each side of the last signature gene to clarify the borders of PK and NRP biosynthetic gene clusters (Medema et al. 2011). NP.searcher annotates pathways as extending 15 kb upstream and downstream from a PKS or NRPS gene. Any extra PKS or NRPS genes in this window are included in the cluster and the gene cluster has elongated another 15 kb from the newly added gene.

7.9 Detection of Unique Gene Clusters

Identification of the gene clusters that can encode novel molecules is thus a basic advantage for genome mining. Biosynthetic gene cluster of the oxytetracycline (broad spectrum antibiotic) identified through BLASTn analysis from Streptomyces rimosus (Zhang et al. 2006), but analysis of the antitumor pederin biosynthetic gene cluster by BLASTn (Piel 2002) cannot identify the onnamide a member of pederin family biosynthetic gene cluster (Piel et al. 2004) although these two biosynthetic compounds have common substantial homology since the biosynthetic pathway of perderin is segmented into three genome zones. Actinomycetes (65–80% GC) and firmicutes (≈35% GC) differ greatly in GC-content, so nucleotide-based searches may not determine linked gene clusters from organisms. Protein-based searches have no limitations. In general, analysis of a single core protein from a PK or NRP gene cluster by BLASTp produces a significant number of strong hits which need to be curated manually to classify associated pathways, the BLASTp analysis of the last PKS protein in erythromycin antibiotic biosynthesis, EryIIIA, from Aeromicrobium erythreum (Brikun et al. 2004), provides more than 100 hits which are statistically the same as the question (*e*-values of 0) with a sequence coverage of more than 90%.

A more automated method to the rapidly related gene cluster identification has been presented by the antiSMASH, ClusterBlast, and Subcluster Blast tools (Blin et al. 2013; Medema et al. 2011). The ClusterBlast algorithm adds the number of preserved genes among routes with bias to PKS and NRPS genes and adds the number of pairs of genes with Synteny among clusters with bias to central PKS and NRPS bias again. Subcluster blast depends on 126 of known protein subclusters which encode the production of starter units.

Identification depends on phylogenetic alignment of PKS and NRPS, for example, phylogenic analyses of anthracycline antitumor antibiotic as a product of aromatic PK. This approach is used in exploring of bacterial genomes searching for new aromatic polyketides (Sun et al. 2012).

7.10 Phylogenetic Binning

The process in which genomic data are clustered into separate taxon, binning, and assembly can be processed in sequence or integrated together; each bin is single genome that assembled separately to remove the errors of assembly of different taxa.

Markov Chain Monte Carlo approach depends on the supposition of homogeneity of the distribution frequency of short sequences throughout the bacterial genome; this approach is used in simple complexity of metagenomic niche (Kislyuk et al. 2009).

For more complex data, PHYSCIMM (Kelley and Salzberg 2010) merges phymm (Brady and Salzberg 2009) and SCIMM are used; PHYSCIMM tends work on the classified genome by phymm while SCIMM works on unclassified genome. PHYSCIMM can give description of microbial samples when species are involved in public database.

MetaWatt (Strous et al. 2012): MetaVelvet, the first step for metagenomic assembly, secondly, the assembled fragments are grouped in clusters depending on observed tetranucleotide frequencies. Thirdly, obtaining the taxonomic data by the use of BLAST and to cover sequence. Taxonomic profile can be obtained by error-free bins through these steps. MetaWatt is considered as an open-source algorithm which is able to be performed in any platform that supports BLAST and Glimmer.

CONCOCT (Alneberg et al. 2014): a binning tool is able to group the metagenomic data into clusters with coverage multiple sequenced fragments. The lengthier contigs are fragmented and the reads data are reverse mapped into contigs to define coverage throughout the sample, The coverage and sequence composition vectors are linked together to compose a collective profile for every contig (Roumpeka et al. 2017).

7.11 Conclusion

Actinobacteria is a unique group of bacteria that still has a hidden ability to produce beneficial secondary metabolites. This ability recently started to be revealed using the novel next-generation technologies for genome sequencing. In silico genome assembly and annotation softwares allow the prediction of smBGCs present in the genome that are responsible for secondary metabolites production. The genomic data can then be assembled by the phylogenetic binning.

7.12 Future Perspectives

Genome mining of *Actinobacteria* will secure new drug candidates in the near future. It will help also to generate original products derivatives as well as non-natural compounds to enhance human health and industry. As long as genome mining and identification of smBGCs' products are proceeding in a positive response cycle, it could eventually be exploited to manipulate synthetic BGCs for the generation of novel bioactive compounds.

References

Alneberg J, Bjarnason BS, De Bruijn I, Schirmer M, Quick J, Ijaz UZ et al (2014) Binning metagenomic contigs by coverage and composition. Nat Methods 11(11):1144–1146. https://doi.org/10.1038/nmeth.3103

Alt S, Wilkinson B (2015) Biosynthesis of the novel macrolide antibiotic anthracimycin. ACS Chem Biol 10(11):2468–2479. https://doi.org/10.1021/acschembio.5b00525

Afiahayati SK, Sakakibara Y (2015) MetaVelvet-SL: an extension of the Velvet assembler to a de novo metagenomic assembler utilizing supervised learning. DNA Res: Int J Rapid Publ Rep Genes Genomes 22(1):69–77. https://doi.org/10.1093/dnares/dsu041

Barka EA, Vatsa P, Sanchez L, Gaveau-Vaillant N, Jacquard C, Klenk HP et al (2016) Taxonomy, physiology, and natural products of Actinobacteria. Microbiol Mol Biol Rev 80(1):1–43. https://doi.org/10.1128/MMBR.00019-15

Basnet DB, Oh TJ, Vu TTH, Sthapit B, Liou K, Lee HC et al (2006) Angucyclines Sch 47554 and Sch 47555 from Streptomyces sp. SCC-2136: cloning, sequencing, and characterization. Mol Cells 22(2):154–162

Belknap KC, Park CJ, Barth BM, Andam CP (2020) Genome mining of biosynthetic and chemotherapeutic gene clusters in Streptomyces bacteria. Sci Rep 10(1):1–9. https://doi.org/10.1038/s41598-020-58904-9

Blin K, Medema MH, Kazempour D et al (2013) antiSMASH 2.0—a versatile platform for genome mining of secondary metabolite producers. Nucleic Acids Res 41:204–212. https://doi.org/10.1093/nar/gkt449

Boisvert S, Raymond F, Godzaridis É, Laviolette F, Corbeil J (2012) Ray meta: scalable de novo metagenome assembly and profiling. Genome boil 13(12):1–13. https://doi.org/10.1186/gb-2012-13-12-r122

Bowe A, Onodera T, Sadakane K, Shibuya T (2012) Succinct de Bruijn graphs. In: International workshop on algorithms in bioinformatics. Springer, Berlin. https://doi.org/10.1007/978-3-642-33122-0_18

Brady A, Salzberg SL (2009) Phymm and PhymmBL: metagenomic phylogenetic classification with interpolated Markov models. Nat Methods 6:673–676. https://doi.org/10.1038/nmeth.1358

Brikun IA, Reeves AR, Cernota WH et al (2004) The erythromycin biosynthetic gene cluster of *Aeromicrobium erythreum*. J Ind Microbiol Biotechnol 31:335–344. https://doi.org/10.1007/s10295-004-0154-5

Cimermancic P, Medema MH, Claesen J, Kurita K, Wieland Brown LC, Mavrommatis K et al (2014) Insights into secondary metabolism from a global analysis of prokaryotic biosynthetic gene clusters. Cell 158:412–421. https://doi.org/10.1016/j.cell.2014.06.034

Cruz-Morales P, Martínez-Guerrero CE, Morales-Escalante MA, Yáñez-Guerra L, Kopp JF, Feldmann J et al (2015) Recapitulation of the evolution of biosynthetic gene clusters reveals hidden chemical diversity on bacterial genomes. BioRxiv 1:020503. https://doi.org/10.1101/020503

De Jong A, van Hijum SA, Bijlsma JJ, Kok J, Kuipers OP (2006) BAGEL: a web-based bacteriocin genome mining tool. Nucleic Acids Res 34(2):273–279. https://doi.org/10.1093/nar/gkl237

Delcher AL, Bratke KA, Powers EC, Salzberg SL (2007) Identifying bacterial genes and endosymbiont DNA with glimmer. J Bioinform 23:673–679. https://doi.org/10.1093/bioinformatics/btm009

Fernandes GL, Shenoy BD, Damare SR (2020) Diversity of bacterial community in the oxygen minimum zones of Arabian Sea and Bay of Bengal as deduced by Illumina sequencing. Front Microbiol 10:3153. https://doi.org/10.3389/fmicb.2019.03153

Hannigan GD, Prihoda D, Palicka A, Soukup J, Klempir O, Rampula L et al (2019) A deep learning genome-mining strategy for biosynthetic gene cluster prediction. Nucleic Acids Res 47 (18):110–110. https://doi.org/10.1093/nar/gkz654

Kelley DR, Salzberg SL (2010) Clustering metagenomic sequences with interpolated Markov models. BMC Bioinform 11:544. https://doi.org/10.1186/1471-2105-11-544

Kelley DR, Liu B, Delcher AL, Pop M, Salzberg SL (2012) Gene prediction with glimmer for metagenomic sequences augmented by classification and clustering. Nucleic Acids Res 40:9. https://doi.org/10.1093/nar/gkr1067

Khater S, Anand S, Mohanty D (2016) In silico methods for linking genes and secondary metabolites: the way forward. Synth Syst Biotechnol 1:80–88. https://doi.org/10.1016/j.synbio.2016.03.001

Kim J, Yi G-S (2012) PKMiner: a database for exploring type II polyketide synthases. BMC Microbiol 12:169. https://doi.org/10.1186/1471-2180-12-169

Kislyuk A, Bhatnagar S, Dushoff J, Weitz JS (2009) Unsupervised statistical clustering of environmental shotgun sequences. BMC Bioinform 10:316. https://doi.org/10.1186/1471-2105-10-316

Koren S, Treangen TJ, Pop M (2011) Bambus 2: scaffolding metagenomes. J Bioinform 27:2964–2971. https://doi.org/10.1093/bioinformatics/btr520

Kurtz S, Phillippy A, Delcher AL, Smoot M, Shumway M, Antonescu C, Salzberg SL (2004) Versatile and open software for comparing large genomes. Genome Biol 5(2):1–9. https://doi.org/10.1186/gb-2004-5-2-r12

Lee N, Hwang S, Lee Y, Cho S, Palsson B, Cho BK (2019) Synthetic biology tools for novel secondary metabolite discovery in Streptomyces. J Microbiol Biotechnol 29:667–686. https://doi.org/10.4014/jmb.1904.04015

Li MHT, Ung PMU, Zajkowski J et al (2009) Automated genome mining for natural products. BMC Bioinform 10:185. https://doi.org/10.1186/1471-2105-10-185

Li D, Liu CM, Luo R, Sadakane K, Lam TW (2015) MEGAHIT: an ultra-fast single-node solution for large and complex metagenomics assembly via succinct de Bruijn graph. J Bioinform 31:1674–1676. https://doi.org/10.1093/bioinformatics/btv033

Lin YS, Kieser HM, Hopwood DA, Chen CW (1994) The chromosomal DNA of Streptomyces lividans 66 is linear. Mol Microbiol 14:1103. https://doi.org/10.1111/j.1365-2958.1993.tb00964.x

Lin LJ, Du FM, Zeng J, Liang ZJ, Zhang XY, Gao XY (2020) Deep insights into fungal diversity in traditional Chinese sour soup by Illumina Miseq sequencing. Food Res Int 137:109439. https://doi.org/10.1016/j.foodres.2020.109439

Medema MH, Blin K, Cimermancic P et al (2011) antiSMASH: rapid identification, annotation and analysis of secondary metabolite biosynthesis gene clusters in bacterial and fungal genome sequences. Nucleic Acids Res 39:339–346. https://doi.org/10.1093/nar/gkr466

Mikheenko A, Saveliev V, Gurevich A (2016) MetaQUAST: evaluation of metagenome assemblies. J Bioinform 32:1088–1090. https://doi.org/10.1093/bioinformatics/btv697

Namiki T, Hachiya T, Tanaka H, Sakakibara Y (2012) MetaVelvet: an extension of velvet assembler to de novo metagenome assembly from short sequence reads. Nucleic Acids Res 40:155. https://doi.org/10.1093/nar/gks678

Nimnoi P, Pongsilp N (2020) Marine bacterial communities in the upper gulf of Thailand assessed by Illumina next-generation sequencing platform. BMC Microbiol 20(1):19. https://doi.org/10.1186/s12866-020-1701-6

Noguchi H, Park J, Takagi T (2006) MetaGene: prokaryotic gene finding from environmental genome shotgun sequences. Nucleic Acids Res 34(19):5623–5630. https://doi.org/10.1093/nar/gkl723

Noguchi H, Taniguchi T, Itoh T (2008) MetaGeneAnnotator: detecting species-specific patterns of ribosomal binding for precise gene prediction in anonymous prokaryotic and phage genomes. DNA Res 15:387–396. https://doi.org/10.1093/dnares/dsn027

Nouioui I, Carro L, García-López M, Meier-Kolthoff JP, Woyke T, Kyrpides NC et al (2018) Genome-based taxonomic classification of the phylum Actinobacteria. Front Microbiol 9:2007. https://doi.org/10.3389/fmicb.2018.02007

Pevzner PA, Tang H, Waterman MS (2001) An Eulerian path approach to DNA fragment assembly. Proc Natl Acad Sci 98(17):9748–9753. https://doi.org/10.1073/pnas.171285098

Peng Y, Leung HC, Yiu SM, Chin FY (2012) IDBA-UD: a de novo assembler for single-cell and metagenomic sequencing data with highly uneven depth. Bioinformatics 28(11):1420–1428. https://doi.org/10.1093/bioinformatics/bts174

Piel J (2002) A polyketide synthase-peptide synthetase gene cluster from an uncultured bacterial symbiont of Paederus beetles. Proc Natl Acad Sci 99(22):14002–14007. https://doi.org/10.1073/pnas.222481399

Piel J, Hui D, Wen G et al (2004) Antitumor polyketide biosynthesis by an uncultivated bacterial symbiont of the marine sponge *Theonella swinhoei*. Proc Natl Acad Sci 101(46):16222–16227. https://doi.org/10.1073/pnas.0405976101

Roumpeka DD, Wallace RJ, Escalettes F, Fotheringham I, Watson M (2017) A review of bioinformatics tools for bio-prospecting from metagenomic sequence data. Front Genet 8:23. https://doi.org/10.3389/fgene.2017.00023

Singh M, Chaudhary S, Sareen D (2017) Non-ribosomal peptide synthetases: identifying the cryptic gene clusters and decoding the natural product. J Biosci 42:175–187. https://doi.org/10.1007/s12038-017-9663-z

Skinnider MA, Dejong CA, Rees PN, Johnston CW, Li H, Webster AL et al (2015) Genomes to natural products prediction informatics for secondary. Nucleic Acids Res 43(20):9645–9662. https://doi.org/10.1093/nar/gkv1012

Sohn JI, Nam JW (2018) The present and future of de novo whole-genome assembly. Brief Bioinformatics 19(1):23–40. https://doi.org/10.1093/bib/bbw096

Starcevic A, Zucko J, Simunkovic J, Long PF, Cullum J, Hranueli D (2008) ClustScan: an integrated program package for the semi-automatic annotation of modular biosynthetic gene clusters and in silico prediction of novel chemical structures. Nucleic Acids Res 36 (21):6882–6892. https://doi.org/10.1093/nar/gkn685

Strous M, Kraft B, Bisdorf R, Tegetmeyer HE (2012) The binning of metagenomic contigs for microbial physiology of mixed cultures. Front Microbiol 3:410. https://doi.org/10.3389/fmicb.2012.00410

Sun W, Peng C, Zhao Y, Li Z (2012) Functional gene-guided discovery of type II polyketides from culturable actinomycetes associated with soft coral *Scleronephthya* sp. PLoS One 7:42847. https://doi.org/10.1371/journal.pone.0042847

Treangen TJ, Koren S, Sommer DD, Liu B, Astrovskaya I, Ondov B et al (2013) MetAMOS: a modular and open source metagenomic assembly and analysis pipeline. Genome Biol 14 (1):1–20. https://doi.org/10.1186/gb-2013-14-1-r2

van Dijk EL, Auger H, Jaszczyszyn Y, Thermes C (2014) Ten years of next-generation sequencing technology. Trends Genet 30:418–426. https://doi.org/10.1016/j.tig.2014.07.001

Wang P, Kim W, Pickens LB et al (2012) Heterologous expression and manipulation of three tetracycline biosynthetic pathways. Angew Chem Int Ed Eng 51:11136–11140. https://doi.org/10.1002/ange.201205426

Weber T, Kim HU (2016) The secondary metabolite bioinformatics portal: computational tools to facilitate synthetic biology of secondary metabolite production. Synth Syst Biotechnol 1 (2):69–79. https://doi.org/10.1016/j.synbio.2015.12.002

Weber T, Rausch C, Lopez P, Hoof I, Gaykova V, Huson DH, Wohlleben W (2009) CLUSEAN: a computer-based framework for the automated analysis of bacterial secondary metabolite biosynthetic gene clusters. J Biotechnol 140(1–2):13–17. https://doi.org/10.1016/j.jbiotec.2009.01.007

Weber T, Blin K, Duddela S, Krug D, Kim HU, Bruccoleri R et al (2015) antiSMASH 3.0—a comprehensive resource for the genome mining of biosynthetic gene clusters. Nucleic Acids Res 43(1):237–243. https://doi.org/10.1093/nar/gkv437

Zazopoulos E, Huang K, Staffa A, Liu W, Bachmann BO, Nonaka K et al (2003) A genomics-guided approach for discovering and expressing cryptic metabolic pathways. Nat Biotechnol 21:187–190. https://doi.org/10.1038/nbt784

Zerbino DR (2010) Using the velvet de novo assembler for short-read sequencing technologies. Curr Protoc Bioinformatics 31(1):11–15. https://doi.org/10.1002/0471250953.bi1105s31

Zhang W, Ames BD, Tsai S-C, Tang Y (2006) Engineered biosynthesis of a novel amidated polyketide, using the malonamylspecific initiation module from the oxytetracycline polyketide synthase. Appl Environ Microbiol 72:2573–2580. https://doi.org/10.1128/AEM.72.4.2573-2580.2006

Ziemert N, Podell S, Penn K et al (2012) The natural product domain seeker NaPDoS: a phylogeny based bioinformatic tool to classify secondary metabolite gene diversity. PLoS One 7:34064. https://doi.org/10.1371/journal.pone.0034064

Chapter 8
Cloning and Heterologous Expression of Natural Products from Actinobacteria

Shabiha Nudrat Hazarika, Pranami Bharadwaj, Aditya Narayan Konwar, and Debajit Thakur

Abstract Biosynthetic gene clusters (BGCs) are imperative in producing natural products in the microbial world which has been regarded as the most important source of antibiotics in the twentieth century. Prospecting of these gene clusters has become critical in recent times owing to the delay in new antibiotic discovery which is compelling deep insights into the biosynthetic machinery. Herein, we focus on the heterologous expression of the BGCs into suitable hosts and its identification through various computational methods. A brief overview of the conventional techniques of cloning and advancements in these techniques has been discussed along with the application of heterologous expression in discovering novel compounds. Genomic approaches have also been discussed which have become a crucial part in exploring the cryptic BGCs that have tremendous potential in delivering us with novel natural products. Metagenomics and various recombination techniques have far-reaching applications in the field of natural product discovery through its exhaustive databases and screening capabilities of unculturable meta-DNA and environmental DNA that has led to discovery of some of the most important antibiotics of recent times. Moreover, these new approaches in cloning and heterologous expression will assist in understanding the intricate mechanisms of secondary metabolism in microorganisms and will promote the exploration of natural products from untapped regions of the world and metagenomes.

S. N. Hazarika
Microbial Biotechnology Laboratory, Life Sciences Division, Institute of Advanced Study in Science and Technology, Guwahati, Assam, India

Department of Molecular Biology and Biotechnology, Cotton University, Guwahati, Assam, India

P. Bharadwaj · A. N. Konwar
Microbial Biotechnology Laboratory, Life Sciences Division, Institute of Advanced Study in Science and Technology, Guwahati, Assam, India

Academy of Scientific and Innovative Research (AcSIR), Ghaziabad, Uttar Pradesh, India

D. Thakur (✉)
Microbial Biotechnology Laboratory, Life Sciences Division, Institute of Advanced Study in Science and Technology, Guwahati, Assam, India

L. Karthik (ed.), *Actinobacteria*, https://doi.org/10.1007/978-981-16-5835-8_8

Keywords Actinobacteria · Antibiotics · Biosynthetic gene clusters · Heterologous expression · Natural products

8.1 Introduction

Natural products (NPs) or secondary metabolites are low molecular mass products that are preserved enormously in nature. They originate from innumerable sources, including terrestrial plants, animals, microorganisms, marine organisms, and invertebrates (Stoesser 2001). These multifarious bioactive molecules have been discovered and reported to play essential role in human life for thousands of years (Koehn and Carter 2005). Natural products and their derivatives act as remarkable source for drug discovery and development to treat various diseases. They are not directly involved in the normal growth and development but serves as defense compounds or signalling molecules that helps improve human health along with nutritional benefits and agricultural productivity (Hassan et al. 2019). At present, approximately 60% of approved medicines are natural products including 61% of anti-cancer drugs and 49% of anti-infection drugs. Till date 64 natural products, 268 natural-product mimics, and 299 natural-product derivatives have been used as antibacterial, antifungal, antiparasitic, antiviral, antitumor agents, antidiabetic, immunosuppressive agents, insecticides as well as herbicides (Newman and Cragg 2020; Patridge et al. 2016).

Microbial natural products take up a notable part in the discovery and development of antibiotic and over the past few years, microbes from all sources are considered factory manufacturing these bioactive natural products. Most natural products antibiotics were detected using “The Waksman Platform”, which is regarded as the golden age of antibiotic discovery (Lewis 2012). This age focused on exploring compounds produced by Actinobacteria, commonly derived from soil. Actinobacteria are filamentous Gram-positive bacteria containing G + C-rich genome that have the biosynthetic potential to produce several family of secondary metabolites (polyketides, alkaloids, terpenoids, peptides, and saccharides) with extensive structural diversity and commercial importance (Hazarika and Thakur 2020; Barka et al. 2016). About two-third of known natural antibiotics are known to be isolated from actinobacteria, of which over 20,000 are bioactive compounds, 7600 are procured from *Streptomyces* and 2500 are from rare actinobacteria species, representing the largest group (45%) of bioactive microbial metabolites (Bérdy 2012; Waksman et al. 2010; Berdy 2005). From these known bioactive microbial compounds only about 200 compounds are directly used in the human, veterinary medicine, and agriculture (Bérdy 2012). Actinobacteria producing bioactive metabolites belongs to the genera *Streptomyces, Micromonospora, Actinoplanes, Amycolatopsis, and Saccharopolyspora*. *Streptomyces* alone accounts for 75% of known antibiotics (de Lima Procópio et al. 2012; Waksman et al. 2010; Demain 2002; Watve et al. 2001). Some of the antibiotics produced by actinobacteria are actinomycin, proactinomycin, streptothricin, and streptomycin, chloramphenicol, actinomycetin, micromonosporin, mycetin, and actinomyces lysozyme and so on (Waksman et al. 2010). Some currently used NPs produced by actinobacteria as antitumor agents are actinomycin D, anthracyclines, bleomycin, mitomycin C, anthracenones, calicheamicin, taxol, and epothilones (Law et al. 2020). These natural products from actinobacteria are synthesized with the help of assembly and regulation of multi-step enzymatic reactions, which are usually known as biosynthetic pathways. Physically organized group of genes called biosynthetic gene clusters (BGCs) encodes these biosynthetic pathways. These gene clusters contain all the genes that encodes for the enzymes required for production, regulation, export of natural products. BGCs are either silent or cryptic in the original actinobacteria; hence, understanding the molecular mechanisms of BGCs is important for redesign of novel NPs (Wright 2017). Natural products have been listed in Table 8.1 that are expressed by heterologous method in a variety of hosts using multiple approaches.

In the past few decades, there is an abrupt decline in the development and inception of new significant drugs medically crucial for the mankind which demands to find new antibiotics due to an increase in public health hardship caused by multidrug resistance (Fair and Tor 2014; Wright 2011). Thus, approaches for identification of BGCs have become a prerequisite for synthesis of natural products. One such possible choice can be the use of genomics which can be used to identify potential drug targets and discover novel gene clusters for the biosynthesis of NPs (Albarano et al. 2020; Ren et al. 2020). Thus, with the advent of genome sequencing techniques, mainly genome mining has enabled to procure new drugs in a rapid and cheaper way. Genome mining has become an attractive tool to produce, reactivate, improve, and modify the pathways for the discovery of novel bioactive compounds from culturable fastidious actinobacteria or metagenomic DNA in heterologous hosts. With the expanding genome sequencing data accumulated through genome mining and heterologous expression, the discovery of NPs has been acquired a new level of pace. Cloning and heterologous expression in suitable heterologous hosts has revealed information about the characterization of biosynthetic pathways and has allowed genetic modifications of these pathways for the synthesis of novel NPs as well as optimization of the yield (Huo et al. 2019; Nah et al. 2017; Hopwood et al. 1985). In this book chapter, we have described the optimized strategies for cloning and heterologous expression of biosynthetic genes for discovery of new natural products antibiotics from actinobacteria to minimize the antibiotic crisis in the twenty-first century. We have also highlighted about different computational tools for the identification and optimization of NP BGCs.

8.2 Identification of BGCs for Known Natural Products

Over the last few decades, genome sequencing has had an immense effect on natural drug discovery. Relentlessly, *Streptomyces coelicolor* is known for the production of actinorhodin (Fig. 8.2a) which has been used as the model for understanding the fundamental aspects of *Streptomyces* genetics and for understanding the genetic

Table 8.1 Some important natural compounds expressed heterologously using different approaches

Sl. no.	Natural product	Original host	Heterologous host	Approach	Reference
1	Chlortetracycline (PKS II)	*Streptomyces aureofaciens* ATCC 10762	*Streptomyces rimosus 461*	Cas9-assisted targeting of chromosome segments (CATCH)	Wang et al. (2019)
2	Divamides (PKS)	Metagenome of symbiotic bacteria of marine tunicates	*Escherichia coli (E. coli)*	Construction of expression vector incorporating whole genetic element for the pathway	Smith et al. (2018)
3	Malacidins (NRPS)	Environmental metagenome	*Streptomyces albus*	TAR cloning	Hover et al. (2018)
4	Grecocyclin (PKS II)	*Streptomyces* sp. Acta 1362	*Streptomyces albus*	TAR integrative	Bilyk et al. (2016)
5	Venemycin (PKSI)	*Streptomyces venezuelae*	*Streptomyces coelicolor*	Cosmid integrative	Thanapipatsiri et al. (2016)
6	Oxytetracyclin (PKS II)	*Streptomyces rimosus* M4018	*Streptomyces venezuelae*	Cosmid integrative	Yin et al. (2016)
7	Metatricycloene (PKS II)	Environmental metagenome	*Streptomyces albus*	Mating *S. albus* with environmental DNA cosmid clone libraries	Iqbal et al. (2016)
8	Actinorhodin (PKS II)	*Streptomyces coelicolor* M145	*E. coli* pKC1139	Phage ΦBT-1 integrase mediated site-specific recombination	Du et al. (2015), Rudd and Hopwood (1979)
9	Salinomycin (PKSI)	*Streptomyces albus*	*Streptomyces coelicolor*	LLHR	Yin et al. (2015)
10	Daptomycin (NRPS)	*Streptomyces roseosporus* NRRL 15998	*E. coli* pKC1139	Phage ΦBT-1 integrase mediated site-specific recombination	Du et al. (2015)
11	Napsamycin (NRPS/PKS)	*Streptomyces roseosporus* NRRL 15998	*E. coli* pKC1139	Phage ΦBT-1 integrase mediated site-specific recombination	Du et al. (2015)
12	Tautomycetin (PKSI)	*Streptomyces* sp. CK4412	*Streptomyces coelicolor* and *Streptomyces lividans*	pSBAC integrative	Nah et al. (2015)

13	Conglobatin (NRPS/PKS)	*Streptomyces conglobatus* ATCC 31005	*Streptomyces coelicolor*	Gibson assembly	Zhou et al. (2015)
14	Arixanthomycins A (PKS II)	Environmental metagenome	*Streptomyces albus*	TAR cloning	Kang and Brady (2014)
15	Streptothricin (NRPS)	*Streptomyces* sp. TP-A0356	*Streptomyces coelicolor*	Cosmid replicative	Li et al. (2013)
16	Chloramphenicol (PKS-NRPS)	*Streptomyces venezuelae*	*Streptomyces coelicolor*	Cosmid integrative	Gomez-Escribano and Bibb (2011)
17	Coelimycin P1 (PKS)	*Streptomyces coelicolor* A3(2)	*Streptomyces coelicolor* M145	Phage ΦBT-1 integrase mediated site-specific recombination	Gomez-Escribano and Bibb (2011)
18	Streptomycin (PKS-NRPS)	*Streptomyces griseus* IFO 13350	*S. avermitilis*, *S. avermitilis* SUKA17	Genome reduced with deletion of 1.4 Mb non-essential genes	Ohnishi et al. (2008)
19	Kanamycin (Aminoglycoside)	*Streptomyces kanamyceticus* ATCC 12853	*Streptomyces venezuelae*	Cosmid replicative	Thapa et al. (2007)
20	Methylenomycin (PKS)	*Streptomyces violaceus-ruber*	*Streptomyces coelicolor* A3(2)	Conjugative transfer of phage P1-derived Artificial Chromosome (PAC) library clones	Hobbs et al. (1992)
21	Undecylprodigiosin (PKS II)	*Streptomyces coelicolor* A3(2)	*Streptomyces lividans 66*	In vivo recombination with a clone carrying two contiguous segments of DNA	Malpartida et al. (1990)

control and regulation of antibiotic synthesis (Belknap et al. 2020). The genes involved in actinorhodin pathway were completely cloned and expressed in a heterologous host at the beginning, which helped in predicting the cloning of numerous other antibiotic gene sets in recent times (Tahlan et al. 2007). *S. coelicolor* was also known to produce three other antibiotics namely, Cda (calcium-dependent antibiotic), Red (undecylprodigiosin) (Fig. 8.2c) and Mmy (methylenomycin A and B) (Fig. 8.2b) (Bentley et al. 2002). A fifth antibiotic named coelimycin P1 (Fig. 8.2d) was discovered in *S. coelicolor* which had been studied in numerous laboratories around the world in a model organism (Gomez-Escribano and Bibb 2011). With the detection of the *Streptomyces coelicolor* and *Streptomyces avermitilis* genomes in the beginning of 2000s, several other genomes of actinobacteria were revealed to encode for diverse cryptic NP BGCs. These cryptic BGCs are poorly expressed at laboratory conditions as they are hidden and cannot be linked to a product. With this discovery of actinobacterial natural product resources a resurgence of NP discovery has been observed leading to the ongoing era of genomics-based drug discovery.

With the advent of genome mining for identification of secondary metabolites, it has been found that actinobacteria are a repository of novel natural products. A large number of biosynthetic gene clusters have been established in actinobacteria involved in the production of natural products with the help of a series of reactions incorporated in a metabolic pathway. It has been delineated that 830 actinobacterial genomes have been identified and studied to produce 11,000 NP BGCs that belong to 4000 chemical families (Doroghazi et al. 2015). The detection of BGCs has directed to a paradigm shift in the natural products research. All the genes obligated for the biosynthesis of natural products are arranged in a contiguous cluster on the bacterial chromosome. For the regulation of expression of the biosynthetic genes, the BGCs incorporate one resistance gene and a pathway-specific regulator. Primitively, the detection of BGCs was based on screening for bioactive metabolites followed by microbial genetics along with the bioassay and chemistry directed classical natural products discovery processes and purification technologies. There were also evidence that identical or similar metabolites being produced by BGCs which were evolutionary distinct. As such with the development of next-generation sequencing, bioinformatics and genomics, genome mining has become a promising source for recognizing the genes which are associated in production of secondary metabolite. This alternative strategy has unveiled a remarkable biosynthetic potential in actinobacteria (Adamek et al. 2017).

The broadening genome sequence data has led to outstanding discovery of natural products BGCs using computational software tools such as antibiotics and secondary metabolite analysis shell (antiSMASH) (Blin et al. 2017, 2019; Medema et al. 2011); CLUster SEquence ANalyzer (CLUSEAN) (Weber et al. 2009); natural solutions to drug discovery (NP.searcher) (Li et al. 2009); the natural product domain seeker (NaPDoS) (Ziemert et al. 2012); genes to natural products (PRISM/GNP) (Johnston et al. 2015; Skinnider et al. 2015) that are adaptable by microbiologists and can be linked to natural product databases such as Minimum Information on Biosynthetic Gene Cluster (MIBiG) (Kautsar et al. 2020), antiSMASH, Database of BIoSynthesis cluster CUrated and InTegrated (DoBISCUIT) (Ichikawa et al. 2013) and Natural Product Atlas (Singh et al. 2019). Using antiSMASH 1110 biosynthetic diversity has been characterized in *Streptomyces* genomes, which is 8–83 BGCs per genome. 34 major classes of BGCs have been detected in other genera of actinobacteria such as *Salinispora* and *Amycolatopsis* (Belknap et al. 2020). These tools are facilitating the discovery of different biosynthetic gene clusters (BGCs), including NRPSs (non-ribosomal peptide synthases; detected in 1062 genomes) (Naughton et al. 2017); type I PKSs (polyketide synthases, present in 981 genomes); type II PKSs (present in 499 genomes) (Weber et al. 2003); other KS (ketide synthases, in 650 genomes) (Selvin et al. 2016); terpenoids (in 697 genomes) (Li et al. 2019), lanthipeptides (in 540 genomes) (Walker et al. 2020), ribosomally synthesized and post-translationally modified peptides (RiPPs) (in 667 genomes) (Zhong et al. 2020) and many others based on their products. The combination of two or more such clusters could be further enhanced to form hybrid BGCs and used to study the diversity of these BGCs (Malik et al. 2020; Fischbach and Walsh 2006).

Even though there is availability of many optimization technologies in native natural products producers such as flourishing the supply of biosynthetic precursors, inducing spontaneous mutations, inserting suitable promoters, overexpression of specific regulators, and disrupting negative regulators a mass of potentially novel natural products is yet to be procured. Thus, mobilization of corresponding BGCs into a viable heterologous host has become a convincing alternative approach to elucidate the hidden secondary metabolites from cryptic BGCs. Heterologous expression of BGCs can synergize to generate enriched chemical diversity of natural products in the antibiotics field (Huo et al. 2019). The schematic representation for the gene editing of BGCs for the development of novel products is shown in Fig. 8.1.

With the onset of genome sequencing or genome engineering and computational tools BGCs can presently be transferred from native strains to discrete heterologous hosts such as *Escherichia coli*, *Bacillus subtilis*, or *Streptomyces lividans* easily and successfully. As most of the BGCs ~90% are partially active or inactive under standard culture conditions, regulatory triggers that are required for the complete activation of the cryptic genes are unknown. Thus, heterologous expression offers a fascinating opportunity to investigate the vast NP diversity in genomes and metagenomes. Thus, for functional expression of natural products from these BGCs, synthetic biology approach can be useful to uncouple BGCs from their natural regulatory hindrance. Approaches such as codon optimization, changing the native DNA constructs or DNA synthesis can be used for unveiling the BGCs, where all the transferred genes must be functionally expressed, with proper folding of all translated products followed by appropriate post-translational modifications. The heterologous hosts must be able to provide all substrates and co-factors, so that BGC expression produces the corresponding product (Ke and Yoshikuni 2020).

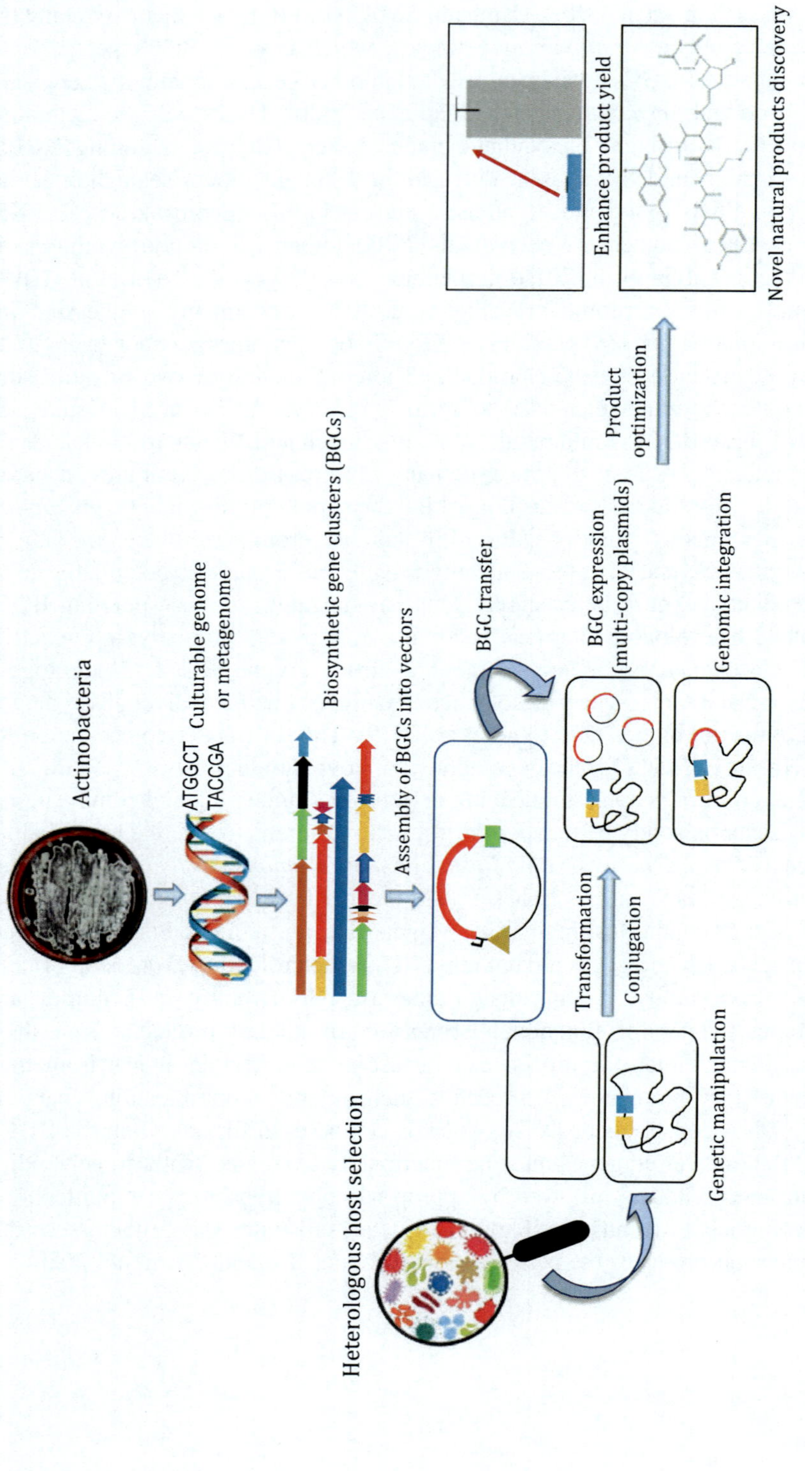

Fig. 8.1 Conventional workflow for identification and heterologous expression of biosynthetic gene clusters

8.3 Conventional Expression Systems

Genome sequencing in actinobacteria is quite challenging due to its high G-C content as well as the presence of highly repetitive homologous sequences in PK-I (Polyketide type I) and NRP (non-ribosomal peptide) containing BGCs. With the advent of time, the conventional systems have evolved to more up-to-date techniques; however, it does not lessen their role and influence in natural product discovery.

8.3.1 Genomic Libraries

The sequence-independent approach for expression of NP BGCs is based on the construction of expression libraries on pure culture genomes or mixed libraries, followed by screening for NPs. The key protocol of sequence-independent methods comprises of high-quality DNAs isolation, followed by DNA fragmentation and finally genomic library is constructed. This conventional protocol is greatly used in scenarios where the genomic information of the native hosts is not well delineated. The advantage of sequence-independent methods lies in the prospect that entire genome can be screened, and it is also possible to capture the BGCs in that genome and discover novel structural natural products (Liu et al. 2021). There are successful instances based on various methods (e.g. bacterial artificial chromosomes (BACs), cosmids, fosmids, phage artificial chromosomes (PACs)) of sequence-independent library cloning, despite the requirement for highly efficient screening assays as the library has very low fragment of positives (Zeng et al. 2001). The steps for library preparation include the following:

8.3.1.1 Isolation of High-Quality DNAs

Since the BGCs are often over a 100 kb, the method used for preparing high-quality DNA is critical for the successful cloning of intact BGCs. The most common methodology is by utilization of CTAB (cetyl trimethyl ammonium ion bromide) extraction buffer for DNA extraction, followed by the addition of phenol-chloroform and/or ethanol precipitation. However, this technique causes shearing of long DNA molecules, and therefore is favourable just for extracting genomic DNAs up to ~10 kb. Before isolating DNA from organisms, cellulase and pectinase are accustomed to hydrolyse the cell wall, so as to organize megabase size genomic DNA. In this technique, not like the standard genomic DNA isolation ways, a low-melting-point agarose gel matrix is employed for embedding the protoplast, nuclei, or cells to safeguard giant DNA fragments from shearing throughout the method of DNA isolation or else, for speedy extraction of top quality and high mass genomic DNA (~20 kb to ~130 kb) from plants, bacteria, animals, etc. Liquid nitrogen is used to grind the cell into a fine powder, followed by cell lysis using SDS buffer and eventually using carboxylated magnetic beads to purify the DNA. Various industrial kits for extracting high-molecular weight DNA are obtainable within the market (e.g. Macherey Nagel, QIAGEN) (Latif and Osman 2017). Different ways were used for the isolation of high-quality and high-molecular weight genomic DNA from the grains of the common cereal crop, *Zea mays*, with a number of minor modifications.

8.3.1.2 DNA Fragmentation

In genomic library construction, numerous ways are out there for DNA fragmentation, e.g. sonication, catalyst digestion, hydraulic shearing (Joneja and Huang 2010). Sonication, by using an ultrasound to get >120 KB fragments, catalyst digestion, through the utilization of site-directed restriction endonuclease Sau3AI to partly digest refined DNA and hydraulic shearing (Hsieh et al. 2000), by repeatedly passing DNA through a syringe needle, are widely used for the development of large-fragment libraries. Sonication and hydraulic shearing (mechanical fragmentation strategies), are way more random as compared to enzymatic shearing which could cause inherently biased and incomplete library, thanks to the uneven restriction sites within the genome, and therefore sonication and hydraulic shearing allows higher control of the size distribution. Once fragmentation is done, DNA samples are further visualised in horizontal agarose gel action to check the yield of fragments or perhaps analysed by fragment analyser. Thus, it paves the approach for separation and extraction of desired size of fragmented DNAs using multiple rounds of pulsed field gel electrophoresis (PFGE) (Sharma-Kuinkel et al. 2016) with completely different ramped pulse times.

8.3.1.3 Cloning

The size-selected fragments from PFGE are end-repaired and ligated to the digestible, dephosphorylated vector, like BAC, cosmid, fosmid, or PAC (Liu et al. 2021) (Fig. 8.1b). The overall ligation product will be reworked into *Escherichia coli* or may also be pre-packaged into a phage for infecting bacterium. Massive sized BGCs are usually split into multiple fragments and reassembled into the whole cluster later, as the insert size of fosmid or cosmid libraries is typically restricted to ~50 kb. As an alternative, PACs will be used to clone inserts of size starting from 60 kb to around 150 kb, whereas BACs have a capability to propagate as well as accommodate DNA fragments with a median size of concerning kb. Several technologies of the lot have been turned into finished product and commercialized by a spread of firms like Agilent, Bio SandT, and epicentre Biotechnologies.

Libraries for both sequenced and unsequenced DNAs can be generated using sequence-independent library cloning, with each clone harbouring around 10 to ~200 kb inserts and promotes the natural product discovery. However, the sequence-independent library cloning technique is sometimes gruelling and time consuming. To dependably cowl the entire genome, around 10–20 folds genome coverage needs to be generated to get the clones harbouring BGCs. These needs improvement of the entire cloning method, as well as the steps like the genome extraction that shouldn't lead to an excessive amount of genomic fragmentation, or the genomic assembly as well as the transformation step ought to be extremely economical to get the specified library size, etc. Moreover, to screen and establish massive sequence clusters, desired BGCs may well be split into totally different clones, particularly using cosmid/fosmid libraries (Lin et al. 2020).

Since there are immense challenges in construction of BAC/PAC genomic libraries for cloning of giant NP BGCs, like recovery of high mass (HMW) genomic DNA, Nah et al. (2017) developed the simplest way. They created the plasmid *Streptomyces* (actinobacteria) bacterial artificial body (pSBAC) system that allowed mobilization of BGC directly from the chromosome of bacterium. There are two rounds of homologous recombination. Firstly, inside the tautomycetin BGC, two distinctive XbaI restriction recognition sites were inserted inside the border regions in actinobacteria sp. CK4412. Inside the second round of homologous recombination, to either finish of the BGC between the XbaI sites and thus the tautomycetin BGC, attP-int deleted pSBAC backbone addition was done. After XbaI digestion of the genomic DNA followed by self-ligation and electro-transformation into *Escherichia coli* (*E. coli*) (host) 80 kb tautomycetin BGC was recovered from the bacterial chromosome. Provided the actinobacterial strain is genetically tractable, this methodology is economical for cloning massive NP BGCs. The main limitation of this methodology lies inside the restricted availability of appropriate restriction sites as a result of the restriction enzyme digestion and self-ligation are elementary to the pSBAC strategy. Significantly improved natural product (NP) production by tautomycetin BGC inside the parental and heterologous expressed strains was reported by the use of various choice markers (apramycin and hygromycin) on the pSBAC plasmids followed by tandem integration of the BGC onto the chromosome.

8.3.2 Integrase-Mediated Site-Specific Recombination (ISR)

Site-specific recombination (SSR) may be described as a bunch of processes among which rearrangement of DNA molecules are done by breaking and rejoining the strands at specific points (Fig. 8.1d, e). SSR involves two short DNA sequences (sites) that might be present in an identical molecule or in many molecules (Olorunniji et al. 2016). For cloning huge NP BGCs from *Streptomyces,* φBT1 ISR was developed by Du et al. (2015) as another strategy to the pSBAC approach. First, at one end of the BGC of interest, a mutagenized attB6 site is introduced through one crossover using a pUC119-based suicide plasmid. Using a pKC1139 (mostly plasmid through single crossover) a self-replicative (attB6) and temperature-sensitive replicon (attP6) is supplemented to the other end of the BGC which inserts the attB and attP integration sites (66) at both the ends of the BGC of interest. Followed by introduction of intφBT1 bearing plasmid, pIJ10500 and integration into the attBφBT1 site among the body of the recombinant strain. To excise the whole BGC, SSR is triggered by the expression of φBT1 integrase between the attB6 and attP6 sites that leaves behind a 42 bp attL and a suicide plasmid backbone among the body and releases a replicative plasmid within the strain containing the BGC (Olorunniji et al. 2019). Exploiting this ISR technique, the authors Du et al. (2015) with success cloned the actinorhodin BGC (23 kb) from *Streptomyces coelicolor* M145 (Hopwood et al. 1985; Rudd and Hopwood 1979) and napsamycin BGC (35 kb) (Fig. 8.2g) and daptomycin BGCs (63 kb) (Fig. 8.2f) from *Streptomyces roseosporus* NRRL15998 (Du et al. 2015). As a result of the self-replicative plasmid bearing the BGC, the potency of ISR system for cloning huge NP BGCs of 80–90% is quite spectacular. For titre improvement, the BGC is reintroduced into the wild-type strain. Conjointly, the system uses totally different mixtures of mutant attB/attP sites to eliminate obviate endogenous NP BGCs for strain/surrogate host development. However, the ISR strategy conjointly needs the target strain to be genetically tractable, just like the pSBAC system.

8.3.3 Linear–Linear Homologous Recombination (LLHR)

This technique exploits the use of the virus Red system that features the virus recombination genes gam, bet, and exo. The gam factor product, gam prevents an *E. coli* enzyme, RecBCD, from degrading linear DNA fragments, permitting preservation of remodelled linear DNA *in vivo*. The bet factor product, Beta (ssDNA-binding supermolecule), promotes annealing of two DNA molecules. The exo factor product, Exo, contains a 5′–3′dsDNA nuclease activity. Operating along these latter two proteins insert linear DNA at the desired target, making genetic recombinants. For dsDNA, Red Exo is believed to degrade from each 5′ ends, exposing ssDNA that is surrounded by Redβ (Sharan et al. 2009). The Red DNA encoding, Redβ, and Redγ proteins from the *E. coli* (λ phage) through short homologous arms (30–50 nt) (Kuzminov 2011) are widely acquainted to promote homologous recombination between linear and circular DNA molecules (LCHR). Datsenko and Wanner (2000) in a very pioneering study, generated 40 single gene deletions on the *E. coli* chromosome using the λ-Red system. For direct cloning of NP BGCs from microbe genomic DNA in 2012, discovered that the full-length RecET proteins increase LLHR. Using a full-length RecET protein-mediated LLHR system they cloned 9 cryptic BGCs ranging from 10 to 37 kb from *Photorhabdus luminescens*. Further, they expressed two of them in *E. coli*, which resulted in the discovery of luminmycin A and luminmide A/B. A two-step LLHR followed by LCHR and is needed to clone the 52 kb plu2670 BGC, demonstrating a potential size limit for LLHR. The identical research group used a three-fragment strategy followed by assembly into one BGC by Red/ET to clone the 106 kb BGC of salinomycin (Fig. 8.2h). Even though, the LLHR strategy is extremely effective, it still remains laborious for larger BGCs.

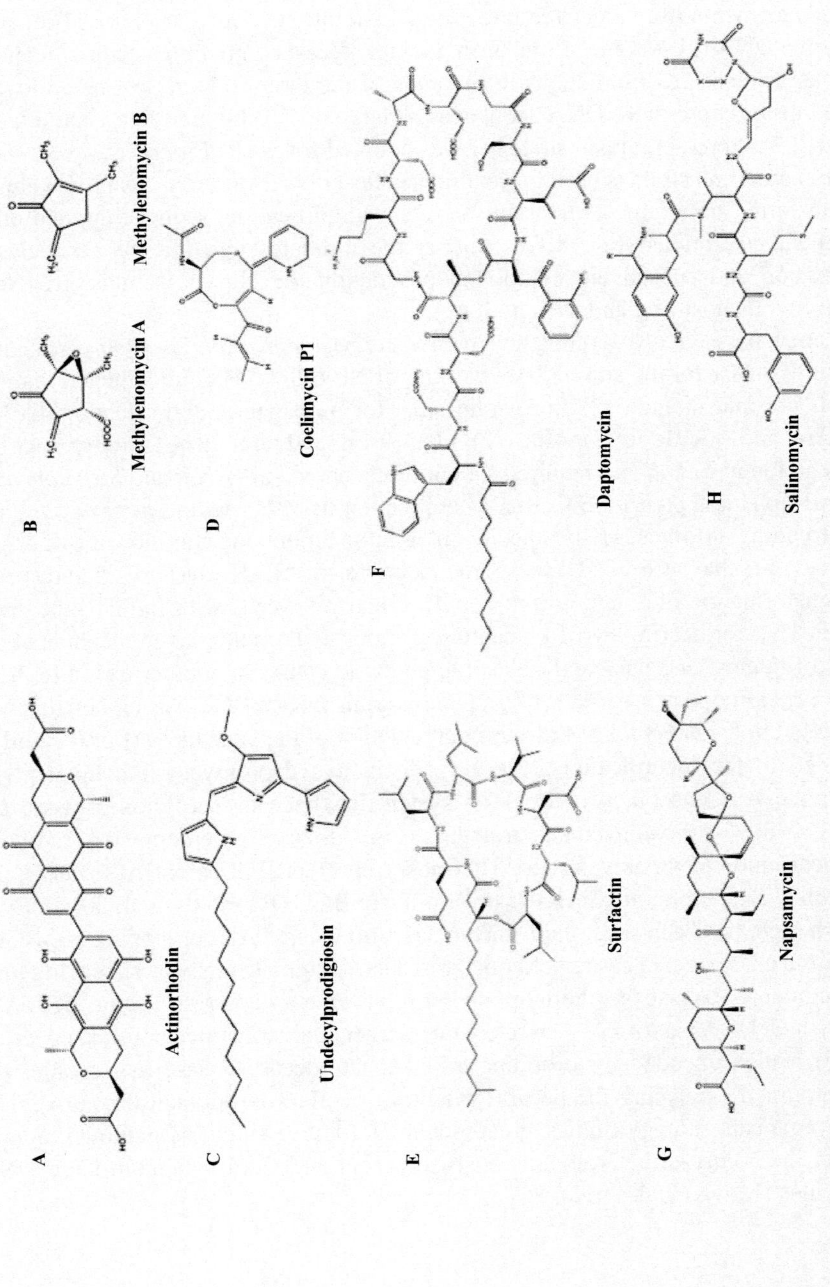

Fig. 8.2 Examples of natural products discovered in actinobacteria facilitated by conventional heterologous expression system

8.3.4 *Transformation-Associated Recombination (TAR): The Solution for Cloning Large NP BGCs*

TAR is based on one transformation step which facilitates the capture of huge genetic elements from a desired target organism into a selectable vector. The target genomic DNA (gDNA) is co-transformed with a yeast compatible vector containing "hooks" which are homologous to the ends of the target region. For heterologous gene cluster expression, DNA sequences as large as 250 kb are often captured and isolated. A "three-organism strategy" was adopted for TAR. Firstly, any *Saccharomyces cerevisiae* strain is used to capture the cluster by TAR and the captured cluster is transferred onto an *E. coli* strain for vector amplification, sequencing, and other beneficial alterations (Fig. 8.1c). And at the ultimate step, for the gene cluster expression and subsequent compound production the cluster is transferred to a heterologous host (Xu and Wright 2019).

Yamanaka et al. (2014) constructed an *S. cerevisiae, E. coli, Streptomyces* shuttle vector pCAP01 for the aim of TAR cloning of NP BGCs. The fundamental features of this plasmid include: (1) In *S. cerevisiae*, for propagation and selection, CEN6/ARSH4 and TRP1 are present; (2) pUC19-ori and aph (3′)-II are present for propagation in *E. coli* and required for antibiotic choice in *E. coli* and *Streptomyces*; (3) traJ-oriTint is present to facilitate conjugation from *E. coli* to *Streptomyces*, and for providing attPintφC31 site-specific integration among the chromosome of *Streptomyces* at the attBφC31 site. An example of TAR system includes the co-transformation of *S. cerevisiae* VL6-48 spheroplast cells with BamHI-linearized pCAP01 vector bearing two 1 kb capturing "hooks", homologous to the ends of the BGC of interest and has XbaI digestible genomic DNA. In another example, from *Saccharomonospora* sp. CNQ-490, a 67 kb cryptic NRPS BGC was cloned that was expressed in *S. coelicolor* M1146 through deletion of the gene tar20 (LuxR-family). This led to the identification of a very distinctive daptomycin like lipopeptide, taromycin A. Consequently, the TAR system has been used to clone several NP BGCs, like amicoumacin, surfactin (Fig. 8.2e), enterocin, napsamycin, alterochromide, cosmomycin, etc. The most important BGC reported to this point through TAR cloning of NP BGCs is Taromycin BGC (Reynolds et al. 2018). Even though there has been several advancements, still using TAR for cloning NP BGCs remains a challenge, in case of cloning large DNA fragments in bacteria having high G-C content. To beat the shortcomings, a replacement capturing vector, pCAP03, which had URA3 as a counter-selection marker was constructed by Tang et al. (2015), which encoded for orotidine 5′-phosphate decarboxylase that synthesizes 5-fluorouracil catalysing the decarboxylation of 5-fluoroorotic acid (5-FOA). The three-organism strategy shows great potential for the study of natural products despite the ability of heterologous hosts to accept and stably maintain large DNA molecules (Kawai et al. 2010).

8.4 Technical Advancements in Heterologous Expression of BGCs

Progress made in the field of cloning and bioengineering has promoted the expression of large BGCs in host systems with ever increasing efficacy and rapidity. The advent of CRISPR/Cas9 along with advancements in synthetic biology has guided the way for cloning multiple fragments of BGCs into a range of host systems. The copy number of genetic elements, i.e. the number of BGCs that are present, the position at which it integrates, and the location of promoter elements, influences the expression capability of the BGCs. All these factors have been considerably looked after in the recent years.

Conventional approaches such as preparation of genomic libraries using vectors including BAC and PAC are most commonly adapted for the expression of large BGCs in various host systems. But recent progress in the field of genome engineering and gene sequencing along with its high-throughput analysis has been imperative for the identification and subsequent extraction of the BGCs from chromosomal DNA which has made the construction of genomic libraries as redundant techniques. Cloning strategies or approaches have been modified tremendously for inserting large fragments of DNA in an efficient way which has resulted in the expression of upto 100 kb sized BGCs in *Streptomyces* or *E. coli*. Recently, a strategy known as Gibson assembly (Fig. 8.1a) was used for obtaining a 41 kb conglobatin BGC from *Streptomyces conglobatus* (Zhou et al. 2015). Gibson assembly utilizes three different enzymes including an exonuclease (5′ exonuclease), a DNA polymerase, and a DNA ligase in a single-step isothermal reaction (Gibson et al. 2009). This technique supports the assemblage of a number of DNA fragments into a concerted recombinant DNA. The rapidity of the technique has well been utilized for DNA library construction of fragments as large as 100 kb (Thomas et al. 2015).

Another technique described as linear plus linear homologous recombination (LLHR) has been used for efficient cloning of bacterial BGCs of sizes upto 52 kb. The expression of a large number of BGCs has been possible with aid of a prophase recombinase RecET-mediated direct cloning of large BGCs (Wang et al. 2016; Fu et al. 2012). Two important proteins, i.e. RecE and RecT from Rac prophage, are utilized for highly efficient LLHR that results in cloning long genes directly from genomic DNA to expression vectors, bypassing library preparation and screening. These two proteins in association with each other facilitate the recombination of large fragments of DNA between two linear strands. Like Redαβ these two proteins also act as 5′–3′ exonuclease and single-stranded annealing protein (SSAP) pair. This has found to be more efficient than the conventional Redαβ-mediated linear plus circular recombination (Fu et al. 2012). This enhanced recombination has helped to clone multiple and large fragments of DNA such as 10 gene clusters from *Photorhabdus luminescens* into expression vectors and express them in heterologous host to identify two important compounds luminmycin A and luminmide A/B (Fu et al. 2012). Recently, a more robust and efficient version of RecET LLHR has been developed known as "ExoCET" where the exonuclease activity of T4 polymerase has been utilized and combined with the RecET-mediated homologous recombination (Wang et al. 2018). This enabled the cloning of fragments larger than 50 kb into expression vectors with nucleotide precision.

Transformation-associated recombination, or TAR, has been most widely used for the cloning of large BGCs into expression vectors such as BAC or YAC. In 1996, Larionov and his team used TAR to clone large human DNA into YAC using two genetic markers M1 and M2 containing the human Alu sequence at one end and a telomerase sequence at the other end (Larionov et al. 1996). This platform has been used to clone numerous BGCs from various host genomic DNAs into expression vectors (Bonet et al. 2015; Li et al. 2015a; b; Ross et al. 2015; Tang et al. 2015). Novel lipopeptide antibiotics, taromycin A and taromycin B, were discovered using TAR-based recombination for direct cloning and refactoring of a "silent" BGC (Reynolds et al. 2018; Yamanaka et al. 2014). The platform has also been used for determining and isolating intermediate compounds of colibactin biosynthetic pathway from *E. coli* using heterologous expression which resulted in increased yield of these compounds (Li et al. 2015b). TAR-based cloning has been associated with CRISPR-Cas9 technology to improve its efficiency by 32%. The Cas9-mediated specific double strand breaks near the desired gene has demonstrated better results and can even be used to gene banks of human genes and its regulatory elements (Lee et al. 2015). In a recent study, CRISPR/Cas12a system along with BAC libraries was used for efficient cloning of large BGCs into *Streptomyces albus* which they named as CAT-FISHING (CRISPR/Cas12a-mediated fast direct biosynthetic gene cluster cloning). The method has been used for capturing of an 87 kb surugamides BGC into *Streptomyces albus* (Liang et al. 2020).

PCR amplification and subsequent assembly of different BGCs have also been implicated to be useful for heterologous expression. TAR-based recombination of the assembled PCR fragments of grecocyclin BGC followed by heterologous expression using new integrative vectors was successfully carried out in *E. coli* and *Streptomyces* (Bilyk et al. 2016). Similar method was used for the discovery of alpiniamide derivatives from *Streptomyces* sp. elucidating a hybrid non-ribosomal peptide synthetase pathway and trans-AT-polyketide synthase pathway (Paulus et al. 2018). A self-resistance-mediated genome mining approach was used along with PCR-based TAR cloning for isolation, characterization, and heterologous expression of novel inhibitors of topoisomerase from myxobacteria (Panter et al. 2018).

Overlap extension PCR yeast homologous recombination (ExRec), direct pathway cloning (DiPac) and sequence- and ligation-independent cloning (SLIC) have been used in recent times for improving the assembly of large number of PCR fragments or chemically synthesized fragments of a number of BGCs that had led to the discovery of some important compounds and scaffolds. ExRec has been used to assemble 15 BGCs (up to 45 kb) from *Photorhabdus* and *Xenorhabdus* (Schimming et al. 2014). DiPac approach, which utilizes long PCR, amplified products and homology-based assembly for construction of vector systems, has been utilized to develop numerous hosts that express large BGCs. The discovery of phenazine and anabaenopeptins along with the transfer of erythromycin BGC into a *Streptomyces* host has been possible using DiPac approach (Greunke et al. 2018). Furthermore, the depsipeptide hapalosin was developed through heterologous expression of cyanobacterial 23 kb BGC using DiPac in combination with SLIC (D'Agostino and Gulder 2018).

These techniques impart great technological advances related to the cloning and assembly of BGCs for heterologous expression. Cas9-mediated cleavage of DNA has introduced precision in cloning techniques that has greatly enhanced the efficiency of genome mining and recombination (Fig. 8.3).

8.5 Application of Heterologous Expression Systems

8.5.1 Novel Compound Discovery Through Heterologous Expression of BGCs

The rise in biosynthetic potentials for novel natural products can be owed to the exponentially increased availability of bacterial genome and metagenome sequences. Microbes generally express a limited number of BGCs that lead to sufficient yields of compounds under laboratory conditions. Majority of biosynthetic pathways mostly tend to stay unexploited given the fact that the majority BGCs exhibit low or maybe undetectable yields of compounds, thereby limiting the evaluation and discovery of novel natural products. Furthermore, although some techniques are adapted to activate cryptic BGCs, there is an absence of an efficient platform to attach genetic information to the biomolecules. Comparatively, to identify and characterize the unknown synthesis pathways expression in heterologous hosts could also be a sturdy approach

8.5.1.1 Cryptic BGC Expression

Genome mining involves the identification of antecedent uncharacterized biosynthetic gene clusters for natural products inside the genomes of antecedent sequenced organisms, followed by the sequence analysis of enzymes encoded by these BGCs, and finally the experimental identification of the products of the gene clusters. Genome sequencing projects throughout the last decade have unconcealed various cryptic synthesis pathways for unknown secondary metabolites in microbes that have in a manner revived drug discovery from microbial metabolites by such distinctive genome mining approaches (Xu et al. 2016). It is noteworthy that despite the very fact that uncovering of notable bioactive compounds remains a problem, the practical genome mining approach is ascendable and even adjustable for the industrial screening of novel antibiotics since the method of bioactivity screening, gene mobilization, and expression is also performed in a very high-throughput fashion similarly. There has been quite a range of flourishing attempts in discovery of BGCs of identified function. The discovery of some novel natural products or their

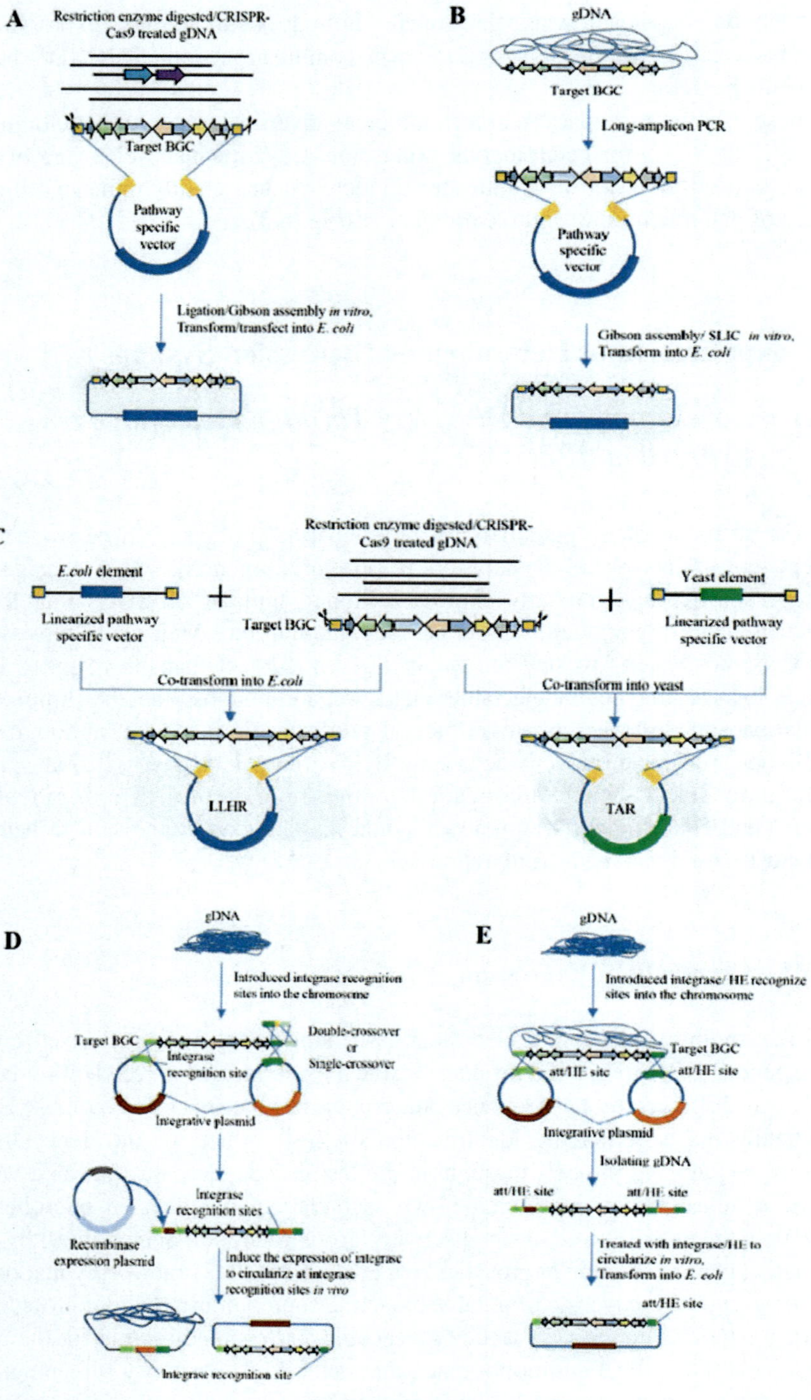

Fig. 8.3 Biosynthetic gene cluster expression for natural products discovery. (**a**) Direct cloning system constructed by enzyme digestion and ligation by ligation or Gibson assembly based cloning of BGCs. (**b**) Direct cloning system constructed by long-amplicon PCR and ligation. (**c**) LLHR (Linear–linear homologous recombination mediated by transformation-associated recombination

derivatives has resulted due to the activation of downregulated BGCs in heterologous systems. The discovery of two novel lipopeptide antibiotics, taromycin A2 and B3, was achieved by cloning a silent NRPS BGC (62 kb) from *Saccharomonospora* sp. CNQ-490 into *S. coelicolor* through chromosomal integration-based expression post-TAR cloning (Yamanaka et al. 2014).

8.5.1.2 Recombineering Techniques

Improved genetic tools like Red/ET recombineering, in addition to cloning massive BGCs, for efficient modification of the biosynthetic genes and corresponding regulatory elements gene manipulation is done by horizontal gene transfer into the heterologous host (Ongley et al. 2013). With the aim of generation of novel NPs and yield improvement conjointly as an in-depth understanding of synthesis mechanisms, a lot of research has been done in regard to recombinant deoxyribonucleic acid technology of synthesis pathways in heterologous hosts with success. These studies embrace feeding of biosynthetic precursors, modification in copy numbers of BGC transcripts, manipulation of transcriptional regulators, promoter engineering, insertion/deletion of tailoring enzymes, and even combined approaches. Because of the versatile and unvaried utilization of NRPS synthesis enzymes, numerous peptides are produced in nature. Scientists have utilized synthetic engineering approaches, as well as domain swaps, module skipping, and cross-talk among others, to change the biosynthetic pathways, sanctioning the identification of even a lot of bioactive rhabdopeptide/xenortide peptides (RXP) derivatives (Cai et al. 2017). Ancient cloning ways or different DNA engineering techniques, like Gibson assembly or CRISPR/Cas, would possibly instead be used in such situations wherever the constraints concerning the recombineering of repeat sequences restrain the appliance of Red/ET technology since repeated sequences often exist in most of the PKS/NRPS biosynthetic genes (Huo et al. 2019).

8.5.2 Novel Compound Discovery Through Heterologous Expression of Metagenomic DNA

Natural product discovery has been a challenging task considering the limitations in screening approaches for the compounds of the microbial world that is still untapped. This is mainly because most of the microorganisms from natural environment are non-culturable and hence impede the screening for natural products. Metagenomics

Fig. 8.3 (continued) (TAR) or RecET in yeast for cloning BGCs (**d**) Site-directed recombination for cloning BGC in vivo (**e**) Site-directed recombination for cloning BGC in vitro. {Adapted from Lin et al. 2020)}(Lin, Nielsen, and Liu, 2020). Copyright © 2020 Lin, Nielsen and Liu of the soil opened up new frontiers in this field that has tremendously expanded the information repositories for natural products from unculturable microorganisms (Handelsman et al. 1998). Efficient transfer of potential metagenomic DNA into cultured hosts and their heterologous expression allows comprehensive evaluation of the environmental DNA and its correlation with their encoded natural product.

Brady and colleagues in The Rockefeller University, USA, were able to discover a number of novel bioactive molecules using metagenomic library screening and subsequent expression of the environment-derived meta-genes in various types of host. In one of the prominent studies, they used functional metagenomics that resulted in the discovery of a novel natural product "metatricycloene" (Fig. 8.4d) (Iqbal et al. 2016). The team screened several species of streptomyces species for high efficiency of heterologous expression of metagenomic DNA, which were cloned into the best naturally privileged species, which was able to express the secondary metabolite with a greater yield. In an another recent study, a new class of antibiotics known as Malacidins (Fig. 8.4c) were discovered using culture-independent screening of calcium-dependent antibiotics from various soil metagenomes (Hover et al. 2018). Here, the team used a sequence-guided approach to screen for calcium-binding motifs from the environmental metagenome. To screen the large metagenome for this particular character, they increased the chances of hits by using natural product (NP) tags through amplifying universal BGCs and

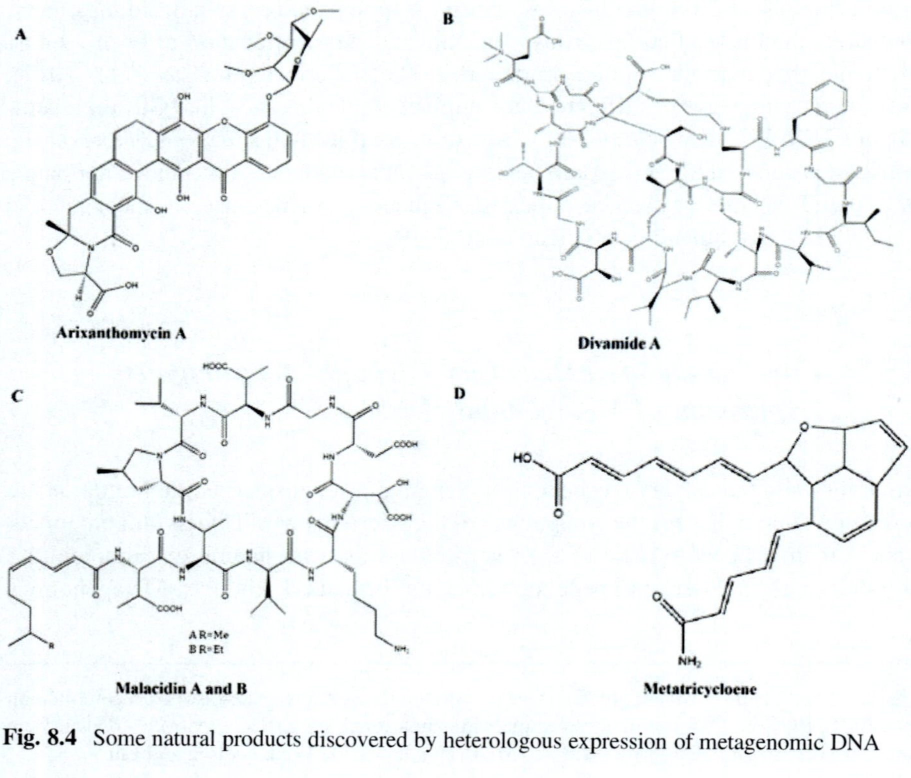

Fig. 8.4 Some natural products discovered by heterologous expression of metagenomic DNA

sequencing these amplicons for further analysis. They developed a database for comparison of these NP sequence tags with characterized BGCs for predicting the secondary metabolite and named it eSNaPD (environmental surveyor of natural product discovery).

Two rare bioactive tryptophan dimers were discovered using a chemical biogeographic survey of the chromopyrrolic acid synthase gene diversity in targeted environmental DNA libraries (Chang et al. 2015). Subsequent heterologous expression and synthetic biology led to the production of hydroxysporine and reductasporine, the two bioactive tryptophan dimers. In another study, a phylogeny-driven analysis of NP sequence tags from environmental DNA led to the discovery of arixanthomycins (Kang and Brady 2014). The ARX gene cluster was heterologously expressed using TAR-based assembly after analysing for ketosynthase beta (KS_{β}) tags and was integrated into the genome of streptomyces albus. The KS_{β} sequence of this gene cluster was phylogenetically distinct from other KS_{β} sequences which led to the discovery of the arixanthomycin A (Fig. 8.4a) having pentangular polyphenolic structure demonstrating anti-proliferative activity against human cancer cell lines. A similar sequence-guided strategy for *arm* cluster of genes was adopted for the discovering a new anthracycline, i.e. arimetamycin using KS_{β} sequences tags, which displayed better activity against multi-drug-resistant cancer cells (Kang and Brady 2013).

In an aim to diversify the microbiomes for the search of natural products, many studies have shifted their focus from commonly exploited soil microbiomes to unconventional niches such as phylloshpere, symbionts, gut and marine microbiomes. Combing for novel compounds in marine symbionts, Schmidt and teammates discovered novel anti-HIV compounds, divamides (Fig. 8.4b), which are a family of lanthipeptides produced by symbiotic bacteria living the tiny marine tunicates (Smith et al. 2018). In a similar approach, Moore and teammates through the heterologous expression of BGCs that encode PBDEs in cyanobacterial hosts from the microbiome of sponge-endosymbionts (Agarwal et al. 2017) discovered polybrominated diphenyl ethers (PBDEs). They employed amino acid-based search instead of the conventional nucleotide-based search for pinpointing the BGCs of PBDEs and subsequent heterologous expression. This example revealed the importance of heterologous expression in amplifying the product to obtain sufficient quantities from limited sources. In addition, this methodology can be used to search for compounds from complex symbionts in the near future. Gut microbiomes were also exploited for the search of novel bioactive compounds. In a prominent study undertaken recently, Fischbach and colleagues adopted heterologous expression, synthetic biology, and computational biology to survey the human gut microbiota for novel bioactive compound, which led to the discovery of novel pyrazinones and dihydropyrazinones (Guo et al. 2017). A number of NRPS BGCs were identified in silico and were integrated into *E. coli* and *B. subtilis*, which yielded a mass of pyrazinones and dihydropyrazinones. Further study revealed that these NRPS gene clusters are actively transcribed under some peculiar host conditions depending on the colonization inside the host by these bacteria. In addition, these metabolites may be released by the microbiota as dipeptide aldehydes, which can serve as potent protease inhibitors and selectively targets a subset of cathepsins in human cell proteomes. This study sets a platform for future studies that aim to explore the metabolic potential of gut microbiomes taking advantage of the human microbiome project.

8.6 Concluding Remarks

Novel natural products have demonstrated themselves to be of extreme importance in confronting infectious diseases and are being investigated thoroughly all around the globe to improve its efficacy and screening process. With the advent of bioinformatics and synthetic biology, heterologous expression of the gene clusters that encodes secondary metabolic products has been used extensively to access the inventory of natural products coded by bacteria and environmental DNA. Advancements in computational capabilities and database managements have largely improved the screening and prediction of natural product pathways allowing interventions that lead to yield improvements or even producing compounds from cryptic BGCs. Furthermore, synthetic biology along with heterologous expression can also be used for reconstituting non-conventional natural product pathways that may, in the future, lead to novel artificial scaffold engineering. Without a doubt, heterologous expression will be imperative in understanding the intricate mechanisms of microbial biosynthesis of natural product and will aid in the discovery of novel scaffolds from diverse environmental sources and microbiomes.

Acknowledgements The authors wish to express sincere gratitude to the Director, IASST, Guwahati, Assam, India, for providing the infrastructure and resources to carry out this work. We would also like to extend thanks to the Department of Science and Technology (DST), Government of India, for the fellowship to SNH and the Department of Biotechnology (DBT), Government of India, for the fellowship to PB and ANK. The authors would like to thank Ms. Gautomi Gogoi, Physical Sciences Division, for rendering help in this work.

References

Adamek M, Spohn M, Stegmann E, Ziemert N (2017) Chapter 2: mining bacterial genomes for secondary metabolite gene clusters 1520. In: Antibiotics. Humana Press, New York, NY, pp 23–47

Agarwal V, Blanton JM, Podell S, Taton A, Schorn MA, Busch J et al (2017) Metagenomic discovery of polybrominated diphenyl ether biosynthesis by marine sponges. Nat Chem Biol 13(5):537–543

Albarano L, Esposito R, Ruocco N, Costantini M (2020) Genome mining as new challenge in natural products discovery. Mar Drugs 18(4):1–17

Barka EA, Vatsa P, Sanchez L, Gaveau-vaillant N, Jacquard C, Klenk H et al (2016) Taxonomy , physiology , and natural products of actinobacteria. Microbiol Mol Biol Rev 80(1):1–43

Belknap KC, Park CJ, Barth BM, Andam CP (2020) Genome mining of biosynthetic and chemotherapeutic gene clusters in Streptomyces bacteria. Sci Rep 10(1):1–9

Bentley SD, Chater KF, Cerdeño-Tárraga AM, Challis GL, Thomson NR, James KD et al (2002) Complete genome sequence of the model actinomycete Streptomyces coelicolor A3(2). Nature 417(6885):141–147

Berdy J (2005) Bioactive microbial metabolites. J Antibiot 58(1):1–26

Bérdy J (2012) Thoughts and facts about antibiotics: Where we are now and where we are heading. J Antibiot 65:385–395

Bilyk O, Sekurova ON, Zotchev SB, Luzhetskyy A (2016) Cloning and heterologous expression of the grecocycline biosynthetic gene cluster. PLoS One 11(7):1–17

Blin K, Wolf T, Chevrette MG, Lu X, Schwalen CJ, Kautsar SA et al (2017) AntiSMASH 4.0 – improvements in chemistry prediction and gene cluster boundary identification. Nucleic Acids Res 45(W1):W36–W41

Blin K, Shaw S, Steinke K, Villebro R, Ziemert N, Lee Y et al (2019) antiSMASH 5.0: updates to the secondary metabolite genome mining pipeline. Nucleic Acids Res 47:81–87

Bonet B, Teufel R, Crüsemann M, Ziemert N, Moore BS (2015) Direct capture and heterologous expression of salinispora natural product genes for the biosynthesis of enterocin. J Nat Prod 78 (3):539–542

Cai X, Nowak S, Wesche F, Bischoff I, Kaiser M, Fürst R, Bode HB (2017) Entomopathogenic bacteria use multiple mechanisms for bioactive peptide library design. Nat Chem 9(4):379–386

Chang FY, Ternei MA, Calle PY, Brady SF (2015) Targeted metagenomics: finding rare tryptophan dimer natural products in the environment. J Am Chem Soc 137(18):6044–6052

D'Agostino PM, Gulder TAM (2018) Direct pathway cloning combined with sequence- and ligation-independent cloning for fast biosynthetic gene cluster refactoring and heterologous expression. ACS Synth Biol 7(7):1702–1708

Datsenko KA, Wanner BL (2000) One-step inactivation of chromosomal genes in *Escherichia coli* K-12 using PCR products. Proc Natl Acad Sci U S A 97(12):6640–6645

de Lima Procópio RE, da Silva IR, Martins MK, de Azevedo JL, de Araújo JM (2012) Antibiotics produced by Streptomyces. Braz J Infect Dis 16(5):466–471

Demain AL (2002) Prescription for an ailing pharmaceutical industry. Nat Publ Group 331

Doroghazi JR, Albright JC, Goering AW, Ju K, Haines R, Tchalukov KA et al (2015) Nat Chem Biol 10(11):963–968

Du D, Wang L, Tian Y, Liu H, Tan H, Niu G (2015) Genome engineering and direct cloning of antibiotic gene clusters via phage φbT1 integrase-mediated site-specific recombination in Streptomyces. Sci Rep 5:1–8

Fair RJ, Tor Y (2014) Perspectives in medicinal chemistry antibiotics and bacterial resistance in the 21st century. Perspect Med Chem 6:25–64

Fischbach MA, Walsh CT (2006) Assembly-line enzymology for polyketide and nonribosomal peptide antibiotics: logic machinery, and mechanisms. Chem Rev 106(8):3468–3496

Fu J, Bian X, Hu S, Wang H, Huang F, Seibert PM et al (2012) Full-length RecE enhances linear-linear homologous recombination and facilitates direct cloning for bioprospecting. Nat Biotechnol 30(5):440–446

Gibson DG, Young L, Chuang RY, Venter JC, Hutchison CA, Smith HO (2009) Enzymatic assembly of DNA molecules up to several hundred kilobases. Nat Methods 6(5):343–345

Gomez-Escribano JP, Bibb MJ (2011) Engineering Streptomyces coelicolor for heterologous expression of secondary metabolite gene clusters. Microb Biotechnol 4(2):207–215

Greunke C, Duell ER, D'Agostino PM, Glöckle A, Lamm K, Gulder TAM (2018) Direct Pathway Cloning (DiPaC) to unlock natural product biosynthetic potential. Metab Eng 47:334–345

Guo CJ, Chang FY, Wyche TP, Backus KM, Acker TM, Funabashi M et al (2017) Discovery of reactive microbiota-derived metabolites that inhibit host proteases. Cell 168(3):517–526. e18

Handelsman J, Rondon MR, Brady SF, Clardy J, Goodman RM (1998) Molecular biological access to the chemistry of unknown soil microbes: a new frontier for natural products. Chem Biol 5(10)

Hassan QP, Bhat AM, Shah AM (2019) Bioprospecting actinobacteria for bioactive secondary metabolites from untapped ecoregions of the northwestern himalayas. In: New and future developments in microbial biotechnology and bioengineering: microbial secondary metabolites biochemistry and applications. Elsevier, pp 77–85

Hazarika SN, Thakur D (2020) Actinobacteria. In: Beneficial microbes in agro-ecology bacteria and fungi. Academic Press, pp 443–476

Hobbs G, Obanye AIC, Petiy J, Mason JC, Barratt E, Gardner DCJ et al (1992) An integrated approach to studying regulation of production of the antibiotic methylenomycin by streptomyces coelicolor A3 (2). J Bacteriol 174(5):1487–1494

Hopwood DA, Malpartida F, Kieser HM, Ikeda H, Duncan J, Fujii I et al (1985) Production of "hybrid" antibiotics by genetic engineering. Nature 314(6012):642–644

Hover BM, Kim SH, Katz M, Charlop-Powers Z, Owen JG, Ternei MA et al (2018) Culture-independent discovery of the malacidins as calcium-dependent antibiotics with activity against multidrug-resistant Gram-positive pathogens. Nat Microbiol 3(4):415–422

Hsieh PC, Xiao JP, O'Loane D, Xu SY (2000) Cloning, expression, and purification of a thermostable nonhomodimeric restriction enzyme, BslI. J Bacteriol 182(4):949–955

Huo L, Hug JJ, Fu C, Bian X, Zhang Y, Müller R (2019) Heterologous expression of bacterial natural product biosynthetic pathways. Nat Prod Rep 36(10):1412–1436

Ichikawa N, Sasagawa M, Yamamoto M, Komaki H, Yoshida Y, Yamazaki S, Fujita N (2013) DoBISCUIT: A database of secondary metabolite biosynthetic gene clusters. Nucleic Acids Res 41(D1):408–414

Iqbal HA, Low-Beinart L, Obiajulu JU, Brady SF (2016) Natural product discovery through improved functional metagenomics in streptomyces. J Am Chem Soc 138(30):9341–9344

Johnston CW, Skinnider MA, Wyatt MA, Li X, Ranieri MRM, Yang L, Zechel DL, Ma B, Magarvey NA (2015) An automated Genomes-to-Natural Products platform (GNP) for the discovery of modular natural products. Nat Commun 6(8421):1–11

Joneja A, Huang X (2010) A device for automated hydrodynamic shearing of genomic DNA. BioTechniques 46(7):553–556

Kang HS, Brady SF (2013) Arimetamycin A: improving clinically relevant families of natural products through sequence-guided screening of soil metagenomes. Angew Chem Int Ed 52 (42):11063–11067

Kang HS, Brady SF (2014) Arixanthomycins A–C: phylogeny-guided discovery of biologically active eDNA-derived pentangular polyphenols. ACS Chem Biol 9(6):1267–1272

Kautsar SA, Blin K, Shaw S, Navarro-mu JC, Terlouw R, Van Der Hooft JJJ et al (2020) MIBiG 2.0: a repository for biosynthetic gene clusters of known function. Nucleic Acids Res 48:454–458

Kawai S, Hashimoto W, Murata K (2010) Transformation of *Saccharomyces cerevisiae* and other fungi: methods and possible underlying mechanism. Bioeng Bugs 1(6):395–403

Ke J, Yoshikuni Y (2020) ScienceDirect Multi-chassis engineering for heterologous production of microbial natural products. Curr Opin Biotechnol 62:88–97

Koehn FE, Carter GT (2005) The evolving role of natural products in drug discovery. Nat Rev Drug Discov 4(3):206–220

Kuzminov A (2011) Homologous recombination-experimental systems, analysis and significance. EcoSal Plus 4(2):1124–1129

Larionov V, Kouprina N, Graves J, Chen XN, Korenberg JR, Resnick MA (1996) Specific cloning of human DNA as yeast artificial chromosomes by transformation-associated recombination. Proc Natl Acad Sci U S A 93(1):491–496

Latif AA, Osman G (2017) Comparison of three genomic DNA extraction methods to obtain high DNA quality from maize. Plant Methods 13(1):1–9

Law JW, Law LN, Letchumanan V, Tan LT, Wong SH, Chan K et al (2020) Anticancer drug discovery from microbial sources: the unique mangrove streptomycetes. Molecules 25(22):5365

Lee NCO, Larionov V, Kouprina N (2015) Highly efficient CRISPR/Cas9-mediated TAR cloning of genes and chromosomal loci from complex genomes in yeast. Nucleic Acids Res 43(8):e55

Lewis K (2012) Recover the lost art of drug discovery. Nature 485:3–4
Li MHT, Ung PMU, Zajkowski J, Garneau S, Sherman DH (2009) Automated genome mining for natural products. BMC Bioinformatics 10:1–10
Li JE, Guo ZY, Huang W, Meng XX, Ai GM, Tang GL, Chen YH (2013) Mining of a streptothricin gene cluster from Streptomyces sp. TP-A0356 genome via heterologous expression. Sci China Life Sci 56(7):619–627
Li Y, Li Z, Yamanaka K, Xu Y, Zhang W, Vlamakis H et al (2015a) Directed natural product biosynthesis gene cluster capture and expression in the model bacterium Bacillus subtilis. Sci Rep 5:1–7
Li ZR, Li Y, Lai JYH, Tang J, Wang B, Lu L et al (2015b) Critical intermediates reveal new biosynthetic events in the enigmatic colibactin pathway. Chembiochem 16(12):1715–1719
Li L, Liu X, Jiang W, Lu Y (2019) Recent advances in synthetic biology approaches to optimize production of bioactive natural products in actinobacteria. Front Microbiol:1–10
Liang M, Liu L, Wang W, Zeng X, Liu J, Karthik L et al (2020) Simple cloning of large natural product biosynthetic gene cluster by CRISPR/Cas12a-mediated fast direct capturing strategy. bioRxiv. preprint
Lin Z, Nielsen J, Liu Z (2020) Bioprospecting through cloning of whole natural product biosynthetic gene clusters. Front Bioeng Biotechnol 8:526
Liu Z, Lin Z, Nielsen J (2021) Expression of fungal biosynthetic gene clusters in S. cerevisiae for natural product discovery. Synth Syst Biotechnol 6(1):20–22
Malik A, Kim YR, Jang IH, Hwang S, Oh DC, Kim SB (2020) Genome-based analysis for the bioactive potential of Streptomyces yeochonensis CN732, an acidophilic filamentous soil actinobacterium. BMC Genomics 21(1):1–16
Malpartida F, Niemi J, Navarrete R, Hopwood DA (1990) Cloning and expression in a heterologous host of the complete set of genes for biosynthesis of the Streptomyces coelicolor antibiotic undecylprodigiosin. Gene 93:91–99
Medema MH, Blin K, Cimermancic P, De Jager V, Zakrzewski P, Fischbach MA et al (2011) antiSMASH : rapid identification , annotation and analysis of secondary metabolite biosynthesis gene clusters in bacterial and fungal genome sequences. Nucleic Acids Res 39:339–346
Nah HJ, Woo MW, Choi SS, Kim ES (2015) Precise cloning and tandem integration of large polyketide biosynthetic gene cluster using Streptomyces artificial chromosome system. Microb Cell Factories 14(1):1–11
Nah HJ, Pyeon HR, Kang SH, Choi SS, Kim ES (2017) Cloning and heterologous expression of a large-sized natural product biosynthetic gene cluster in Streptomyces species. Front Microbiol 8:394
Naughton LM, Romano S, Gara FO, Dobson ADW (2017) Identification of secondary metabolite gene clusters in the pseudovibrio genus reveals encouraging biosynthetic potential toward the production of novel bioactive compounds. Front Microbiol 8:1–15
Newman DJ, Cragg GM (2020) Natural products as sources of new drugs over the nearly four decades from 01/1981 to 09/2019. J Nat Prod 83(3):770–803
Ohnishi Y, Ishikawa J, Hara H, Suzuki H, Ikenoya M, Ikeda H et al (2008) Genome sequence of the streptomycin-producing microorganism *Streptomyces griseus* IFO 13350. J Bacteriol 190 (11):4050–4060
Olorunniji FJ, Rosser SJ, Stark WM (2016) Site-specific recombinases: molecular machines for the genetic revolution. Biochem J 473(6):673–684
Olorunniji FJ, Lawson-Williams M, McPherson AL, Paget JE, Stark WM, Rosser SJ (2019) Control of ϕC31 integrase-mediated site-specific recombination by protein trans-splicing. Nucleic Acids Res 47(21):11452–11460
Ongley SE, Bian X, Neilan BA, Müller R (2013) Recent advances in the heterologous expression of microbial natural product biosynthetic pathways. Nat Prod Rep 30(8):1121–1138
Panter F, Krug D, Baumann S, Müller R (2018) Self-resistance guided genome mining uncovers new topoisomerase inhibitors from myxobacteria. Chem Sci 9(21):4898–4908

Patridge E, Gareiss P, Kinch MS, Hoyer D (2016) An analysis of FDA-approved drugs: natural products and their derivatives. Drug Discov Today 21(2):204–207

Paulus C, Rebets Y, Zapp J, Rückert C, Kalinowski J, Luzhetskyy A (2018) New alpiniamides from Streptomyces sp. IB2014/011-12 assembled by an unusual hybrid non-ribosomal peptide synthetase trans-AT polyketide synthase enzyme. Front Microbiol 9:1–15

Ren H, Shi C, Zhao H (2020) Computational tools for discovering and engineering natural product biosynthetic pathways. ISCIENCE 23(1):100795

Reynolds KA, Luhavaya H, Li J, Dahesh S, Nizet V, Yamanaka K, Moore BS (2018) Isolation and structure elucidation of lipopeptide antibiotic taromycin B from the activated taromycin biosynthetic gene cluster. J Antibiot 71(2):333–338

Ross AC, Gulland LES, Dorrestein PC, Moore BS (2015) Targeted capture and heterologous expression of the pseudoalteromonas alterochromide gene cluster in escherichia coli represents a promising natural product exploratory platform. ACS Synth Biol 4(4):414–420

Rudd BAM, Hopwood DA (1979) Genetics of actinorhodin biosynthesis by *Streptomyces coelicolor* A3(2). J Gen Microbiol 114(1):35–43

Schimming O, Fleischhacker F, Nollmann FI, Bode HB (2014) Yeast homologous recombination cloning leading to the novel peptides ambactin and xenolindicin. Chembiochem 15 (9):1290–1294

Sharan, SK, Thomason, LC, Kuznetsov, SG, Court, DL (2009) Recombineering: a homologous recombination-based method of genetic engineering. Nat Prot 4(2): 206-223

Selvin J, Sathiyanarayanan G, Lipton AN, Al-Dhabi NA, Valan Arasu M, Kiran GS (2016) Ketide Synthase (KS) domain prediction and analysis of iterative Type II PKS gene in marine sponge-associated actinobacteria producing biosurfactants and antimicrobial agents. Front Microbiol 7: 1–12

Sharma-Kuinkel BK, Rude TH, Fowler VG Jr (2016) Pulse field gel electrophoresis. Methods Mol Biol 1373:117–130

Singh AL, Aniebok V, Balunas MJ, Van Santen A, Bunsko D, Neto FC et al (2019) The natural products Atlas: an open access knowledge base for microbial natural products discovery. ACS Central Sci 5(11):1824–1833

Skinnider MA, Dejong CA, Rees PN, Johnston CW, Li H, Webster ALH et al (2015) Genomes to natural products. PRediction Inform Second Metab (PRISM) 43(20):9645–9662

Smith TE, Pond CD, Pierce E, Harmer ZP, Kwan J, Zachariah MM et al (2018) Accessing chemical diversity from the uncultivated symbionts of small marine animals. Nat Chem Biol 14 (2):179–185

Stoesser G (2001) The EMBL nucleotide sequence database. Nucleic Acids Res 29(1):17–21

Tahlan K, Ahn SK, Sing A, Bodnaruk TD, Willems AR, Davidson AR, Nodwell JR (2007) Initiation of actinorhodin export in Streptomyces coelicolor. Mol Microbiol 63(4):951–961

Tang X, Li J, Millán-Aguiñaga N, Zhang JJ, O'Neill EC, Ugalde JA et al (2015) Identification of thiotetronic acid antibiotic biosynthetic pathways by target-directed genome mining. ACS Chem Biol 10(12):2841–2849

Thanapipatsiri A, Gomez-Escribano JP, Song L, Bibb MJ, Al-Bassam M, Chandra G et al (2016) Discovery of unusual biaryl polyketides by activation of a silent streptomyces venezuelae biosynthetic gene cluster. Chembiochem 17(22):2189–2198

Thapa LP, Oh TJ, Lee HC, Liou K, Park JW, Yoon YJ, Sohng JK (2007) Heterologous expression of the kanamycin biosynthetic gene cluster (pSKC2) in Streptomyces venezuelae YJ003. Appl Microbiol Biotechnol 76(6):1357–1364

Thomas S, Maynard ND, Gill J (2015) DNA library construction using Gibson assembly. Nat Methods 12(11):i–ii

Waksman SA, Schatz A, Reynolds DM (2010) Production of antibiotic substances by actinomycetes. Ann N Y Acad Sci 1213(1):112–124

Walker MC, Eslami SM, Hetrick KJ, Ackenhusen SE, Mitchell DA, Van Der Donk WA (2020) Precursor peptide-targeted mining of more than one hundred thousand genomes expands the lanthipeptide natural product family. BMC Genomics 21:1–17

Wang H, Li Z, Jia R, Hou Y, Yin J, Bian X et al (2016) RecET direct cloning and Redαβ recombineering of biosynthetic gene clusters, large operons or single genes for heterologous expression. Nat Protoc 11(7):1175–1190

Wang H, Li Z, Jia R, Yin J, Li A, Xia L et al (2018) ExoCET: exonuclease in vitro assembly combined with RecET recombination for highly efficient direct DNA cloning from complex genomes. Nucleic Acids Res 46(5):e28

Wang X, Yin S, Bai J, Liu Y, Fan K, Wang H et al (2019) Heterologous production of chlortetracycline in an industrial grade *Streptomyces rimosus* host. Appl Microbiol Biotechnol 103 (16):6645–6655

Watve MG, Tickoo R, Jog MM, Bhole BD (2001) How many antibiotics are produced by the genus Streptomyces? Arch Microbiol 176(5):386–390

Weber T, Welzel K, Pelzer S, Vente A, Wohlleben W (2003) Exploiting the genetic potential of polyketide producing streptomycetes. J Biotechnol 106(2–3):221–232

Weber T, Rausch C, Lopez P, Hoof I, Gaykova V, Huson DH, Wohlleben W (2009) CLUSEAN : a computer-based framework for the automated analysis of bacterial secondary metabolite biosynthetic gene clusters. J Biotechnol 140(1–2):13–17

Wright GD (2011) Molecular mechanisms of antibiotic resistance. Chem Commun 47 (14):4055–4061

Wright GD (2017) Opportunities for natural products in 21st century antibiotic discovery. Nat Prod Rep 34(7):694–701

Xu M, Wright GD (2019) Heterologous expression-facilitated natural products' discovery in actinomycetes. J Ind Microbiol Biotechnol 46(3–4):415–431

Xu M, Wang Y, Zhao Z, Gao G, Huang SX, Kang Q et al (2016) Functional genome mining for metabolites encoded by large gene clusters through heterologous expression of a whole-genome bacterial artificial chromosome library in Streptomyces spp. Appl Environ Microbiol 82 (19):5795–5805

Yamanaka K, Reynolds KA, Kersten RD, Ryan KS, Gonzalez DJ, Nizet V et al (2014) Direct cloning and refactoring of a silent lipopeptide biosynthetic gene cluster yields the antibiotic taromycin A. Proc Natl Acad Sci U S A 111(5):1957–1962

Yin J, Hoffmann M, Bian X, Tu Q, Yan F, Xia L et al (2015) Direct cloning and heterologous expression of the salinomycin biosynthetic gene cluster from Streptomyces albus DSM41398 in Streptomyces coelicolor A3(2). Sci Rep 5(1):15081

Yin S, Li Z, Wang X, Wang H, Jia X, Ai G et al (2016) Heterologous expression of oxytetracycline biosynthetic gene cluster in *Streptomyces venezuela*e WVR2006 to improve production level and to alter fermentation process. Appl Microbiol Biotechnol 100(24):10563–10572

Zeng C, Kouprina N, Zhu B, Cairo A, Hoek M, Cross G et al (2001) Large-insert BAC/YAC libraries for selective re-isolation of genomic regions by homologous recombination in yeast. Genomics 77(1–2):27–34

Zhong Z, He B, Li J, Li Y (2020) Challenges and advances in genome mining of ribosomally synthesized and post-translationally modified peptides (RiPPs). Synth Syst Biotechnol 5 (3):155–172

Zhou Y, Murphy AC, Samborskyy M, Prediger P, Dias LC, Leadlay PF (2015) Iterative mechanism of macrodiolide formation in the anticancer compound conglobatin. Chem Biol 22(6):745–754

Ziemert N, Podell S, Penn K, Badger JH, Allen E, Jensen PR (2012) The Natural product domain seeker napdos : a phylogeny based bioinformatic tool to classify secondary metabolite gene diversity. PLoS One 7(3):1–9

Chapter 9
Synthetic Biology and Metabolic Engineering in Actinobacteria for Natural Product Production

Pachara Sattayawat and Wasu Pathom-aree

Abstract Synthetic biology has been employed mostly in *Escherichia coli* and yeasts for almost 20 years, not until recently that this field has been expanded to actinobacteria. Metabolic engineering is considered as an overlapping field with synthetic biology and has been a pioneer synthetic biology-based strategy adapted to natural product (NP) production in actinobacteria. This group of microorganisms is still thought of as challenging hosts for engineering because of their distinct biology. Yet, the number of studies on such topics has increased over the last decades indicating underexplored potential of actinobacteria as an engineering host. In this chapter, fundamental understandings on synthetic biology, recent works, and developments on synthetic biology-based strategies (including the ground-breaking CRISPR) in actinobacteria are discussed. We also highlight the potential of actinobacteria as a host for metabolic engineering to enhance NP production and future perspectives on this topic are additionally provided.

Keywords Actinobacteria · Natural products · Synthetic biology · Metabolic engineering · Overproduction · Streptomyces · Rare actinobacteria

9.1 Introduction

Synthetic biology is a rapidly emerging discipline, which has demonstrated significant achievements over the last 20 years since its birth. The number of publications on synthetic biology has greatly increased suggesting the rapid growth of research

P. Sattayawat
Department of Biology, Faculty of Science, Chiang Mai University, Chiang Mai, Thailand

W. Pathom-aree (✉)
Research Center of Microbial Diversity and Sustainable Utilization, Department of Biology, Faculty of Science, Chiang Mai University, Chiang Mai, Thailand
e-mail: wasu.p@cmu.ac.th

L. Karthik (ed.), *Actinobacteria*, https://doi.org/10.1007/978-981-16-5835-8_9

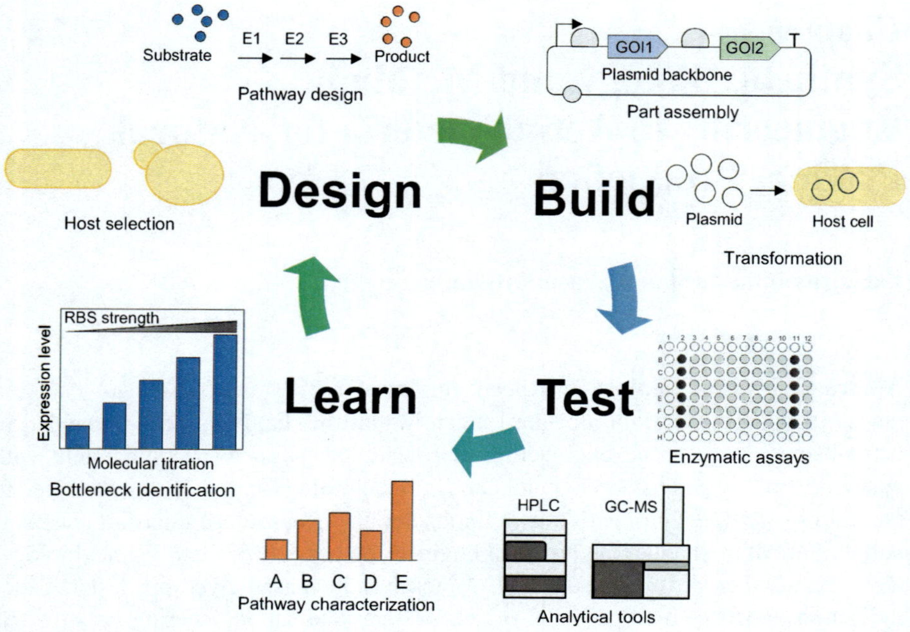

Fig. 9.1 Synthetic biology's "Design-Build-Test-Learn (DBTL)" cycle adapted from Petzold et al. (2015)

interest in this field (Raimbault et al. 2016). The definition of synthetic biology can be interpreted directly as a study of artificial or unnatural biology. But to be more specific, it can generally be defined as a field of study that aims to construct novel biological systems or redesign existing systems in nature (Endy 2005; Roberts et al. 2013) through a wide range of engineering approaches. With this principle, synthetic biology allows biologists to make a shortcut through evolution to achieve biological parts, devices, or systems for desirable applications.

In this sense, one may ask then what the differences between synthetic biology and other bioengineering disciplines are. To answer this question, it was proposed that synthetic biology focuses on the design and construction of the whole system unlike other fields where the focus may only be on particular genes or proteins (Keasling 2008). Moreover, it is also viewed that synthetic biology is a more complex engineering as all known factors are often considered in the designing step. Research on synthetic biology often follows a unique pattern of its general workflow called "Design-Build-Test-Learn (DBTL) cycle." To begin the workflow (Fig. 9.1), the system is initially designed based on the possibilities of occurring when all essential components are available. The design is then implemented, or all components are assembled. To give an example, if a novel synthetic pathway is designed to be implemented in a selected microbial host, all genes responsible for the pathways are put together commonly in plasmid systems and transformed to the host. Subsequently, the newly created system is investigated whether the engineering has taken place and the designed system operates or not. The final step is to learn or comprehend the factors that may prevent the system to function optimally. Prior to the start of the second round, knowledge learnt is contemplated to further optimize the system for better performances. As mentioned, synthetic biology requires efforts on design and construction of the systems. This suggests that often the careful design of the work is an important factor to achieve final goals in this field of research.

The works on synthetic biology have been performed mainly in microorganisms as they are the relatively simplest chassis to engineer (Kim et al. 2016) including actinobacteria. Metabolic engineering is considered at some level an overlapping field with synthetic biology (Nielsen and Keasling 2011). In this respect, one of the most common practices adapting synthetic biology to actinobacteria is to metabolically engineer actinobacteria for natural products (NPs) overproduction or to transfer the biosynthetic gene clusters (BGCs) from native hosts to a more suitable microbial chassis. Host cell is viewed as a factory for natural product production and metabolic engineering is a tool to improve cell performances.

In this chapter, we then summarize recent works and developments on metabolic engineering in actinobacteria and highlight the potential of actinobacteria as a host for metabolic engineering to enhance NP production. Future perspectives on metabolic engineering for NP production are also discussed.

9.2 An Overview of Engineering in Actinobacteria for Natural Products (NPs) Overproduction

As mentioned in previous chapters, actinobacteria are a large group of Gram-positive bacteria with unique characteristics such as the high GC-content genomes and their filamentous morphology in some genera (Barka et al. 2016; Nouioui et al. 2018). They are best known for their ability to produce biologically active natural products (NPs), which benefit human as a source of bioactive compounds for almost a century (Demain and Sanchez 2009; Scheffler et al. 2013). Traditional discovery of NP relies on time-consuming and laborious workflow, but recent technologies such as next-generation sequencing (NGS) open new opportunities to novel NP discovery from actinobacteria (Gomez-Escribano et al. 2016). With these progresses, genome mining has revealed great potential of "hidden" biosynthetic gene clusters (BGCs) (Foulston 2019; Ziemert et al. 2016). However, most of these newly discovered BGCs remain silent or "cryptic" under standard laboratory conditions and need to be activated prior to characterization (Luo et al. 2013; Olano et al. 2014). Moreover, even though some bioactive compounds from actinobacteria are commercially available such as gentamicin and lipstatin, the production of a major number of NPs is still insufficient for commercial production. This situation emphasizes an urgent need to improve the production yield. Metabolic engineering offers a powerful tool to tackle such challenges. Although this state-of-the-art technique has been employed to several microorganisms to achieve commercialization of both natural and non-natural products, the work on actinobacteria has not much been investigated. Recent achievements in the field of metabolic engineering in actinobacteria are outlined in this chapter together with the potential of actinobacteria as a host microorganism for metabolic engineering.

Metabolic engineering is generally defined as an intentional modification of cellular metabolisms for the production of targeted chemicals (Yu et al. 2014). This approach relies heavily on genetic engineering techniques and tools. However, genetic engineering in actinobacteria has been known to be more difficult as compared to other model microorganisms, e.g., *E. coli* and *Saccharomyces cerevisiae* mainly due to their idiosyncratic genetic, physiological and morphological traits (Musiol-Kroll et al. 2019). Nevertheless, a number of recent studies on metabolic engineering in actinobacteria for NP production are now available as summarized in Table 9.1. These efforts highlight the potential of actinobacteria as indispensable NP producers. Metabolic engineering in actinobacteria is classified broadly as (i) manipulation of native metabolisms to improve the yield of native NPs and (ii) heterologous expression of non-native BGCs in actinobacterial heterologous hosts.

Several strategies are employed to manipulate native metabolisms of actinobacteria. A combination of several metabolic engineering strategies is often used to maximize the efficiency of metabolic engineering. These strategies and examples of recent works are elucidated.

Carbon flux redirection is one of the most well-studied strategies. Deletion of competitive pathways to increase targeted precursor supply is often used to increase the metabolic flux toward final products of interest. In one study, deletion of gene clusters encoding enzymes for competitive pathways in *Streptomyces albus* DSM 41398 had shown to enhance the production of salinomycin up to 11 folds (Lu et al. 2016). Similar to the aforementioned strategy, overexpression of bottleneck enzymes to enhance the flux, which leads to a more efficient reaction of the limiting step, has also been studied. Overexpression of an insufficiently active enzyme in chlortetracycline biosynthetic pathway in *S. aureofaciens* increased the production around 1.75-fold compared to the parental strain (Zhu et al. 2013). Moreover, a comprehensive review on recombinant protein production in actinobacteria was recently published highlighting the capability of actinobacteria as hosts for heterologous enzyme expression (Berini et al. 2020). Superior efflux pump system has been known to contribute to cellular tolerance toward native or non-native bio-chemicals (Dunlop et al. 2011). Overexpression of transporters, thus holds great potential for production improvements and has been shown previously in actinobacteria (Xie et al. 2017).

Manipulation of BGC regulatory networks (i.e., activator and repressor) is another useful strategy. For instance, deletion of homologs (*shbR1* and *shbR3*) that repress the transcription of a central transcriptional regulator and validamycin biosynthetic genes resulted in an increased yield of 55% (Tan et al. 2015; Tan et al. 2013).

Moreover, amplification of BGCs or structural genes involved in NP production has been proven to be another effective strategy. Introduction of an extra copy of moenomycin gene cluster led to approximately 2.2-fold increase in nosokomycin A production. It is also worth mentioning that the introduction of regulatory genes was also investigated in the same study but only a marginal increase was observed (Kuzhyk et al. 2019).

BGC refactoring is a strategy that dissects BGCs into small modular units in order to independently optimize the function of each unit (Smanski et al. 2016). Hence, this can be implemented in many ways including replacement of natural promoters (Li et al. 2018), or even the whole operon with optimized artificial ones (Song et al. 2019), which is often seen in heterologous expression where BGCs cannot maximally express in homologous hosts. Heterologous hosts for NP production are often actinobacterial strains with well-known genomic and metabolic backgrounds and/or have been constructed and optimized for such expressions. *Streptomyces* is considered the first choice as a heterologous chassis, *S. albus* (Myronovskyi et al. 2018), *S. coelicolor* (Gomez-Escribano and Bibb 2011), *S. avermitilis* (Komatsu et al. 2010), and *S. lividans* (Ahmed et al. 2020) have been previously developed. Myronovskyi and Luzhetskyy (2019) comprehensively reviewed these model heterologous strains with recent developments on genetic parts to facilitate high success rates for future engineering. Reduced genome is arguably the most common modified characteristic to increase the strain performance as this could lower metabolic backgrounds of producing strains and, in turn, increase native precursor availability, thus heterologous hosts often have their genome minimized. For example, deletion of 0.7 Mb of non-essential regions and an undesirable native NP gene cluster on *S. chattanoogensis* resulted in cleaner metabolite profiles and improved heterologous NP production (Bu et al. 2019).

As mentioned, the implementation of several strategies to enhance the production of one final product is often seen. To give an instance, combining inactivation of repressor, deletion of competitive pathways and amplification of a chosen BGC resulted in 12-fold improved production of prodigiosins (Liu et al. 2017). Table 9.1 lists examples of recent metabolic engineering achievements in actinobacteria for NP production.

Table 9.1 Examples of natural products from engineered *Streptomyces* strains and yield improvements from 2010 to 2020.

Natural product	Host	Metabolic manipulation	Yield improvement	Reference
Salinomycin	*S. albus* DSM 41398	Carbon flux redirection	11-fold	Lu et al. (2016)
Tylactone	*S. venezuelae*	Carbon flux redirection	10-fold	Jung et al. (2014)
Mithramycins	*S. argillaceus* ATCC12956	Carbon flux redirection	229%	Zabala et al. (2013)
Chlortetracycline	*S. aureofaciens*	Overexpression of bottleneck enzyme(s)	~1.75-fold	Zhu et al. (2013)
Griseoviridin	*S. griseoviridis* NRRL 2427	Overexpression of transporter(s)	~3-fold	Xie et al. (2017)
Viridogrisein	*S. griseoviridis* NRRL 2427	Overexpression of transporter(s)	~3-fold	Xie et al. (2017)
Amphotericin B	*S. nodosus*	Overexpression of bottleneck enzyme(s) and transporter(s)	28%	Zhang et al. (2020a, b)
Nosokomycin A	*S. ghanaensis* ATCC14672	Amplification of BGC (s)	3-fold	Kuzhyk et al. (2019)
FK506	*S. tsukubaensis*	Amplification of structural gene(s)	146%	Huang et al. (2013)
Validamycin	*S. hygroscopicus* 5008	Manipulation of regulatory network(s)	55%	Tan et al. (2015)
Erythromycin A	*Sa. erythraea* A226	Manipulation of regulatory network(s)	41%	Wu et al. (2016)
Nystatin A1	*S. ahygroscopicus*	Manipulation of regulatory network(s)	2.1-fold	Cui et al. (2015)
Tiancimycins	*Streptomyces* sp. CB03234	Manipulation of regulatory network(s)	13.9-fold	Zhang et al. (2020a, b)
Actinorhodin	*S. coelicolor*	BGC refactoring	1.3-fold	Li et al. (2018)
Oxytetracycline	*S. coelicolor*	BGC refactoring	9.1-fold	Li et al. (2018)
Tunicamycin	*S. albus* Del14	Heterologous expression with reduced-genome host	20%	Ahmed et al. (2020)
Deoxycinnamycin	*S. lividans* TK24	Heterologous expression with reduced-genome host	4.5-fold	Ahmed et al. (2020)
Actinorhodin	*S. chattanoogensis*	Heterologous expression with reduced-genome host	2-3 times	Bu et al. (2019)

(continued)

Table 9.1 (continued)

Natural product	Host	Metabolic manipulation	Yield improvement	Reference
Spinosad	*S. albus* J1074	Heterologous expression with BGC refactoring	328-fold	Song et al. (2019)
Bottromycin	*S. lividans* TK24	Heterologous expression with BGC refactoring	50-fold	Horbal et al. (2018a, b)
Pristinamycin II	*S. pristinaespiralis* HCCB10218	Amplification of BGC (s) and manipulation of regulatory networks	1.5-fold	Li et al. (2015)
Prodigiosins	*S. coelicolor* SBJ106	Carbon flux redirection, manipulation of regulatory networks, and amplification of BGC(s)	12-fold	Liu et al. (2017)

Streptomyces is a model actinobacterial genus for metabolic engineering as it is considered the most thorough studied genus with solid databases for genomes, metabolisms, and cultivation protocols, which all facilitate metabolic engineering (Bekker et al. 2014; Robertsen et al. 2018). Due to a vast difference within the group of actinobacteria, genetic engineering protocols for actinobacteria is considered genus-dependent. However, with recent developments, other genera of actinobacteria have started to gain attention as potential hosts for metabolic engineering including *Amycolatopsis* and *Micromonospora*.

Amycolatopsis is a rare actinobacterial genus and best known for its ability to naturally produce vancomycin (Jung et al. 2007) and synthesize vanillin from a cheap substrate, ferulic acid, on an industrial scale (Fleige et al. 2013; Fleige et al. 2016; Fleige and Steinbüchel 2014; Ma and Daugulis 2014). It has also been reported to show high tolerance level to vanillin (Fleige et al. 2016). Recently, *A. jarponicum* was metabolically engineered to optimize the production of a naturally produced chelator called [S,S]-ethylenediamine-disuccinic acid (EDDS) by amplification of a [S,S]-EDDS biosynthetic BGC, replacement of a native promoter to enhance the expression of biosynthetic genes, and increasing a bottleneck precursor supply (Edenhart et al. 2020). With this potential, the interest in metabolic engineering of *Amycolatopsis* has been expanding and genetic background knowledge for this actinobacteria has started to build up (Fleige and Steinbüchel 2014; Gao et al. 2019). Currently, NCBI database offers whole genome sequences of at least 65 *Amycolatopsis* strains (data access: January 2021).

Micromonospora is another actinobacterial genus most regarded for their ability to naturally synthesize gentamycin C (Weinstein et al. 1963). Engineered *Micromonospora* strain is presently used in industrial production of this antibiotic. With such recognitions, NCBI database contains at least 81 whole genome sequences from members of this genus (data access: January 2021) and metabolic engineering in *Micromonospora* has gained increasing attention. To give some instances, overexpression of *genR* and *genS* (genes encoding enzymes for the final reaction steps for gentamycin B biosynthesis) resulting in a 64% increase in gentamycin B production in *M. echinospora* CCTCC M 2018898 was recently reported (Chang et al. 2019). Similarly, overexpression of *kanM1* and *genM2* which are responsible for glycosylation of intermediates in gentamycin B and C1 pathway, significantly improved the production of both NPs in *M. echinospora* (Wu et al. 2017).

9.3 Recent Advances in Metabolic Engineering in Actinobacteria

Though general biology of actinobacteria complicates genetic and metabolic engineering, developments of molecular techniques and tools have overcome some limitations and/or facilitated the work. Recent advances in metabolic engineering have been comprehensively reviewed in previous works (Deng et al. 2017; Li et al. 2019; Palazzotto et al. 2019). To demonstrate this, a few developments of techniques and tools for metabolic engineering in actinobacteria in the time period of 2015–2020 are discussed in this section.

9.3.1 CRISPR Technology

First and foremost, with recent breakthroughs in Clustered Regulatory Interspaced Short Palindromic Repeats (CRISPR) genome editing, metabolic engineering has benefited substantially from such advances. In actinobacteria, the use of CRISPR has started in 2015 when the first few CRISPR systems were developed (Cobb et al. 2015; Tong et al. 2015). These achievements were marked and followed by many studies. CRISPR system was discovered as an adaptive immune system in bacteria and later biotechnological functions of CRISPR system in genome editing was first described in 2012 (Jinek et al. 2012). A simplified scheme for CRISPR/Cas9 is illustrated in Fig. 9.2. In brief, this technology relies on enzymes that cut DNA at particular sites called DNA endonucleases (most notably Cas9) associated with a single guide RNA (sgRNA). The sgRNA is a short RNA that helps DNA endonucleases to precisely identify and cut specific regions on chromosomal DNA according to complementary sequences on the sgRNA. Protospacer adjacent motif or PAM site is a three-base pair region also playing a role in helping Cas9 recognizes the cut sites. The cut then leads to double-stranded breaks (DSBs) on the target DNA. After cutting, the cells utilize natural repair mechanisms to rejoin the cut sites; 1) non-homologous end joining (NHEJ) and 2) homologous directed repair (HDR) mechanisms. Non-homologous end joining (NHEJ) mechanism to repair the breaks results in random mutations (insertions or deletions (InDels)). Therefore, the

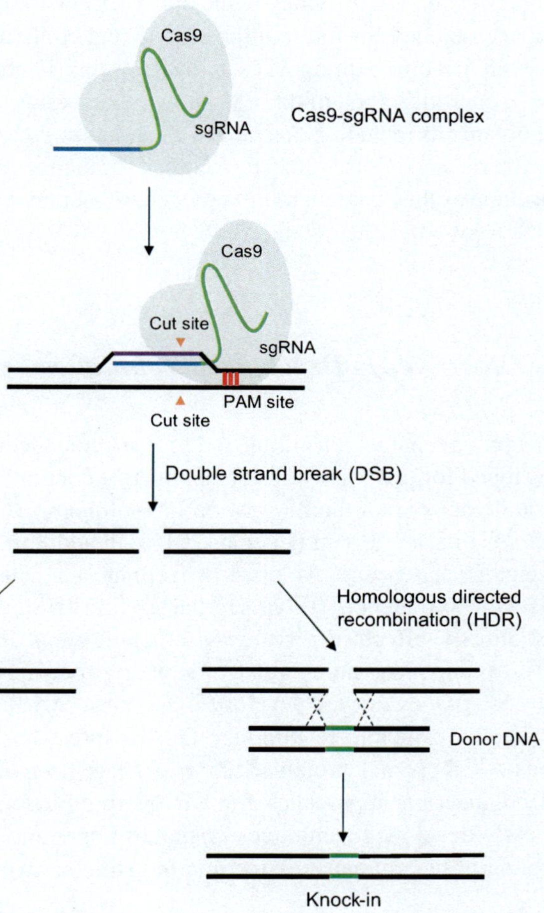

Fig. 9.2 CRISPR/Cas9 genome editing adapted from Hsu et al. (2014), Bortesi and Fischer (2015)

repaired genes are inactivated or knocked out because the gene sequence has been changed. In the case of homologous directed recombination (HDR), donor DNA is also provided to the reaction and the cells use this donor DNA as a template to repair digested DNA. This leads to an insertion of the donor sequence to the cut site. With its potentials, this technology is creatively used for several applications (Hsu et al. 2014) to knock-out, knock-in, and to introduce precise point mutations to the genome. Moreover, with CRISPR, markerless engineered strains are created per design which is one of the unique traits compared with other engineering methods (Zhang et al. 2018; Alberti and Corre 2019; Ji et al. 2019).

The recent CRISPR/Cas9 systems related to natural product discovery, characterization, and production have been reviewed (Tao et al. 2018). Several CRISPR toolkits and CRISPR-derived technologies have been developed for actinobacteria over the years since the first use of CRISPR in actinobacteria (Alberti and Corre

2019; Zhao et al. 2020). One of the most recent toolkits for *Streptomyces* was reported and detailed for future utilization (Tong et al. 2020).

However, genome editing via CRISPR system is challenging and needs to be considered carefully. First of all, it is known that expression of Cas9 protein is toxic in some organisms including *Streptomyces* and its expression is poor in heterologous hosts (Alberti and Corre 2019). Another challenge is due to the genetics of actinobacteria as they contain multicopy genes that may complicate the manipulation via CRISPR system (Zhao et al. 2020).

9.3.2 Other Newly Developed Techniques and Tools

A new technique, namely, Multiplexed Site-specific Genome Engineering (MSGE), was developed for pristinamycin II (PII) overproduction in *S. pristinaespiralis*. This technique demonstrated the integration of multicopy BGCs to the chromosome of *S. coelicolor* in a single step (Li et al. 2017). In addition, a genetic tool, "an *in vivo* RBS selector," has been developed to facilitate the selection of RBSs in order to maximize the production (Lilya Horbal et al. 2018a, b) as expression level of proteins directly affects the pathway efficiency and determines the final yield. Similarly, a universal autoregulated strategy for fine-tuning the expression of BGCs in *Streptomyces* was developed (Li et al. 2018). Ribosome engineering is another known principle to optimize BGC expression. Recently, introduction of mutations to ribosomal protein S12 (*rpsL*) has been investigated and presented rational engineering approaches for NP overproduction (Lopatniuk et al. 2019). These newly developed techniques potentially open the doors for high-throughput optimization of heterologous expression in actinobacteria.

9.4 Advantages of Actinobacteria as a Host for Metabolic Engineering

As stated earlier, *E. coli* and *S. cerevisiae* are the most common model microorganisms for metabolic engineering with well-studied metabolisms and widely available genetic/molecular tools, yet they fail to efficiently produce NPs or often require significant further modifications (Myronovskyi and Luzhetskyy 2019; Yang et al. 2019). The reasons lie in their biology compared with the biology of actinobacteria which, in turn, affects NP production. This variation among different groups of microorganisms allows the production of desired products to be optimized, if suitable hosts are considered. In this chapter, we address the advantages of actinobacteria as a host for employing metabolic engineering in order to enhance NP production. Points are discussed exclusively on heterologous production as to compare with other host microorganisms. The advantages are elucidated separately on physiological and molecular perspectives.

Physiologically, native tolerance of actinobacteria toward NPs is one of the most useful features in NP production. It is generally known that the overproduction of chemicals in native hosts has benefited from their evolution to tolerate the native metabolites compared to heterologous hosts. Enhancement of host tolerance has shown to result in an improvement of productivity compared to the parental strains (Estévez et al. 2020; Horbal et al. 2018a, b; Hosoya et al. 1998; Hu and Ochi 2001). It has also been demonstrated that, in some cases, the production has not been commercially satisfied in *E. coli* and *S. cerevisiae* despite their well-developed metabolic engineering techniques and tools (Lian et al. 2018; Yang et al. 2019). Arguably, the challenge is the heterologous host tolerance level toward non-native compounds. Erythromycin A was produced in *E. coli* by transferring the BGCs from their native producers to *E. coli*. The yield was first reported at only 10 mg/L (Zhang et al. 2010) and since then the production has not reached commercial stages. Certainly, several factors can be further optimized, the question is whether tolerance level is one of them. Furthermore, NP biosynthetic pathways often comprise many reaction steps, thus involve several and, sometimes, unique precursors and cofactors. The supplementation of these components is therefore needed either intracellularly or extracellularly for the heterologous hosts. Native precursors and cofactors in actinobacteria are the factors that can significantly facilitate the production as these precursors in heterologous hosts may not meet the metabolic demands or do not naturally exist. Efforts have been put into the production of polyketides in *E. coli*, yet relatively unsuccessful without further manipulations (Fang et al. 2018; Stevens et al. 2013; Zhang et al. 2010, 2012). The same principle can be addressed also for native transporters (Méndez and Salas 2001), which, as mentioned, could influence overall production. Lastly from physiological point of view, as only a few heterologous productions of actinobacterial NPs have been reported in *E. coli*. One influencing factor is the fundamental properties of non-native proteins. Expression of proteins involving in NP biosynthesis in *E. coli* was challenging due to their distinctive properties (e.g., insolubility or extensively long polypeptide) (Cummings et al. 2019; Pfeifer et al. 2001; Watanabe et al. 2003).

From molecular perspective, one unique molecular trait of actinobacteria is the high GC content (50-70%) in their genomes (Ventura et al. 2007), which makes them codon-biased microorganisms (Lal et al. 2016). This may be viewed as a downside for practicing genetic engineering; however, optimistically, this is considered an advantage as a host for the expression of high GC-content genes. Since most BGCs in questions are natively from actinobacteria, thus it is highly likely that they are high in GC content. One obvious example is a tailoring gene in erythromycin biosynthetic pathway, *eryG*, which contains around 72% GC and the attempt to express this gene in *E. coli* resulted in a relatively low enzyme activity (Paulus et al. 1990). Moreover, as BGCs are sophisticated systems often comprising multiple genes, this makes it challenging for the heterologous hosts. The expression of each enzyme in the pathway needs to be validated (Nielsen 2019). For example, in order to heterologously produce erythromycin A, a 55 kb gene cluster needed to be expressed in *E. coli* (Zhang et al. 2010) excluding further modifications that also needed to optimize the production. This may create molecular burdens and result in a low production in *E. coli*. Finally, regulatory networks in actinobacteria are tightly controlled and complex (Liu et al. 2013) and they also have complicated post-modification processes during biosynthesis. In this regard, the use of naturally existing systems often lessens the efforts in constructing the entirely synthetic ones.

9.5 Concluding Remarks and Future Perspectives

Overall, this chapter does not only demonstrate that actinobacteria are considered indispensable hosts for NP production, but it is also evident that metabolic engineering is a powerful tool for NP production improvement. With recent breakthroughs in genetic and metabolic engineering field, the number of publications on actinobacterial metabolic engineering is rising. Undoubtedly, better understandings on BGCs and metabolic networks of actinobacteria are needed to further maximize the efficiency of employing metabolic engineering for the improvement of NP production. Still, this does not compromise the growing interest in this field. Hence, we anticipate that the golden age of metabolic engineering in actinobacteria has yet to come. Additionally, actinobacteria have long been used in industries for NP production, they have been optimized for such purposes and their industrial systems have been developed. The production processes ranging from compound characterization to commercialization could, therefore, be bypassed accordingly to the well-established knowledge where possible. Nevertheless, it is worth noting that apart from metabolic engineering approaches to increase the yield and productivity, midstream and downstream processes can also be optimized to further maximize the production (Lee and Kim 2015). Fermentation conditions are among the most well-studied factors and have been shown to improve the final production (Reeves et al. 2007; Zhu et al. 2014). In conclusion, this chapter summarizes recent works on metabolic engineering in actinobacteria and recent advances in relevant technique and tool developments. We also address the advantages of actinobacteria as hosts for metabolic engineering. With evidences provided, actinobacteria potentially outrun other hosts as NP producers and their biology further assists the implementation of metabolic engineering to improve the production.

References

Ahmed Y, Rebets Y, Estévez MR, Zapp J, Myronovskyi M, Luzhetskyy A (2020) Engineering of *Streptomyces lividans* for heterologous expression of secondary metabolite gene clusters. Microb Cell Fact 19:1–16. https://doi.org/10.1186/s12934-020-1277-8

Alberti F, Corre C (2019) Editing streptomycete genomes in the CRISPR/Cas9 age. Nat Prod Rep 36:1237–1248. https://doi.org/10.1039/c8np00081f

Barka EA, Vatsa P, Sanchez L, Nathalie Gaveau-Vaillant CJ, Klenk H-P, Clément C, Ouhdouch Y, van Wezeld GP (2016) Taxonomy, physiology, and natural products of actinobacteria. Microbiol Mol Biol Rev 80:1–43. https://doi.org/10.1128/MMBR.00019-15.Address

Bekker V, Dodd A, Brady D, Rumbold K (2014) Tools for metabolic engineering in *Streptomyces*. Bioengineered 5(5):293–299. https://doi.org/10.4161/bioe.29935

Berini F, Marinelli F, Binda E (2020) Streptomycetes: attractive hosts for recombinant protein production. Front Microbiol 11:1958. https://doi.org/10.3389/fmicb.2020.01958

Bortesi L, Fischer R (2015) The CRISPR/Cas9 system for plant genome editing and beyond. Biotechnol Adv 33:41–52. https://doi.org/10.1016/j.biotechadv.2014.12.006

Bu QT, Yu P, Wang J, Li ZY, Chen XA, Mao XM, Li YQ (2019) Rational construction of genome-reduced and high-efficient industrial *Streptomyces* chassis based on multiple comparative genomic approaches. Microb Cell Fact 18:1–17. https://doi.org/10.1186/s12934-019-1055-7

Chang Y, Chai B, Ding Y, He M, Zheng L, Teng Y, Deng Z, Yu Y, Liu T (2019) Overproduction of gentamicin B in industrial strain *Micromonospora echinospora* CCTCC M 2018898 by cloning of the missing genes *genR* and *genS*. Metab Eng Commun 9:1–10. https://doi.org/10.1016/j.mec.2019.e00096

Cobb RE, Wang Y, Zhao H (2015) High-efficiency multiplex genome editing of *Streptomyces* species using an engineered CRISPR/Cas system. ACS Synth Biol 4:723–728. https://doi.org/10.1021/sb500351f

Cui H, Ni X, Shao W, Su J, Su J, Ren J, Xia H (2015) Functional manipulations of the tetramycin positive regulatory gene *ttmRIV* to enhance the production of tetramycin A and nystatin A1 in *Streptomyces ahygroscopicus*. J Ind Microbiol Biotechnol 42(9):1273–1282. https://doi.org/10.1007/s10295-015-1660-3

Cummings M, Peters AD, Whitehead GFS, Menon BRK, Micklefield J, Webb SJ, Takano E (2019) Assembling a plug-and-play production line for combinatorial biosynthesis of aromatic polyketides in *Escherichia coli*. PLoS Biol 17(7):1–34. https://doi.org/10.1371/journal.pbio.3000347

Demain AL, Sanchez S (2009) Microbial drug discovery: 80 years of progress. J Antibiot (Tokyo) 62:5–16. https://doi.org/10.1038/ja.2008.16

Deng Y, Zhang X, Zhang X (2017) Recent advances in genetic modification systems for actinobacteria. Appl Microbiol Biotechnol 101(6):2217–2226. https://doi.org/10.1007/s00253-017-8156-1

Dunlop MJ, Dossani ZY, Szmidt HL, Chu HC, Lee TS, Keasling JD, Hadi MZ, Mukhopadhyay A (2011) Engineering microbial biofuel tolerance and export using efflux pumps. Mol Syst Biol 7 (487):1–7. https://doi.org/10.1038/msb.2011.21

Edenhart S, Denneler M, Spohn M, Doskocil E, Kavšček M, Amon T, Kosec G, Smole J, Bardl B, Biermann M, Roth M, Wohlleben W, Stegmann E (2020) Metabolic engineering of *Amycolatopsis japonicum* for optimized production of [S,S]-EDDS, a biodegradable chelator. Metab Eng 60:148–156. https://doi.org/10.1016/j.ymben.2020.04.003

Endy D (2005) Foundations for engineering biology. Nature 438:449–453. https://doi.org/10.1038/nature04342

Estévez MR, Myronovskyi M, Rosenkränzer B, Paululat T, Petzke L, Ristau J, Luzhetskyy A (2020) Novel fredericamycin variant overproduced by a streptomycin-resistant *Streptomyces albus* subsp. *chlorinus* strain. Mar Drugs 18(284):1–12. https://doi.org/10.3390/md18060284

Fang L, Guell M, Church GM, Pfeifer BA (2018) Heterologous erythromycin production across strain and plasmid construction. Biotechnol Prog 34:271–276. https://doi.org/10.1002/btpr.2567

Fleige C, Steinbüchel A (2014) Construction of expression vectors for metabolic engineering of the vanillin-producing actinomycete *Amycolatopsis* sp. ATCC 39116. Appl Microbiol Biotechnol 98:6387–6395. https://doi.org/10.1007/s00253-014-5724-5

Fleige C, Hansen G, Kroll J, Steinbüchel A (2013) Investigation of the *Amycolatopsis* sp. strain ATCC 39116 vanillin dehydrogenase and its impact on the biotechnical production of vanillin. Appl Environ Microbiol 79:81–90. https://doi.org/10.1128/AEM.02358-12

Fleige C, Meyer F, Steinbüchel A (2016) Metabolic engineering of the actinomycete *Amycolatopsis* sp. strain ATCC 39116 towards enhanced production of natural vanillin. Appl Environ Microbiol 82:3410–3419. https://doi.org/10.1128/AEM.00802-16

Foulston L (2019) Genome mining and prospects for antibiotic discovery. Curr Opin Microbiol 51:1–8. https://doi.org/10.1016/j.mib.2019.01.001

Gao H, Murugesan B, Hoßbach J, Evans SK, Stark WM, Smith MCM (2019) Integrating vectors for genetic studies in the rare actinomycete *Amycolatopsis marina*. BMC Biotechnol 19:1–10. https://doi.org/10.1186/s12896-019-0521-y

Gomez-Escribano JP, Bibb MJ (2011) Engineering *Streptomyces coelicolor* for heterologous expression of secondary metabolite gene clusters. J Microbial Biotechnol 4(2):207–215. https://doi.org/10.1111/j.1751-7915.2010.00219.x

Gomez-Escribano JP, Alt S, Bibb MJ (2016) Next generation sequencing of actinobacteria for the discovery of novel natural products. Mar Drugs 14(4):6–8. https://doi.org/10.3390/md14040078

Horbal L, Marques F, Nadmid S, Mendes MV, Luzhetskyy A (2018a) Secondary metabolites overproduction through transcriptional gene cluster refactoring. Metab Eng 49:299–315. https://doi.org/10.1016/j.ymben.2018.09.010

Horbal L, Siegl T, Luzhetskyy A (2018b) A set of synthetic versatile genetic control elements for the efficient expression of genes in actinobacteria. Sci Rep 8:1–13. https://doi.org/10.1038/s41598-017-18846-1

Hosoya Y, Okamoto S, Muramatsu H, Ochi K (1998) Acquisition of certain streptomycin-resistant (str) mutations enhances antibiotic production in bacteria. Antimicrob Agents Chemother 42:2041–2047. https://doi.org/10.1128/aac.42.8.2041

Hsu PD, Lander ES, Zhang F (2014) Development and applications of CRISPR-Cas9 for genome engineering. Cell 157:1262–1278. https://doi.org/10.1016/j.cell.2014.05.010

Hu H, Ochi K (2001) Novel approach for improving the productivity of antibiotic-producing strains by inducing combined resistant mutations. Appl Environ Microbiol 67:1885–1892. https://doi.org/10.1128/AEM.67.4.1885-1892.2001

Huang D, Xia M, Li S, Wen J, Jia X (2013) Enhancement of FK506 production by engineering secondary pathways of *Streptomyces tsukubaensis* and exogenous feeding strategies. J Ind Microbiol Biotechnol 40(9):1023–1037. https://doi.org/10.1007/s10295-013-1301-7

Ji CH, Kim H, Kang HS (2019) Synthetic inducible regulatory systems optimized for the modulation of secondary metabolite production in *Streptomyces*. ACS Synth Biol 8(3):577–586. https://doi.org/10.1021/acssynbio.9b00001

Jinek, M., Chylinski, K., Fonfara, I., Hauer, M., et al. (2012) A programmable dual-RNA – guided DNA endonuclease in adaptive bacterial immunity. Science 337, 816–822. Available from: doi: https://doi.org/10.1126/science.1225829.

Jung HM, Kim SY, Moon HJ, Oh DK, Lee JK (2007) Optimization of culture conditions and scale-up to pilot and plant scales for vancomycin production by *Amycolatopsis orientalis*. Appl Microbiol Biotechnol 77(4):789–795. https://doi.org/10.1007/s00253-007-1221-4

Jung WS, Kim E, Yoo YJ, Ban YH, Kim EJ, Yoon YJ (2014) Characterization and engineering of the ethylmalonyl-CoA pathway towards the improved heterologous production of polyketides in *Streptomyces venezuelae*. Appl Microbiol Biotechnol 98(8):3701–3713. https://doi.org/10.1007/s00253-013-5503-8

Keasling JD (2008) Synthetic biology for synthetic chemistry. ACS Chem Biol 3:64–76. https://doi.org/10.1021/cb7002434

Kim J, Salvador M, Saunders E, Gonz J, Avignone-rossa C, Ignacio J (2016) Properties of alternative microbial hosts used in synthetic biology: towards the design of a modular. Essays Biochem 60(4):303–313. https://doi.org/10.1042/EBC20160015

Komatsu M, Uchiyama T, Omura S, Cane DE, Ikeda H (2010) Genome-minimized *Streptomyces* host for the heterologous expression of secondary metabolism. Proc Natl Acad Sci U S A 107 (6):2646–2651. https://doi.org/10.1073/pnas.0914833107

Kuzhyk Y, Lopatniuk M, Luzhetskyy A, Fedorenko V, Ostash B (2019) Genome engineering approaches to improve nosokomycin A production by *Streptomyces ghanaensis* B38.3. Indian J. Microbiology 59(1):109–111. https://doi.org/10.1007/s12088-018-0761-x

Lal D, Verma M, Behura SK, Lal R (2016) Codon usage bias in phylum actinobacteria: relevance to environmental adaptation and host pathogenicity. Res Microbiol 167(8):669–677. https://doi.org/10.1016/j.resmic.2016.06.003

Lee SY, Kim HU (2015) Systems strategies for developing industrial microbial strains. Nat Biotechnol 33:1061–1072. https://doi.org/10.1038/nbt.3365

Li L, Zhao Y, Ruan L, Yang S, Ge M, Jiang W, Lu Y (2015) A stepwise increase in pristinamycin II biosynthesis by *Streptomyces pristinaespiralis* through combinatorial metabolic engineering. Metab Eng 29:12–25. https://doi.org/10.1016/j.ymben.2015.02.001

Li L, Zheng G, Chen J, Ge M, Jiang W, Lu Y (2017) Multiplexed site-specific genome engineering for overproducing bioactive secondary metabolites in actinomycetes. Metab Eng 40:80–92. https://doi.org/10.1016/j.ymben.2017.01.004

Li S, Wang J, Xiang W, Yang K, Li Z, Wang W (2018) An autoregulated fine-tuning strategy for titer improvement of secondary metabolites using native promoters in *Streptomyces*. ACS Synth Biol 7(2):522–530. https://doi.org/10.1021/acssynbio.7b00318

Li L, Liu X, Jiang W, Lu Y (2019) Recent advances in synthetic biology approaches to optimize production of bioactive natural products in actinobacteria. Front Microbiol 10:1–10. https://doi.org/10.3389/fmicb.2019.02467

Lian J, Mishra S, Zhao H (2018) Recent advances in metabolic engineering of *Saccharomyces cerevisiae*: new tools and their applications. Metab Eng 50:85–108. https://doi.org/10.1016/j.ymben.2018.04.011

Liu G, Chater KF, Chandra G, Niu G, Tan H (2013) Molecular regulation of antibiotic biosynthesis in *Streptomyces*. Microbiol Mol Biol Rev 77:112–143. https://doi.org/10.1128/mmbr.00054-12

Liu P, Zhu H, Zheng G, Jiang W, Lu Y (2017) Metabolic engineering of *Streptomyces coelicolor* for enhanced prodigiosins (RED) production. Sci China Life Sci 60:948–957. https://doi.org/10.1007/s11427-017-9117-x

Lopatniuk M, Myronovskyi M, Nottebrock A, Busche T, Kalinowski J, Ostash B, Fedorenko V, Luzhetskyy A (2019) Effect of "ribosome engineering" on the transcription level and production of *S. albus* indigenous secondary metabolites. Appl Microbiol Biotechnol 103(17):7097–7110. https://doi.org/10.1007/s00253-019-10005-y

Lu C, Zhang X, Jiang M, Bai L (2016) Enhanced salinomycin production by adjusting the supply of polyketide extender units in *Streptomyces albus*. Metab Eng 35:129–137. https://doi.org/10.1016/j.ymben.2016.02.012

Luo Y, Huang H, Liang J, Wang M, Lu L, Shao Z, Cobb RE, Zhao H (2013) Activation and characterization of a cryptic polycyclic tetramate macrolactam biosynthetic gene cluster. Nat Commun 4:1–8. https://doi.org/10.1038/ncomms3894

Ma XK, Daugulis AJ (2014) Effect of bioconversion conditions on vanillin production by *Amycolatopsis* sp. ATCC 39116 through an analysis of competing by-product formation. Bioprocess Biosyst Eng 37(5):891–899. https://doi.org/10.1007/s00449-013-1060-x

Méndez C, Salas JA (2001) The role of ABC transporters in antibiotic-producing organisms: Drug secretion and resistance mechanisms. Res Microbiol 152:341–350. https://doi.org/10.1016/S0923-2508(01)01205-0

Musiol-Kroll EM, Tocchetti A, Sosio M, Stegmann E (2019) Challenges and advances in genetic manipulation of filamentous actinomycetes-the remarkable producers of specialized metabolites. Nat Prod Rep 36:1351–1369. https://doi.org/10.1039/c9np00029a

Myronovskyi M, Luzhetskyy A (2019) Heterologous production of small molecules in the optimized: *Streptomyces* hosts. Nat Prod Rep 36:1281–1294. https://doi.org/10.1039/c9np00023b

Myronovskyi M, Rosenkränzer B, Nadmid S, Pujic P, Normand P, Luzhetskyy A (2018) Generation of a cluster-free *Streptomyces albus* chassis strains for improved heterologous expression of secondary metabolite clusters. Metab Eng 49:316–324. https://doi.org/10.1016/j.ymben.2018.09.004

Nielsen J (2019) Cell factory engineering for improved production of natural products. Nat Prod Rep 36:1233–1236. https://doi.org/10.1039/c9np00005d

Nielsen J, Keasling JD (2011) Synergies between synthetic biology and metabolic engineering. Nat Biotechnol 29:693–695. https://doi.org/10.1038/nbt.1937

Nouioui I, Carro L, García-López M, Meier-Kolthoff JP, Woyke T, Kyrpides NC, Pukall R, Klenk HP, Goodfellow M, Göker M (2018) Genome-based taxonomic classification of the phylum actinobacteria. Front Microbiol 9:1–119. https://doi.org/10.3389/fmicb.2018.02007

Olano C, García I, González A, Rodriguez M, Rozas D, Rubio J, Sánchez-Hidalgo M, Braña AF, Méndez C, Salas JA (2014) Activation and identification of five clusters for secondary metabolites in *Streptomyces albus* J1074. J Microbial Biotechnol 7:242–256. https://doi.org/10.1111/1751-7915.12116

Palazzotto E, Tong Y, Lee SY, Weber T (2019) Synthetic biology and metabolic engineering of actinomycetes for natural product discovery. Biotechnol Adv 37:107366. https://doi.org/10.1016/j.biotechadv.2019.03.005

Paulus TJ, Tuan JS, Luebke VE, Maine GT, DeWitt JP, Katz L (1990) Mutation and cloning of *eryG*, the structural gene for erythromycin O-methyltransferase from *Saccharopolyspora erythraea*, and expression of *eryG* in *Escherichia coli*. J Bacteriol 172:2541–2546. https://doi.org/10.1128/jb.172.5.2541-2546.1990

Petzold CJ, Chan LJG, Nhan M, Adams PD (2015) Analytics for metabolic engineering. Front Bioeng Biotechnol 3:1–11. https://doi.org/10.3389/fbioe.2015.00135

Pfeifer BA, Admiraal SJ, Gramajo H, Cane DE, Khosla C (2001) Biosynthesis of complex polyketides in a metabolically engineered strain of *E. coli*. Science 291:1790–1792. https://doi.org/10.1126/science.1058092

Raimbault B, Cointet JP, Joly PB (2016) Mapping the emergence of synthetic biology. PLoS One 11(9):1–19. https://doi.org/10.1371/journal.pone.0161522

Reeves AR, Brikun IA, Cernota WH, Leach BI, Gonzalez MC, Mark Weber J (2007) Engineering of the methylmalonyl-CoA metabolite node of *Saccharopolyspora erythraea* for increased erythromycin production. Metab Eng 9:293–303. https://doi.org/10.1016/j.ymben.2007.02.001

Roberts MAJ, Cranenburgh RM, Stevens MP, Oyston PCF (2013) Synthetic biology: biology by design. Microbiology 159:1219–1220. https://doi.org/10.1099/mic.0.069724-0

Robertsen HL, Weber T, Kim HU, Lee SY (2018) Toward systems metabolic engineering of streptomycetes for secondary metabolites production. Biotechnol J 13:1–12. https://doi.org/10.1002/biot.201700465

Scheffler RJ, Colmer S, Tynan H, Demain AL, Gullo VP (2013) Antimicrobials, drug discovery, and genome mining. Appl Microbiol Biotechnol 97(3):969–978. https://doi.org/10.1007/s00253-012-4609-8

Smanski MJ, Zhou H, Claesen J, Shen B, Fischbach MA, Voigt CA (2016) Synthetic biology to access and expand nature's chemical diversity. Nat Rev Microbiol 14:135–149. https://doi.org/10.1038/nrmicro.2015.24

Song C, Luan J, Cui Q, Duan Q, Li Z, Gao Y, Li R, Li A, Shen Y, Li Y, Stewart AF, Zhang Y, Fu J, Wang H (2019) Enhanced heterologous spinosad production from a 79-kb synthetic multioperon assembly. ACS Synth Biol 8:137–147. https://doi.org/10.1021/acssynbio.8b00402

Stevens DC, Conway KR, Pearce N, Villegas-Peñaranda LR, Garza AG, Boddy CN (2013) Alternative sigma factor over-expression enables heterologous expression of a type II polyketide biosynthetic pathway in *Escherichia coli*. PLoS One 8(5). https://doi.org/10.1371/journal.pone.0064858

Tan GY, Bai L, Zhong JJ (2013) Exogenous 1,4-butyrolactone stimulates a-factor-like cascade and validamycin biosynthesis in *Streptomyces hygroscopicus* 5008. Biotechnol Bioeng 110(11):2984–2993. https://doi.org/10.1002/bit.24965

Tan GY, Peng Y, Lu C, Bai L, Zhong JJ (2015) Engineering validamycin production by tandem deletion of γ-butyrolactone receptor genes in *Streptomyces hygroscopicus* 5008. Metab Eng 28:74–81. https://doi.org/10.1016/j.ymben.2014.12.003

Tao W, Yang A, Deng Z, Sun Y (2018) CRISPR/Cas9-based editing of *Streptomyces* for discovery, characterization, and production of natural products. Front Microbiol 9:1660. https://doi.org/10.3389/fmicb.2018.01660

Tong Y, Charusanti P, Zhang L, Weber T, Lee SY (2015) CRISPR-Cas9 based engineering of actinomycetal genomes. ACS Synth Biol 4(9):1020–1029. https://doi.org/10.1021/acssynbio.5b00038

Tong Y, Whitford CM, Blin K, Jørgensen TS, Weber T, Lee SY (2020) CRISPR–Cas9, CRISPRi and CRISPR-BEST-mediated genetic manipulation in streptomycetes. Nat Protoc 15:2470–2502. https://doi.org/10.1038/s41596-020-0339-z

Ventura M, Canchaya C, Tauch A, Chandra G, Fitzgerald GF, Chater KF, van Sinderen D (2007) Genomics of actinobacteria: tracing the evolutionary history of an ancient phylum. Microbiol Mol Biol Rev 71:495–548. https://doi.org/10.1128/mmbr.00005-07

Watanabe K, Rude MA, Walsh CT, Khosla C (2003) Engineered biosynthesis of an ansamycin polyketide precursor in *Escherichia coli*. Proc Natl Acad Sci U S A 100:9774–9778. https://doi.org/10.1073/pnas.1632167100

Weinstein MJ, Luedemann GM, Oden EM, Wagman GH, Rosselet JP, Marquez JA, Coniglio CT, Charney W, Herzog HL, Black J (1963) Gentamicin, a new antibiotic complex from *Micromonospora*. J Med Chem 6:463–464. https://doi.org/10.1021/jm00340a034

Wu H, Wang Y, Yuan L, Mao Y, Wang W, Zhu L, Wu P, Fu C, Müller R, Weaver DT, Zhang L, Zhang B (2016) Inactivation of SACE_3446, a TetR family transcriptional regulator, stimulates erythromycin production in *Saccharopolyspora erythraea*. Synth Syst Biotechnol 1:39–46. https://doi.org/10.1016/j.synbio.2016.01.004

Wu Z, Gao W, Zhou S, Wen Z, Ni X, Xia H (2017) Improving gentamicin B and gentamicin C1a production by engineering the glycosyltransferases that transfer primary metabolites into secondary metabolites biosynthesis. Microbiol Res 203:40–46. https://doi.org/10.1016/j.micres.2017.06.006

Xie Y, Ma J, Qin X, Li Q, Ju J (2017) Identification and utilization of two important transporters: SgvT1 and SgvT2, for griseoviridin and viridogrisein biosynthesis in *Streptomyces griseoviridis*. Microb Cell Fact 16:177. https://doi.org/10.1186/s12934-017-0792-8

Yang D, Park SY, Park YS, Eun H, Lee SY (2019) Metabolic engineering of *Escherichia coli* for natural product biosynthesis. Trends Biotechnol 38:745–765. https://doi.org/10.1016/j.tibtech.2019.11.007

Yu AQ, Pratomo Juwono NK, Leong SSJ, Chang MW (2014) Production of fatty acid-derived valuable chemicals in synthetic microbes. Front Bioeng Biotechnol 2:1–12. https://doi.org/10.3389/fbioe.2014.00078

Zabala D, Braña AF, Flórez AB, Salas JA, Méndez C (2013) Engineering precursor metabolite pools for increasing production of antitumor mithramycins in *Streptomyces argillaceus*. Metab Eng 20:187–197. https://doi.org/10.1016/j.ymben.2013.10.002

Zhang H, Wang Y, Wu J, Skalina K, Pfeifer BA (2010) Complete biosynthesis of erythromycin A and designed analogs using *E. coli* as a heterologous host. Chem Biol 17:1232–1240. https://doi.org/10.1016/j.chembiol.2010.09.013

Zhang H, Skalina K, Jiang M, Pfeifer BA (2012) Improved *E. coli* erythromycin a production through the application of metabolic and bioprocess engineering. Biotechnol Prog 28:292–296. https://doi.org/10.1002/btpr.702

Zhang J, Hong W, Zong W, Wang P, Wang Y (2018) Markerless genome editing in *Clostridium beijerinckii* using the CRISPR-Cpf1 system. J Biotechnol 284:27–30. https://doi.org/10.1016/j.jbiotec.2018.07.040

Zhang B, Zhou YT, Jiang SX, Zhang YH, Huang K, Liu ZQ, Zheng YG (2020a) Amphotericin B biosynthesis in *Streptomyces nodosus*: quantitative analysis of metabolism via LC-MS/MS based metabolomics for rational design. Microb Cell Fact 19(7):1–12. https://doi.org/10.1186/s12934-020-1290-y

Zhang F, Gao D, Lin J, Zhu M, Zhuang Z, Duan Y, Zhu X (2020b) Construction of inducible genetic switch for the global regulator WblA to sustain both overproduction of tiancimycins and on-demand sporulation in *Streptomyces* sp. CB03234. ACS Synth Biol 9(6):1460–1467. https://doi.org/10.1021/acssynbio.0c00114

Zhao Y, Li G, Chen Y, Lu Y (2020) Challenges and advances in genome editing technologies in *Streptomyces*. Biomolecules 10(5):1–14. https://doi.org/10.3390/biom10050734

Zhu T, Cheng X, Liu Y, Deng Z, You D (2013) Deciphering and engineering of the final step halogenase for improved chlortetracycline biosynthesis in industrial *Streptomyces aureofaciens*. Metab Eng 19:69–78. https://doi.org/10.1016/j.ymben.2013.06.003

Zhu T, Wang L, Wang W, Hu Z, Yu M, Wang K, Cui Z (2014) Enhanced production of lipstatin from *Streptomyces toxytricini* by optimizing fermentation conditions and medium. J Gen Appl Microbiol 60(3):106–111. https://doi.org/10.2323/jgam.60.106

Ziemert N, Alanjary M, Weber T (2016) The evolution of genome mining in microbes-a review. Nat Prod Rep 33:988–1005. https://doi.org/10.1039/c6np00025h

Chapter 10
CRISPR ERA: Current Applications and Future Perspectives on Actinobacteria

Loganathan Karthik, Vishnu Kirthi Arivarasan, Mahendran Ruckmani Vignesh, and Palakshi Anitha

Abstract Genome editing technologies have been proven to be an important strategy for drug discovery in the recent years. Among them, clustered regularly interspaced short palindromic repeats (CRISPR) and associated proteins (CRISPR/Cas) is an efficient genome editing tool, which helps mainly in gene regulation and activation of biosynthetic gene clusters (BGC) for the discovery of novel bioactive natural products (NP) apart from its other applications. Amidst wide variety of organisms, CRISPR/Cas system aids extensively in activating silent BGCs in actinobacteria as they are an important source of antibiotics and other pharmaceutical drugs. Actinomycetes are the source of two third of the available antibiotics derived from microbial fermentation. Majority of these antibiotics are derived from *Streptomyces* genus, as its codes for a high number of NP BGCs. This chapter recapitulates the recent advances in CRISPR/Cas-based genome editing approaches in actinobacteria, mining of Cas3, and future perspectives in the discovery of NPs from novel species of actinomycetes of different origins.

Keywords CRISPR/Cas · Actinomycete · Streptomyces · BGC · Genome editing · Natural product

L. Karthik (✉)
Synthetic Biology, R&D Center, Salem Microbes Private Limited, Salem, India

V. K. Arivarasan
Department of Microbiology, Lovely Professional University, Punjab, India

M. R. Vignesh
Society of Chemical and Synthetic Biology, Vellore, Tamil Nadu, India

P. Anitha
Madurai Kamaraj University, Madurai, Tamil Nadu, India

L. Karthik (ed.), *Actinobacteria*, https://doi.org/10.1007/978-981-16-5835-8_10

10.1 Introduction

10.1.1 Biosynthetic Gene Clusters and Natural Products

Natural products (NPs) also known as secondary metabolites are produced in plants, microbes, and invertebrates. They help the producing organisms to tolerate stresses, to surpass nearby microbes, and play essential role at specific physiological and developmental stages of that organism (Davies 2013; Challis and Hopwood 2003). NPs derived from microorganisms are much diverse in structure and are rich sources for novel drug discovery which aid in treatment of several human diseases. Especially, a valuable source of bioactive NPs, which further developed into pharmaceutical drugs to treat cancer, infections, and diseases related to immune system, is bacteria (Sekurova et al. 2019). Though the role of NPs is important for survival of microorganisms in a particular habitat, molecules emerging from the secondary metabolism are of utmost agro-industrial and therapeutic importance (Loureiro et al. 2018; Walsh and Fischbach 2010). These bioactive NPs are produced by a kind of machinery which is encoded by a group of genes called as biosynthetic gene cluster which includes biosynthetic genes and genes for self-resistance, export, and expression control (Walsh and Fischbach 2010; Medema et al. 2015; Tenconi and Rigali 2018).

A biosynthetic gene cluster (BGC) can be defined as group of genes clustered together in a specific genome to encode a specialized secondary metabolite (Medema et al. 2015). Moreover, BGCs contain not only genes which encodes all the enzymes that are responsible for the production of a secondary metabolite, but also pathway-specific regulatory genes (Keller et al. 2005). A single BGC is capable of synthesizing several NPs which are related to each other in many aspects and whose bioactivities vary only in terms of specificity and/or strength. It has also been proved that a single BGC can be responsible for synthesis of two NPs, namely ferroverdins and bagremycins, which have different chemical compositions and structural organizations and are synthesized under different culture conditions (Martinet et al. 2019). Also, a lone BGC from a deep-sea derived microorganism accounts for four NPs, with diverse biological activities, representing two different anthracene scaffold variants (Alali et al. 2021). Polyketide synthases (PKS) and non-ribosomal peptide synthases (NRPS) are the most popular targets in genome mining for novel NPs and they are well known to synthesize diverse range of secondary metabolites which are used as antibiotics, immunosuppressants, and antifungals (Ayuso-Sacido and Genilloud 2005).

10.1.2 Actinomycetes Phyla as a Source of Bioactive Compound

Among the phylum of bacteria, Actinomycetes, apart from its role in maintaining soil microbiota, it is also a major source of pharmaceutically important natural bioactive compounds. Natural products (NPs) of microbes especially of the actinomycetes have been explored as an excellent resource in modern drugs discovery. Moreover, the genomic era has given many recent techniques to further strengthen the identification of several NP BGCs from the genomes of actinomycetes (Tao et al. 2018). Although different approaches like RNA polymerase engineering (Shima et al. 1996; Hosaka et al. 2009), one strain many compounds co-culturing technique (Bode et al. 2002), equipping with chemical elicitors (Romero et al. 2011), and transcription factor decoy approach (Wang et al. 2019; Nguyen et al. 2020) have been in use for decades, all these methods have its own draw back in activating the low expression BGCs. It is evident from earlier studies, streptomyces genome provide more than 30 BGCs and provide as a rich source of antibiotics (Culp et al. 2019). Till now only few BGCs of only few strains have been explored for its chemical importance and majority of the BGCs are overlooked due to its decreased or untraceable expression levels in the genome (Tong et al. 2015). The time-consuming homologous recombination methods with a suicide or temperature sensitive or plasmid self-replication were used earlier in editing BGCs in streptomyces. Nevertheless, for the discovery of NP drugs, streptomyces are of the utmost importance, of which the studying process depends on efficient genome editing methods. Therefore, it is necessary to have the accurate and the rapid techniques to identify the cryptic BGCs of distinct metabolic pathways in the host system to uncover the novel compounds that were hidden by the already existing NPs in actinomycetes. From the culture of marine *Streptomyces olivaceus* SCSIO T05, a single rsd BGC is identified responsible for the production of rishirilide B, rishirilide C, lupinacidin A, and galvaquinone B belonging to two different anthracene scaffolds, which have antithrombolytic activity, inhibitory effects against cancer cell invasion, epigenetic modulatory activity, and cytotoxicity against lung cancer cells (Zhang et al. 2018; Igarashi et al. 2011; Sottorff et al. 2019).

10.2 CRISPR/CAS-Based Gene Editing Strategies in Streptomyces

In 2015, the genome editing of streptomyces was first shown by cobb et al., followed by various in vivo and in vitro methods by distinct research group to identify the unidentified BGCs using CRISPR/Cas9 technique. There are four different methods that are mostly used in editing the streptomyces genome (Zhao et al. 2020a). (1) The gene deletion, point mutation, and promoter replacement can be performed, respectively, by *Homology directed repair (HDR)* editing in Streptomyces. This method

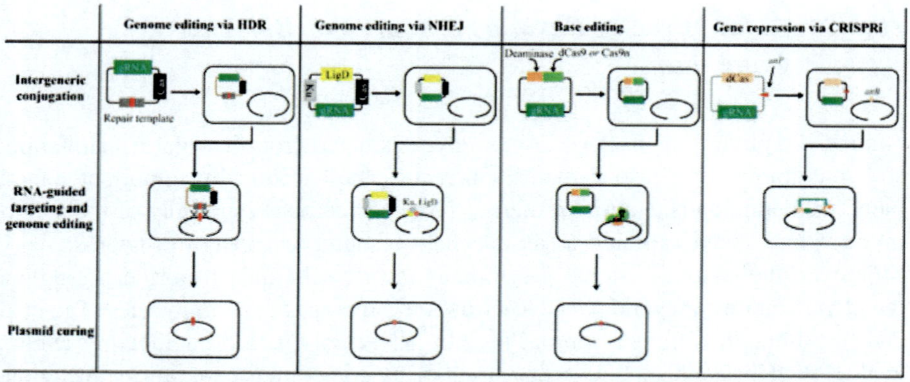

Fig. 10.1 Interpretation of the CRISPR/Cas system derived genome editing technologies in *Streptomyces* (Source: Zhao et al. 2020a)

employs the transfer of editing plasmid containing the expression cassettes of cas nuclease, a small segment of guide RNA (gRNA), and a homologous repair template into streptomyces. This would result in the cleavage of target site by the gRNA-mediated cas9 nuclease which would result in the double-stranded break (DSB) that will further be repaired by the HDR homologous repair template to provide the specific mutations in the genome. (2) *Non-homologous end joining (NHEJ)*-mediated editing in Streptomyces. This technique can be achieved by the successful transfer of editing plasmid containing the cas endonuclease segment, gRNA sequence, and the ku/ligD genes into streptomyces which would result in the DSB through intergenic conjugation. Small-sized deletion, insertion, or substitution (mostly 1–3 bp) close to the target site can be achieved by using a reconstituted NHEJ system co-expressing the scalig D. (3) *Base editor*. The target site editing can be done through intergenic conjugation while introducing the editor plasmid that would code for the gRNA and deaminase fused with cas9/cas9n to induce the base substitution in the streptomyces. In all these methods the final step of plasmid curing will be performed at the temperature of about 37 °C to obtain plasmid-free mutants and continue multiple rounds of genome editing, (4) in *CIRPSRi mediated gene editing*, gRNA is introduced into Streptomyces by conjugation for recombination at the target location for the plasmid to integrate into the genome. An inhibition of transcription of the specific genes can be done by dCas/gRNA complex by interfering with transcriptional initiation (Fig. 10.1).

On various studies, the CRISPR/Cas system has been shown as an aggressive technology for genome editing across plant, animal, and microbial kingdom. Further, these techniques employ better manipulation of genomes with many positive aspects which helps in modern drug discovery and other pharmaceutical studies (Arazoe et al. 2018). However, the toxicity and the genome instability and toxicity by the cas9 nuclease/SgRNA protein that exist in the technique has still need to be explored in depth. Multiple researches unraveled distinct approaches of using CRISPR/Cas-based gene editing technique to engineer the metabolic pathways for its downstream applications. Recent, researches shown that the genetic engineering in actinomycetes genome solves some of the major problems that exist in the classical CRISPR techniques. A publication that appeared in 2020 by Ye et al. demonstrated that fine tuning the expression level may be useful to reduce the toxicity of Cas9 in the host strain in gene editing. High Cas9 expression level may be toxic to the cells and the level of toxicity varies between actinomycete strain. Based on their findings, the study has also introduced five set of plasmids (pCM (−cas9), pCMU, pCMUtuf, pCMU-4, and pCMU-4tuf) which have showed the distinct level of resistance for Cas9 protein in *S. coelicolor* and *S. lividans*. In addition, the plasmid pCMU 4.4 was developed by the same group to use the plasmid in various other strains of actinomycetes that are very sensitive to the Cas9 expression by using a weaker constitutive promoter.

10.2.1 Nuclease-Dependent Editing System

In the nuclease-based CRISPR editing systems, the type II-A CRISPR-Cas9 and type V-A CRISPR-Cas12a (previously referred as Cpf1) are the majorly used editing methods in prokaryotes and have showed a promising effect by their natural ability to develop immunity against the invading foreign DNA (Swarts and Jinek 2018). The defense mechanism occurs in three steps (spacer acquisition, crRNA biogenesis, and target interference) for Cas9/Cas12a (For information on the mechanism of action of Cas9/Cas12a system ref. Yao et al. 2018). Moreover, the Cas12a-based systems alleviated some of the issues present while using Cas9, like requirement of G-rich protospacer-adjacent motifs to recognize the target sequence by Cas9. Further, the Cas12a system has the RNase activity for pre-crRNA and single promoter is required for the expression of multiple guide CrRNA. Thus, these features of CRISPR/Cas12a facilitate the genome editing by targeting to multiple genomic loci through a single transcript (Abudayyeh et al. 2016).

10.2.2 CRISPRi and CRISPRa-Based Genome Editing

The very recent development in CRISPR/Cas-based genome editing is the introduction of nuclease defective cas nuclease (dcas) instead of the earlier cas9 version. The method is known as CRISPRi and the technique employs repression of genes at the transcriptional (initiation and elongation) level (Zheng et al. 2019). This simple and strong technology for repression of gene offered by dCas, which is aided by single guide RNA, can almost target any DNA sequence (Yao et al. 2018). dCas9 (D10A and H841A) of nuclease defective *S. pyogenes* are most widely used for various purposes. By blocking RNA polymerase sterically, CRISPRi inhibits transcription (Yao et al. 2018), which works like a transcriptional repressor of bacteria that have limitations such as recognizing specific DNA sequence (Xiao et al. 2016). CRISPRi was first developed in bacteria as a method for controlling gene expression widely on a genome scale using an RNA guided dCas9 system which could also be used for gene repression in mammalian cells (Qi et al. 2013). In eukaryotes, for strong gene repression, auxiliary inhibitors are essential to bind with dCas, whereas in bacteria, dCas/sgRNA complex is single-handedly capable of repressing genes strongly (Gilbert et al. 2013; Qi et al. 2013). In multiplexed gene repression, dCas12a CRISPRi requires only a lone CRISPR array expression (Zhang et al. 2017b), whereas dCas9 needs multiple sgRNAs which express independently (Qi et al. 2013).

Three different CRISPRi gene editing tools in Streptomyces have been introduced by Tong et al., in 2015, replacing Cas9 with dCas9, which enabled efficient repression of single genes upon induction. The first method relies on the replicative plasmid pGM1190 harboring temperature-sensitive replicon pSG5, which will have the thiostrepton-inducible promoter (tipAp) driven expression of the dCas9/sgRNA complex. Later, two methods were developed in *S. coelicolor* based on dCas9 and ddCpf1 whose expression is driven by thiostrepton-inducible promoter (tipAp). Using these two CRISPRi systems, the repression of up to four genes at high efficiency was demonstrated simultaneously. The later methods have two advantages over the earlier method of using replicative plasmid (e.g., pGM1190). First, their repression effects are likely to be stable as they are integrated into the genome. Further, it is important to note that simultaneous repression of multiple targets using the dCas9-based system involves a time-consuming procedure to construct multiple sgRNA expression cassettes with independent promoters and terminators. In contrast, only a single customized CRISPR array with one promoter is required owing to the pre-crRNA processing ability of ddCpf1, which is time-saving and convenient. Therefore, the ddCpf1-based CRISPRi system has an advantage over dCas9-based systems for multiplexed gene repression (Li et al. 2018a, b, c; Zhao et al. 2018, 2020a, b). In these systems, repression of a maximum of up to four genes was achieved simultaneously at high efficiency. Also, they have some advantages such as stable repression effects and higher efficiency of transformation of pSET152 plasmid comparing to the inducible CRISPRi tool. In *S. bingchenggensis*, transcription level of the gene *sbi_04868* was down-regulated which is involved in the primary metabolism of milbemycins production by using the ddCpf-based integrative CRIPSRi system (Liu et al. 2021). Furthermore, in a recent study by Tian et al. in 2020, a novel control strategy which is pathway independent named EQCi has been developed integrating the CRISPRi tool with an endogenous quorum sensing (QS) system in the industrially important rapamycin producing strain *S. rapamycinicus*, and found that transcription repression degree in multiple targets by EQCi was equal to that of single target based on CRISPRi. Application of this EQCi genetic circuit, having advantages of both the QS system and CRISPRi, maximizes biosynthesis of the target product by aiding down-regulation of pathways which leads to deflection in metabolic fluxes.

Another recent adaptation technique in CRISPR/Cas-based genome editing for gene expression variation is CRISPR activation (CRISPRa). CRISPR-based transcriptional activation (CRISPRa), which has an activator domain, fuses RNA polymerase from dCas9/sgRNA complex with the promoter of a target gene (Alok et al. 2020). In bacteria, especially in *E. coli*, CRISPRa is achieved by fusing dCas9 system with the ω subunit of the RNA polymerase; this complex then binds to the region upstream of the promoter, thus activating transcription (Bikard et al. 2013). In a study by Hu et al. (2018), over a 100-fold increase in transcriptional activation has been achieved by using ω-subunit ω-(I12N) bound with dCas9 system. CRISPRa is not much used in prokaryotes due to the lack of transcriptional activators suitable for the technique (Zhang et al. 2018). CRISRa needs to target a narrow region of around 90 bp upstream the transcriptional start site (TSS) for optimal activation effect. Thus, application of CRISPRa is limited to genes having no noticeable PAM sites around the 90 bp upstream the TSS (Dong et al. 2018). Therefore, study on CRISPRa genome editing in *Streptomyces* is lacking, although it has potential application in production of novel NPs through gene cluster activation.

10.2.3 CRISPR/Cas9: Knock-in Strategy

Here, knock-in of promoters is performed strategically to activate silent BGCs using CRISPR/Cas9 systems in *Streptomyces* sp. Zhang et al. 2017a used this strategy with the help of pCRISPomyces-2 plasmid (Cobb et al. 2015) to efficiently replaced the regions of native promoters with constitutive promoters in *Streptomyces* species. As a result, three uncharacterized and two known BGCs in *Streptomyces sp.* are activated using this knock-in strategy (Table 10.1). In addition to this, by knocking-in bidirectional promoters using CRISPR/Cas9 system, erythromycin production was increased by 58.3% in *Saccharopolyspora erythraea* (Liu et al. 2019).

10.2.4 CAPTURE Method

The cloning of BGCs which are larger in size and having high GC content is highly challenging in large scale. To overcome this, a new, robust, and highly efficient

Table 10.1 CRISPR/Cas9 Knock-in strategy in *Streptomyces*

BGCs	Promoter	Species	Reference
Indigoidine	*kasO*p*	*S. albus*	Zhang et al. (2017a, b)
Actinorhodin (ACT), undecylprodigiosin (RED)		*S. lividans*	
Alteramide A, Polycyclic tetramate macrolactam, FR-900098, type I polyketides		*S. roseosporus*	
Type III polyketide		*S. venezuelae*	
Pentangular type II polyketide		*S. viridochromogenes*	

direct cloning method called Cas12a-assisted precise targeted cloning using in vivo Cre-lox recombination (CAPTURE) has been developed recently by combining T4 polymerase exo + fill-in DNA assembly, Cas12a digestion and circularization of in vivo Cre-*lox* DNA (Fig. 10.2) (Enghiad et al. 2021). Regardless of their DNA sequence repetition and GC content, BGCs ranging from 10 to 113 kb can be cloned directly by this method. In this method, 43 uncharacterized NP BGCs from different *Streptomyces and Bacillus* species were cloned and heterologously expressed in *S. avermitilis*, *S. lividans,* and *B. subtilis*, resulting in identification of 7 BGC capable of producing compounds which have antibacterial activity. In addition to other direct cloning methods, CAPTURE also gives a high-level efficiency of approximately 100% and robustness.

10.2.5 CRISPR-BEST

To address the limitations of CRISPR-Cas9 system such as genome instability due to the DNA double-strand breaks (DSBs) and effects of off-target genome editing, Tong et al. (2019) have developed a novel genome editing system named CRISPR-Base Editing SysTem (CRISPR-BEST) which is a high-efficient, DSB-free, single nucleotide editing system based on pSG5 (Muth 2018) in different *Streptomyces* species. Also, CRISP-BEST reduces Cas9 toxicity and stress on chromosome by DSBs compared to the other CRISPR-based editing system. Furthermore, two variants of this system, namely CRISPR-cBEST and CRISPR-aBEST, are capable of converting the base pairs C:G to T:A and A:T to G:C, respectively (Fig. 10.3). In *Streptomyces collinus* Tü365, a non-model actinomycete, CRISPR-cBEST simultaneously inactivated two copies of a target gene upon introduction of a STOP codon. Also, Tong et al. 2020 developed a CRISPR toolkit which includes CRIPR-BEST system for Streptomycetes, where its function is inactivation of coding genes, which also can be applied in in vivo protein engineering by customizing amino acid substitution.

10.2.6 CAT-Fishing

In an attempt to simplify the process of discovery of NPs, our team developed the CRISPR/Cas12a-mediated fast direct biosynthetic gene cluster cloning (CAT-FISHING) strategy by direct cloning of large fragments of DNA from high GC *Streptomyces* genomic DNA or bacterial artificial chromosome efficiently, making it a direct and simple method. This strategy has the potential to facilitate the novel bioactive small molecules discovery from microorganisms for drug development (Liang et al. 2020).

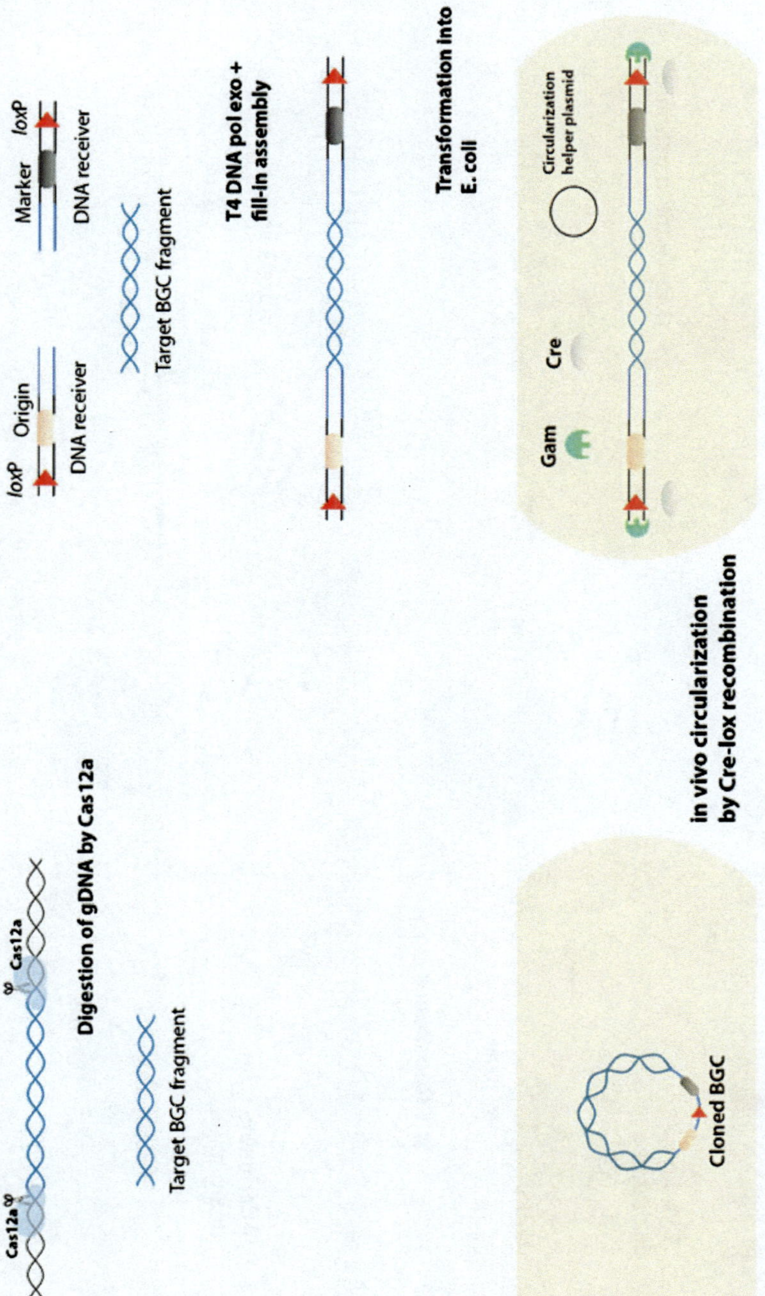

Fig. 10.2 Overview of CAPTURE method (Source: Enghiad et al. 2021)

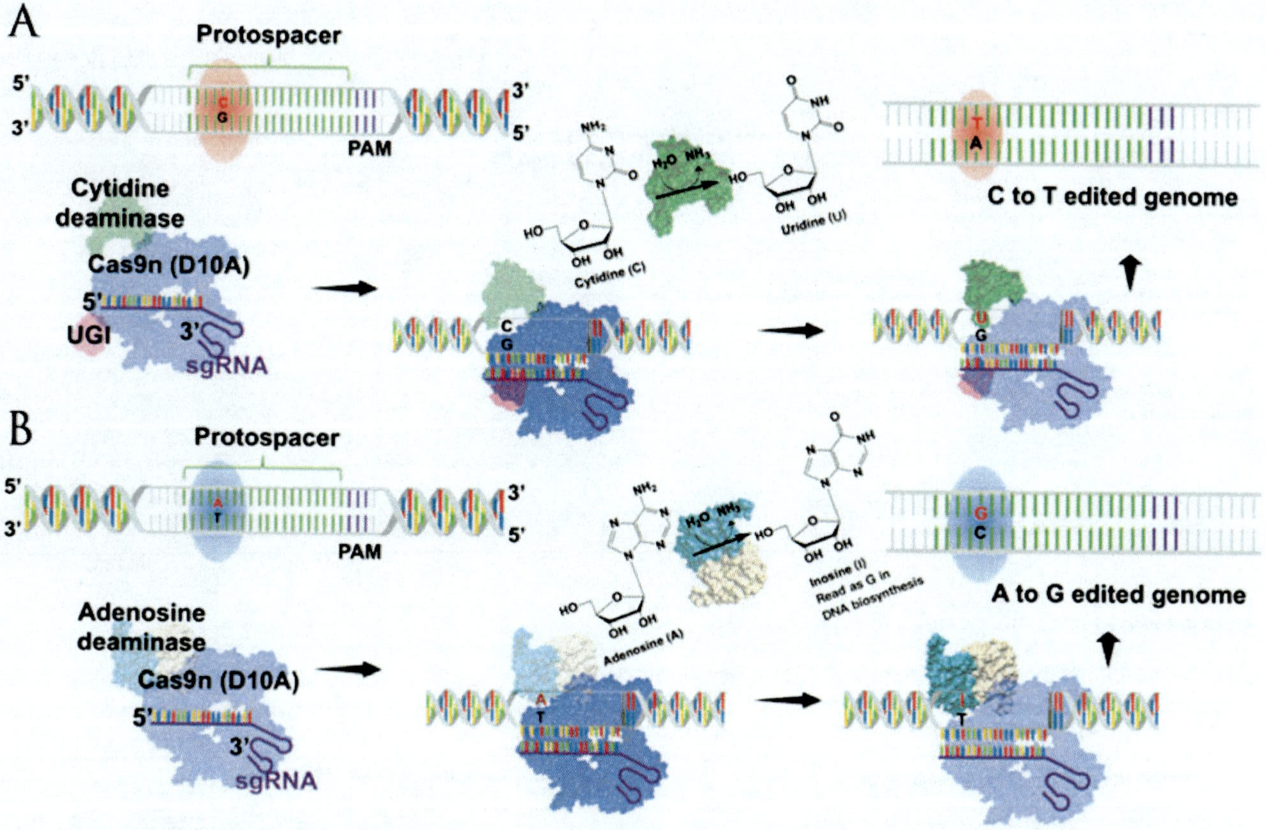

Fig. 10.3 Workflow of CRISPR-BEST. (A and B) Overview of base editing strategy for CRISPR-cBEST and CRISPR-aBEST, respectively. (Source: Tong et al. 2019)

10.3 CRISPR/CAS Systems in Other Actinomycetes

CRISPR/Cas systems and CRISPR-associated proteins are not only used as genome editing tool, but also as a defense mechanism against mobile genetic elements (MGE) in prokaryotes through an adaptive immunity by introducing short MGE sequences, called as protospacers, into the CRISPR locus (Choi and Lee 2016). In prokaryotes, CRISPRs are considered to be means of adaptive immunity against bacteriophages (Sorek et al. 2013). The variable spacers in CRISPRs have viral or plasmid DNA fragments that provide immunity upon succeeding encounters with the virus (Barrangou et al. 2007).

10.3.1 Gardnerella *spp.*

In bacterial vaginosis (BV), a most common infection which results in white gray discharge in vagina and a fishy smell among women mostly of reproductive age (Ruiz-Hernández et al. 2020), *Gardnerella* spp., another member of phylum actinobacteria, play a pivotal role in the development of the infection (Schwebke et al. 2014; Muzny et al. 2019; Morrill et al. 2020; Rosca et al. 2020). Among other bacteria which are associated with BV, *Gardnerella vaginalis* has been proved to have higher potential for virulence throughout the process of this infection (Castro et al. 2019). The presence of CRISPR/Cas systems in *G. vaginalis* strains and the possibility of genetic material transfer between the strains regulated by CRISPR/Cas system was first described by Pleckaityte et al. in 2012. An analysis of CRISPR/Cas system in *Gardnerella* spp. revealed the presence of CRISPR-mediated immunity which provides a mechanism against the effect of the phages infecting the *Gardnerella* species (Ruiz-Hernández et al. 2020).

10.3.2 Salinispora

An important emerging source for bioactive NPs is marine actinobacteria enclosing peculiar structural classes of natural compounds (Fenical and Jensen 2006). Actinobacteria are extensively distributed in the marine ecosystem, and native marine actinobacteria have also been described (Bull et al. 2005). A pan-tropical distribution of *Salinispora* genus, under phylum actinobacteria, can be found in marine sediments (Jensen and Mafnas 2006; Freel et al. 2012). In a study by Wietz et al. (2014), three previously undefined CRISPR subtypes were identified in *Salinispora* strains, and it has also been concluded that a key force in the ecology and evolution of *Salinispora* is characterized by the CRISPR-based interaction with the MGEs.

10.3.3 Nonomuraea *sp.*

Su et al. (2020) reported that for the first time, an enhanced ecumicin production was achieved through promoter engineering and CRISPR/Cas9-based genome editing of *Nonomuraea sp.*, a rare actinomycete. By combining the ribosome-binding site of *kasO*P* with that of the ecumicin gene cluster *ecuE*, increased production of active ecumicin component EcuH14 was attained. This study indicates that the *Nonomuraea* sp. might be a potential source of novel antibiotics and other pharmaceutical compounds.

10.4 Application of CRISPR/Cas Systems In Biotechnology

Industrial biotechnology is reliant on these techniques to meet the growing demands for, and expand the catalog of, chemicals, metabolites, and biomolecules that can be produced by microbial fermentation. To date, efficient genomic engineering in microbes has relied on the use of DNA donors in combination with endogenous DNA repair machinery, exogenous recombination systems, selectable markers, site-specific recombinases, group II intron retro transposition, and the use of artificial chromosomes (Esvelt and Wang 2013; David and Siewers 2014). CRISPR-Cas technologies have greatly reshaped the field of biology. In general, CRISPR-Cas systems are composed of a CRISPR RNA (crRNA) and Cas proteins. The genetic modification can then be introduced by either the error-prone NHEJ or HDR that creates precise genomic modifications. Most prokaryotes employ HDR. The initial genome engineering was done by natural CRISPR-Cas systems and later the technique was developed for gene regulation (Yao et al. 2018). These technologies have great contribution in the field of DNA imaging, bacterial immunization, virome tracking, and gene cloning (Cho et al. 2018; Donohoue et al. 2018). In 2019, Culp et al. reported the identification of various hidden antibiotics in actinomycetes by using the CRISPR/Cas9 genome engineering. Further, they have recovered distinct known variants of the antibiotics, such as thiolactomycin, amicetin, phenanthroviridin, and 5-chloro-3formalindone. The following table gives the brief overview of CRISPR/Cas-mediated plasmids that are developed by various research groups for genome editing in streptomyces (Table 10.2). Further, the industrially relevant streptomyces species is also give in Table 10.3.

In the recent years, CRISPR/Cas systems have provided a powerful tool for silent gene cluster activation, gene expression control, single or double site mutations, and in-frame gene deletion in different Streptomyces strains (Dong et al. 2017; Huang et al. 2015; Cobb et al. 2015; Zeng et al. 2015; Li et al. 2018a, b, c; Jia et al. 2017; Tong et al. 2015; Zhang et al. 2017a, b). spCas9 was used in several studies for genome editing which is codon-optimized (Huang et al. 2015; Cobb et al. 2015; Zeng et al. 2015; Tong et al. 2015; Bao et al. 2015). High efficiency in genome editing of about 66–100% has been achieved in *S. ablus* and *S. viridochromogenes*

Table 10.2 Brief list of CRISPR/Cas-based gene editing plasmids in *Streptomyces*

Editing plasmid	Endonuclease	origin	Editing efficiency (%)	Fusion sequence	Target gene	Reference
pCRISPomyces-1	Cas9	*Streptococcus pyogenes*	21–25	–	redN, *actVA-ORF5* *redN, actVA-ORF5/phpM, phpD/ redN/actVA-ORF5, shg_05713, sshg_00040/ sshg_00050*	Cobb et al. (2015)
pCRISPomyces-2	Cas9	*Streptococcus pyogenes*	67–100	–		
pKCcas9dO	Cas9	*Streptococcus pyogenes*	29–100	–	*actII-orf4*, *redD*, and *glnR and large gene clusters like,* actinorhodin (ACT), undecylprodigiosin (RED), and Ca2 + −dependent antibiotic	Huang et al. (2015)
pCRISPR-Cas9	Cas9	*Streptococcus pyogenes*	3–100	–	actIORF1, actVB actIORF1, actVB actIORF1, actVB	Tong et al. (2015)
pCRISPR-Cas9-ScaligD	Cas9	*Streptococcus pyogenes*	69–77	LigD		
pCRISPR-dCas9	Cas9	*Streptococcus pyogenes*	ND	–		
pWHU2653	Cas9	*Streptococcus pyogenes*	93–99	CodA(sm)	Actinorhodin polyketide chain length factor gene actI-ORF2	Zeng et al. (2015)
pMWCas9	Cas9	*Streptococcus pyogenes*	ND	Cod(sm)	Polyketide synthase (PKS) genes	Mo et al. (2019)
pQS-*gusA*	Cas9	*Streptococcus pyogenes*	100	GusA	actIORFI, orange-pigmented carotenoid gene cluster, abyssomicin gene cluster	Wang et al. (2019)
pQS-*idgS*	Cas9	*Streptococcus pyogenes*	100	IdgS		
pWHU2653-TRMA	Cas9	*Streptococcus pyogenes*	8.3–80	AtpD	actII-ORF4, redD	Wang et al. (2019)
pKC1139-TRMA	Cas9	*Streptococcus pyogenes*	8.3–80	AtpD		

(continued)

Table 10.2 (continued)

Editing plasmid	Endonuclease	origin	Editing efficiency (%)	Fusion sequence	Target gene	Reference
pKCCpf1	Cpf1	*Francisella novicida*	75–95	–	actI-orf1	Li et al. (2018a, b, c)
pKCCpf1-MsmE	Cpf1	*Francisella novicida*	10–56.7	LigD, Ku		
pSETddCpf1	dd Cpf1	*Francisella novicida*	11.8–95.2	–		
pCRISPomyces-Sth1Cas9	Cas9	*Streptococcus thermophilus*	100	–	?	Yeo et al. (2019)
pCRISPomyces-SaCas9	Cas9	*Staphylococcus aureus*	87–100	–		
pCRISPomyces-FnCpf1	Cpf1	*Francisella novicida*	87–100	–		
pSET-dCas9	dCas9	*Streptococcus pyogenes*	ND	–	actII4, actI1, redQ, cdaPS1 and cpkA	Zhao et al. (2018)
pSET-dCas9-actII-4-NT-S1	dCas9	*Streptococcus pyogenes*	68–99	–		
pCRISPR-cBEST	Cas9n	*Streptococcus pyogenes*	0–100	rAPOBEC1	?	Tong et al. (2019)
pCRISPR-aBEST	Cas9n	*Streptococcus pyogenes*	0–100	ecTadA		
pKC-dCas9-CDA-UL*str*	dCas9	*Streptococcus pyogenes*	15–100	PmCDA1	actVI-ORF3, actVI-ORF2 redX, redL, redX, actIII	Zhao et al. (2020b)

Table 10.3 *Streptomyces* used in CRISPR technology with industrial applications

Species	Industrial relevance	Type of modifications
Streptomyces albus	Heterologous secondary metabolites production	Recombination
Streptomyces coelicolor	Pharmacologically active and industrially relevant secondary metabolites isolation	Recombination and CRISPRi
Streptomyces lividans	Active secondary metabolites with pharmacological and industrial relevance	Recombination
Streptomyces viridochromogenes	Active secondary metabolites with pharmacological and industrial relevance	Recombination

and 70–100% in *S. coelicolor* (Cobb et al. 2015; Huang et al. 2015; Tong et al. 2015). In a study by Cobb et al. in 2015, an efficient genome deletion was reported ranging from 20 to 30 bp by the use of sgRNA CRISPR system by comparing dual tracr-crRNA expression cassette with a sgRNA expression cassette. In *S. coelicolor*, a fnCas12a assisted genome editing system has been developed so that spCas9 can be introduced into industrial Streptomyces, particularly economically important strains such as *S. pristinaespiralis* HCCB10218 and *S. hygroscopicus* SIPI-KF (Li et al. 2018a, b, c). With the help of NHEJ and HDR, gene deletion efficiency in *S. coelicolor* has reached 0–50% and 75–95%, respectively, and also CRISPRi system based on dCas12a has been developed in Streptomyces for transcriptional repression resulting in 95% repression (Li et al. 2018a, b, c). High-efficient knock-in strategies based on CRISPR/Cas9 and CRISPR/Cas12a has been reported in *S. albus*, *S. lividans*, *S. roseosporus*, *S. venezuelae*, and *S. viridochromogenes* in order to activate silent BGCs (Li et al. 2017; Zhang et al. 2017a, b). Zhang et al. (2017a, b) also reported an enhanced biosynthesis of bioactive compounds and a novel type II polyketide by replacing native promoters with the constitutive kasO*p promoter using CRISPR/Cas9 system. This strategy contributes to an enhanced technique for silent BGCs activation and to the discovery of novel uncharacterized compounds. Using CRISPR/Cas9 system, mutation in single and double site, and target gene disruption are also successfully achieved in S. rimosus, where single site mutation rate is 100% (Jia et al. 2017). Though CRISPR/Cas9 system has proven to be a powerful genome editing tool in actinomycetes, it is important to adjust the expression levels of Cas9 because in untested strains, the system was proved to be toxic by exhibiting absence or delayed growth of the organisms, but upon lowering levels of Cas9, the toxicity is reduced (Ye et al. 2020).

Furthermore, in the recent years, CRISPR/Cas systems have provided a powerful tool for silent gene cluster activation, gene expression control, single or double site mutations, and in-frame gene deletion in different Streptomyces strains (Dong et al. 2017; Huang et al. 2015; Cobb et al. 2015; Zeng et al. 2015; Li et al. 2018a, b, c; Jia et al. 2017; Tong et al. 2015; Zhang et al. 2017a, b). spCas9 was used in several studies for genome editing which is codon-optimized (Huang et al. 2015; Cobb et al. 2015; Zeng et al. 2015; Tong et al. 2015; Bao et al. 2015). High efficiency in genome editing of about 66–100% has been achieved in *S. albus* and *S. viridochromogenes* and 70–100% in *S. coelicolor* (Cobb et al. 2015; Huang et al. 2015; Tong et al.

2015). In a study by Cobb et al. in 2015, an efficient genome deletion was reported ranging from 20 to 30 bp by the use of sgRNA CRISPR system by comparing dual tracr-crRNA expression cassette with a sgRNA expression cassette. In *S. coelicolor*, a fnCas12a assisted genome editing system has been developed so that spCas9 can be introduced into industrial Streptomyces, particularly economically important strains such as *S. pristinaespiralis* HCCB10218 and *S. hygroscopicus* SIPI-KF (Li et al. 2018a, b, c). With the help of NHEJ and HDR, gene deletion efficiency in *S. coelicolor* has reached 0–50% and 75–95%, respectively, and also CRISPRi system based on dCas12a has been developed in Streptomyces for transcriptional repression resulting in 95% repression (Li et al. 2018a, b, c). High-efficient knock-in strategies based on CRISPR/Cas9 and CRISPR/Cas12a has been reported in *S. albus, S. lividans, S. roseosporus, S. venezuelae*, and S. *viridochromogenes* in order to activate silent BGCs (Li et al. 2017; Zhang et al. 2017a, b). Zhang et al. (2017a, b) also reported an enhanced biosynthesis of bioactive compounds and a novel type II polyketide by replacing native promoters with the constitutive KasO*p promoter using CRISPR/Cas9 system. This strategy contributes to an enhanced technique for silent BGCs activation and to the discovery of novel uncharacterized compounds. Using CRISPR/Cas9 system, mutation in single and double site and target gene disruption are also successfully achieved in *S. rimosus*, where single site mutation rate is 100% (Jia et al. 2017).

10.5 CRISPR Cas 3 Genes from Actinobacteria

We verified and confirmed full length sequence of CRISPR-dependent cas3 gene of *Streptomyces hygroscopicus* (A6F03_RS20145) was retrieved from NCBI gene database. The putative cas3 genes were mined from genomes of all microbes available at IMG/M database using BLASTP. The gene equivalents including the hypothetical and putative protein entries were retrieved as an amino acid FASTA file. It resulted in about 807 entries comprising cas3 genes of phylum *Actinobacteria* (Fig. 10.4). The amino acid sequences were clustered with 90% similarity using **uclust** algorithm in **USEARCH** command line tool (V- 11.0.667). Whole cas3 three genes retrieved and representative sequence from each cluster was taken and aligned using CLUSTALW algorithm in MEGA X. The phylogenetic tree data was supplied to Interactive Tree of Life webserver to visualize the phylogenetic tree. In future, if we reveal the function of these novel Cas3 genes, it can be used for actinobacteria gene editing and other applications.

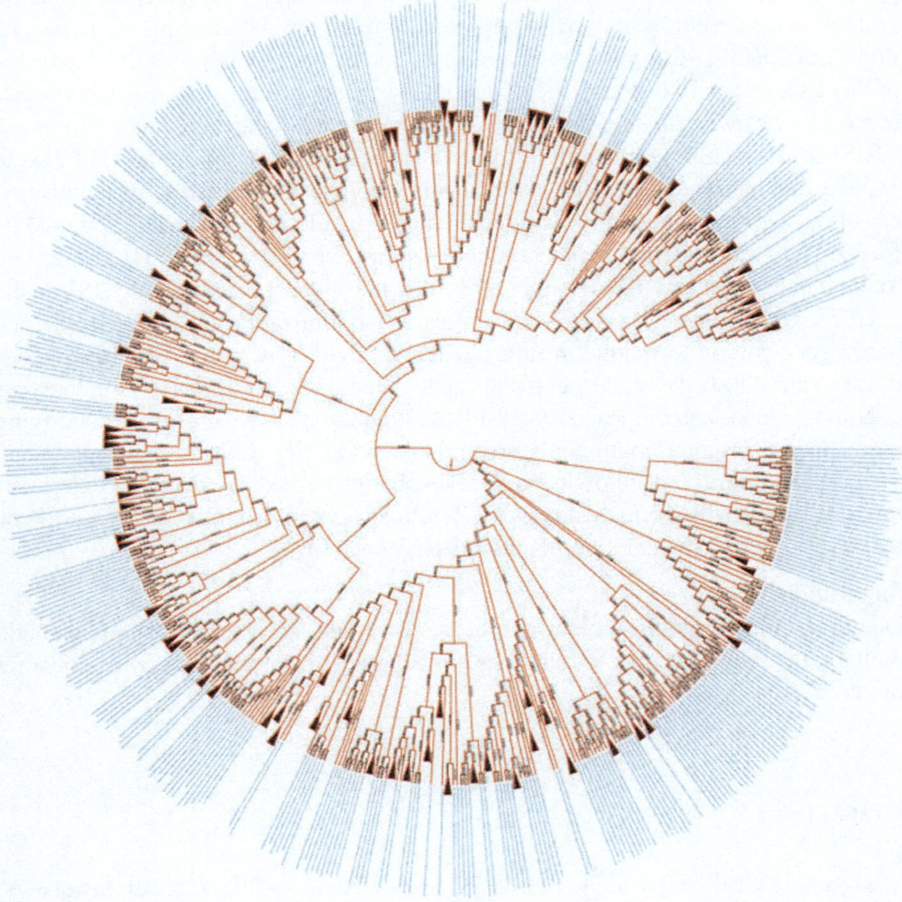

Fig. 10.4 Mining of Cas3 gene from Actinobacteria

10.6 Future Perspectives

There is no doubt that the CRISPR/Cas technology-based genome editing has revolutionized the genome editing process across organisms. The technique has been shown efficient, accurate, and less time and labor consuming. Although, in the past decade, the technique has attracted several researchers to overwhelm with CRISPR/Cas-based gene editing, use of these techniques for real application is still at the nascent stage. Number of several research group has demonstrated by using various CRISPR-based techniques to edit the single or multiple genes, small of huge BGCs in various *Streptomyces* species. However, mostly these testing level works were shown for the efficacy and the type of mutation in the selected organism. It is time and necessary to take up the approach to next level from proof of concept to application to explore the feasibility of editing the beneficial genes in metabolic pathways. Further improvement is still needed, including the application for larger scale genome editing in rare actinomycetes. In addition, maximizing the beneficial impact of CRISPR-Cas systems on industrial biotechnology requires the expansion of this technology into strains specific to industrial applications. As all technologies come with its own setbacks, it is necessary to continue the standardization process of CRISPR/Cas technique to apply it in various organisms not just to test but also to explore the maximum possibility of real-time application especially for the discovery of the active compound for various therapies. In addition, the research should be focused to understand the major bottle necks that exist with Cas9/Cas12a nuclease toxicity in the host and the off-target effect that introduce the nonspecific DSB in the CRISPR techniques. Therefore, more tools and optimization in the CRISPR/Cas-based gene editing technique in future will shed light on several unexplored genes those were masked by the existing gene candidates in distinct streptomyces genomes. Upcoming improved CRISPR techniques may be useful in identifying more natural product compounds not only in prokaryotes but also in eukaryotic systems. In future, the following questions should be addressed such as How the microorganism will react when newer CRISPR system model introduces? And How the native microorganism controls the off-target activity?

Suggestion
Due to the higher intelligence of microorganisms and latest revolution in synthetic biology, the time has come to frame the ethical guidelines for microbial engineering research.

References

Abudayyeh OO, Gootenberg JS, Konermann S, Joung J, Slaymaker IM, Cox DB, Shmakov S, Makarova KS, Semenova E, Minakhin L, Severinov K, Regev A, Lander ES, Koonin EV, Zhang F (2016) C2c2 is a singlecomponent programmable RNA-guided RNA-targeting CRISPR effector. Science 353(6299):aaf5573. https://doi.org/10.1126/science.aaf5573

Alali A, Zhang L, Li J, Zuo C, Wassouf D, Yan X, Schwarzer P, Günther S, Einsle O, Bechthold A (2021) Biosynthesis of the tricyclic aromatic Type II polyketide rishirilide: new potential third ring oxygenation after three cyclization steps. Mol Biotechnol 63:502–514. https://doi.org/10.1007/s12033-021-00314-x

Alok A, Tiwari S, Kaur J (2020) CRISPR/Cas9 mediated genome engineering in microbes and its application in plant beneficial effects. In: Molecular aspects of plant beneficial microbes in agriculture. Elsevier, pp 351–359

Arazoe T, Kondo A, Nishida K (2018) Targeted nucleotide editing technologies for microbial metabolic engineering. Biotechnol J 13:1700596. https://doi.org/10.1002/biot.201700596

Ayuso-Sacido A, Genilloud O (2005) New PCR primers for the screening of NRPS and PKS-I systems in actinomycetes: detection and distribution of these biosynthetic gene sequences in major taxonomic groups. Microb Ecol 49:10–24. https://doi.org/10.1007/s00248-004-0249-6

Bao Z, Xiao H, Liang J, Zhang L, Xiong X, Sun N, Si T, Zhao H (2015) Homology-integrated CRISPR–Cas (HI-CRISPR) system for one-step multigene disruption in *Saccharomyces cerevisiae*. ACS Synth Biol 4:585–594. https://doi.org/10.1021/sb500255k

Bull AT, Stach JE, Ward AC, Goodfellow M (2005) Marine actinobacteria: perspectives, challenges, future directions. Antonie Van Leeuwenhoek 87(3):65–79

Barrangou R, Fremaux C, Deveau H, Richards M, Boyaval P, Moineau S, Romero DA, Horvath P (2007) CRISPR provides acquired resistance against viruses in prokaryotes. Science 315 (5819):1709–1712. https://doi.org/10.1126/science.1138140

Bikard D, Jiang W, Samai P, Hochschild A, Zhang F, Marraffini LA (2013) Programmable repression and activation of bacterial gene expression using an engineered CRISPR-Cas system. Nucleic Acids Res 41:7429–7437. https://doi.org/10.1093/nar/gkt520

Bode HB, Bethe B, Höfs R, Zeeck A (2002) Big effects from small changes: possible ways to explore nature's chemical diversity. Chembiochem 3:619–627. https://doi.org/10.1002/1439-7633(20020703)3:7<619::AID-CBIC619>3.0.CO;2-9

Castro J, Machado D, Cerca N (2019) Unveiling the role of Gardnerella vaginalis in polymicrobial bacterial vaginosis biofilms: the impact of other vaginal pathogens living as neighbors. ISME J 13(5):1306–1317. https://doi.org/10.1038/s41396-018-0337-0

Challis GL, Hopwood DA (2003) Synergy and contingency as driving forces for the evolution of multiple secondary metabolite production by Streptomyces species. Proc Natl Acad Sci 100: 14555–14561. https://doi.org/10.1073/pnas.1934677100

Cho S, Shin J, Cho B-K (2018) Applications of CRISPR/Cas system to bacterial metabolic engineering. IJMS 19:1089. https://doi.org/10.3390/ijms19041089

Choi KR, Lee SY (2016) CRISPR technologies for bacterial systems: current achievements and future directions. Biotechnol Adv 34:1180–1209. https://doi.org/10.1016/j.biotechadv.2016.08.002

Cobb RE, Wang Y, Zhao H (2015) High-efficiency multiplex genome editing of *Streptomyces* species using an engineered CRISPR/Cas system. ACS Synth Biol 4:723–728. https://doi.org/10.1021/sb500351f

Culp EJ, Yim G, Waglechner N, Wang W, Pawlowski AC, Wright GD (2019) Hidden antibiotics in actinomycetes can be identified by inactivation of gene clusters for common antibiotics. Nat Biotechnol 37:1149–1154. https://doi.org/10.1038/s41587-019-0241-9

David F, Siewers V (2014) Advances in yeast genome engineering. FEMS Yeast Res. https://doi.org/10.1111/1567-1364.12200

Davies J (2013) Specialized microbial metabolites: functions and origins. J Antibiot 66:361–364. https://doi.org/10.1038/ja.2013.61

Dong D, Guo M, Wang S, Zhu Y, Wang S, Xiong Z, Yang J, Xu Z, Huang Z (2017) Structural basis of CRISPR–SpyCas9 inhibition by an anti-CRISPR protein. Nature 546:436–439. https://doi.org/10.1038/nature22377

Dong C, Fontana J, Patel A, Carothers JM, Zalatan JG (2018) Synthetic CRISPR-Cas gene activators for transcriptional reprogramming in bacteria. Nat Commun 9:2489. https://doi.org/10.1038/s41467-018-04901-6

Donohoue PD, Barrangou R, May AP (2018) Advances in industrial biotechnology using CRISPR-Cas systems. Trends Biotechnol 36(2):134–146. https://doi.org/10.1016/j.tibtech.2017.07.007

Enghiad B, Huang C, Guo F, Jiang G, Wang B, Tabatabaei SK, Martin TA, Zhao H (2021) Cas12a-assisted precise targeted cloning using in vivo Cre-lox recombination. Nat Commun 12:1171. https://doi.org/10.1038/s41467-021-21275-4

Esvelt KM, Wang HH (2013) Genome-scale engineering for systems and synthetic biology. Mol Syst Biol 9:641. https://doi.org/10.1038/msb.2012.66

Fenical W, Jensen PR (2006) Developing a new resource for drug discovery: marine actinomycete bacteria. Nat Chem Biol 2(12):666–673. https://doi.org/10.1038/nchembio841

Freel KC, Edlund A, Jensen PR (2012) Microdiversity and evidence for high dispersal rates in the marine actinomycete 'Salinispora pacifica'. Environ Microbiol 14(2):480–493. https://doi.org/10.1111/j.1462-2920.2011.02641.x

Gilbert LA, Larson MH, Morsut L, Liu Z, Brar GA, Torres SE, Stern-Ginossar N, Brandman O, Whitehead EH, Doudna JA, Lim WA, Weissman JS, Qi LS (2013) CRISPR-mediated modular RNA-guided regulation of transcription in eukaryotes. Cell 154:442–451. https://doi.org/10.1016/j.cell.2013.06.044

Hosaka T, Ohnishi-Kameyama M, Muramatsu H, Murakami K, Tsurumi Y, Kodani S, Yoshida M, Fujie A, Ochi K (2009) Antibacterial discovery in actinomycetes strains with mutations in RNA polymerase or ribosomal protein S12. Nat Biotechnol 27:462–464. https://doi.org/10.1038/nbt.1538

Hu JH, Miller SM, Geurts MH, Tang W, Chen L, Sun N, Zeina CM, Gao X, Rees HA, Lin Z, Liu DR (2018) Evolved Cas9 variants with broad PAM compatibility and high DNA specificity. Nature 556:57–63. https://doi.org/10.1038/nature26155

Huang H, Zheng G, Jiang W, Hu H, Lu Y (2015) One-step high-efficiency CRISPR/Cas9-mediated genome editing in Streptomyces. Acta Biochim Biophys Sin 47:231–243. https://doi.org/10.1093/abbs/gmv007

Igarashi Y, Yanase S, Sugimoto K, Enomoto M, Miyanaga S, Trujillo ME, Saiki I, Kuwahara S (2011) Lupinacidin C, an inhibitor of tumor cell invasion from *Micromonospora lupini*. J Nat Prod 74:862–865. https://doi.org/10.1021/np100779t

Jensen PR, Mafnas C (2006) Biogeography of the marine actinomycete Salinispora. Environ Microbiol 8(11):1881–1888. https://doi.org/10.1111/j.1462-2920.2006.01093.x

Jia H, Zhang L, Wang T, Han J, Tang H, Zhang L (2017) Development of a CRISPR/Cas9-mediated gene-editing tool in Streptomyces rimosus. Microbiology 163:1148–1155. https://doi.org/10.1099/mic.0.000501

Keller NP, Turner G, Bennett JW (2005) Fungal secondary metabolism—from biochemistry to genomics. Nat Rev Microbiol 3:937–947. https://doi.org/10.1038/nrmicro1286

Li L, Jiang W, Lu Y (2017) New strategies and approaches for engineering biosynthetic gene clusters of microbial natural products. Biotechnol Adv 35(8):936–949. https://doi.org/10.1016/j.biotechadv.2017.03.007

Li L, Wei K, Zheng G, Liu X, Chen S, Jiang W, Lu Y (2018a) CRISPR-Cpf1-assisted multiplex genome editing and transcriptional repression in streptomyces. Appl Environ Microbiol:84. https://doi.org/10.1128/AEM.00827-18

Li S-Y, Cheng Q-X, Liu J-K, Nie X-Q, Zhao G-P, Wang J (2018b) CRISPR-Cas12a has both cis- and trans-cleavage activities on single-stranded DNA. Cell Res 28:491–493. https://doi.org/10.1038/s41422-018-0022-x

Li S-Y, Cheng Q-X, Wang J-M, Li X-Y, Zhang Z-L, Gao S, Cao R-B, Zhao G-P, Wang J (2018c) CRISPR-Cas12a-assisted nucleic acid detection. Cell Discov 4:20. https://doi.org/10.1038/s41421-018-0028-z

Liang M, Liu L, Wang W, Zeng X, Liu J, Karthik L, Zhu G, Bai L, Hou C, Chen X, Ouyang L, Liu X, Hu B, Xia X, Tong Y, Lou C, Alterovitz G, Tan G-Y, Zhang L-X (2020) Simple cloning of large natural product biosynthetic gene clusters from Streptomyces by an engineered CRISPR/Cas12a system. Synth Biol

Liu Y, Ren C-Y, Wei W-P, You D, Yin B-C, Ye B-C (2019) A CRISPR-Cas9 strategy for activating the *Saccharopolyspora erythraea* erythromycin biosynthetic gene cluster with knock-in bidirectional promoters. ACS Synth Biol 8:1134–1143. https://doi.org/10.1021/acssynbio.9b00024

Liu Y, Wang H, Li S, Zhang Y, Cheng X, Xiang W, Wang X (2021) Engineering of primary metabolic pathways for titer improvement of milbemycins in Streptomyces bingchenggensis. Appl Microbiol Biotechnol 105:1875–1887. https://doi.org/10.1007/s00253-021-11164-7

Loureiro C, Medema MH, van der Oost J, Sipkema D (2018) Exploration and exploitation of the environment for novel specialized metabolites. Curr Opin Biotechnol 50:206–213. https://doi.org/10.1016/j.copbio.2018.01.017

Martinet L, Naômé A, Deflandre B, Maciejewska M, Tellatin D, Tenconi E, Smargiasso N, de Pauw E, van Wezel GP, Rigali S (2019) A single biosynthetic gene cluster is responsible for the production of bagremycin antibiotics and ferroverdin iron chelators. mBio:10. https://doi.org/10.1128/mBio.01230-19

Medema MH, Kottmann R, Yilmaz P et al (2015) Minimum information about a biosynthetic gene cluster. Nat Chem Biol 11:625–631. https://doi.org/10.1038/nchembio.1890

Mo J, Wang S, Zhang W, Li C, Deng Z, Zhang L, Qu X (2019) Efficient editing DNA regions with high sequence identity in actinomycetal genomes by a CRISPR-Cas9 system. Synth Syst Biotechnol 4:86–91. https://doi.org/10.1016/j.synbio.2019.02.004

Morrill S, Gilbert NM, Lewis AL (2020) Gardnerella vaginalis as a cause of bacterial vaginosis: appraisal of the evidence from in vivo models. Front Cell Infect Microbiol 10:168. https://doi.org/10.3389/fcimb.2020.00168

Muth G (2018) The pSG5-based thermosensitive vector family for genome editing and gene expression in actinomycetes. Appl Microbiol Biotechnol 102:9067–9080. https://doi.org/10.1007/s00253-018-9334-5

Muzny CA, Taylor CM, Swords WE, Tamhane A, Chattopadhyay D, Cerca N, Schwebke JR (2019) An updated conceptual model on the pathogenesis of bacterial vaginosis. J Infect Dis 220 (9):1399–1405. https://doi.org/10.1093/infdis/jiz342

Nguyen CT, Dhakal D, Pham VTT, Nguyen HT, Sohng J-K (2020) Recent advances in strategies for activation and discovery/characterization of cryptic biosynthetic gene clusters in Streptomyces. Microorganisms 8(4):616

Qi LS, Larson MH, Gilbert LA, Doudna JA, Weissman JS, Arkin AP, Lim WA (2013) Repurposing CRISPR as an RNA-guided platform for sequence-specific control of gene expression. Cell 152: 1173–1183. https://doi.org/10.1016/j.cell.2013.02.022

Romero D, Traxler MF, López D, Kolter R (2011) Antibiotics as signal molecules. Chem Rev 111: 5492–5505. https://doi.org/10.1021/cr2000509

Ruiz-Hernández UE, Pelcastre-Rodriguez LI, Cabrero-Martínez OA, Hernández-Cortez C, Castro-Escarpulli G (2020) Analysis of CRISPR-Cas systems in Gardnerella suggests its potential role in the mechanisms of bacterial vaginosis. Comput Biol Chem 89:107381. https://doi.org/10.1016/j.compbiolchem.2020.107381

Schwebke JR, Muzny CA, Josey WE (2014) Role of Gardnerella vaginalis in the pathogenesis of bacterial vaginosis: a conceptual model. J Infect Dis 210(3):338–343. https://doi.org/10.1093/infdis/jiu089

Sekurova ON, Schneider O, Zotchev SB (2019) Novel bioactive natural products from bacteria via bioprospecting, genome mining and metabolic engineering. Microb Biotechnol 12:828–844. https://doi.org/10.1111/1751-7915.13398

Shima J, Hesketh A, Okamoto S, Kawamoto S, Ochi K (1996) Induction of actinorhodin production by rpsL (encoding ribosomal protein S12) mutations that confer streptomycin resistance in Streptomyces lividans and Streptomyces coelicolor A3(2). J Bacteriol 178:7276–7284. https://doi.org/10.1128/jb.178.24.7276-7284.1996

Sottorff I, Künzel S, Wiese J, Lipfert M, Preußke N, Sönnichsen F, Imhoff J (2019) Antitumor anthraquinones from an Easter Island Sea Anemone: animal or bacterial origin? Mar Drugs 17: 154. https://doi.org/10.3390/md17030154

Su C, Tuan N-Q, Lee M-J, Zhang X-Y, Cheng J-H, Jin Y-Y, Zhao X-Q, Suh J-W (2020) Enhanced production of active ecumicin component with higher antituberculosis activity by the rare Actinomycete *Nonomuraea* sp. MJM5123 using a novel promoter-engineering strategy. ACS Synth Biol 9:3019–3029. https://doi.org/10.1021/acssynbio.0c00248

Swarts DC, Jinek M (2018) Cas9 versus Cas12a/Cpf1: structure-function comparisons and implications for genome editing. Wiley Interdiscip Rev RNA:e1481. https://doi.org/10.1002/wrna.1481

Sorek R, Lawrence CM, Wiedenheft B (2013) CRISPR-mediated adaptive immune systems in bacteria and archaea. Annu Rev Biochem 82:237–266. https://doi.org/10.1146/annurev-biochem-072911-172315

Rosca AS, Castro J, Sousa LGV, Cerca N (2020) Gardnerella and vaginal health: the truth is out there. FEMS Microbiol Rev 44(1):73–105. https://doi.org/10.1093/femsre/fuz027

Tao W, Yang A, Deng Z, Sun Y (2018) CRISPR/Cas9-based editing of streptomyces for discovery, characterization, and production of natural products. Front Microbiol 9:1660. https://doi.org/10.3389/fmicb.2018.01660

Tenconi E, Rigali S (2018) Self-resistance mechanisms to DNA-damaging antitumor antibiotics in actinobacteria. Curr Opin Microbiol 45:100–108. https://doi.org/10.1016/j.mib.2018.03.003

Tian J, Yang G, Gu Y, Sun X, Lu Y, Jiang W (2020) Developing an endogenous quorum-sensing based CRISPRi circuit for autonomous and tunable dynamic regulation of multiple targets in Streptomyces. Nucleic Acids Res 48:8188–8202. https://doi.org/10.1093/nar/gkaa602

Tong Y, Charusanti P, Zhang L, Weber T, Lee SY (2015) CRISPR-Cas9 based engineering of actinomycetal genomes. ACS Synth Biol 4:1020–1029. https://doi.org/10.1021/acssynbio.5b00038

Tong Y, Whitford CM, Robertsen HL, Blin K, Jørgensen TS, Klitgaard AK, Gren T, Jiang X, Weber T, Lee SY (2019) Highly efficient DSB-free base editing for streptomycetes with CRISPR-BEST. Proc Natl Acad Sci U S A 116:20366–20375. https://doi.org/10.1073/pnas.1913493116

Tong Y, Whitford CM, Blin K, Jørgensen TS, Weber T, Lee SY (2020) CRISPR–Cas9, CRISPRi and CRISPR-BEST-mediated genetic manipulation in streptomycetes. Nat Protoc 15:2470–2502. https://doi.org/10.1038/s41596-020-0339-z

Walsh CT, Fischbach MA (2010) Natural products version 2.0: connecting genes to molecules. J Am Chem Soc 132:2469–2493. https://doi.org/10.1021/ja909118a

Wang K, Zhao Q-W, Liu Y-F, Sun C-F, Chen X-A, Burchmore R, Burgess K, Li Y-Q, Mao X-M (2019) Multi-layer controls of Cas9 activity coupled with ATP synthase over-expression for efficient genome editing in streptomyces. Front Bioeng Biotechnol 7:304. https://doi.org/10.3389/fbioe.2019.00304

Wietz M, Millan-Aguinaga N, Jensen PR (2014) CRISPR-Cas systems in the marine actinomycete Salinispora: linkages with phage defense, microdiversity and biogeography. BMC Genomics 15:936. https://doi.org/10.1186/1471-2164-15-936

Xiao Y, Bowen CH, Liu D, Zhang F (2016) Exploiting nongenetic cell-to-cell variation for enhanced biosynthesis. Nat Chem Biol 12:339–344. https://doi.org/10.1038/nchembio.2046

Yao R, Liu D, Jia X, Zheng Y, Liu W, Xiao Y (2018) CRISPR-Cas9/Cas12a biotechnology and application in bacteria. Synth Syst Biotechnol 3:135–149. https://doi.org/10.1016/j.synbio.2018.09.004

Ye S, Enghiad B, Zhao H, Takano E (2020) Fine-tuning the regulation of Cas9 expression levels for efficient CRISPR-Cas9 mediated recombination in *Streptomyces*. J Ind Microbiol Biotechnol 47:413–423. https://doi.org/10.1007/s10295-020-02277-5

Yeo WL, Heng E, Tan LL, Lim YW, Lim YH, Hoon S, Zhao H, Zhang MM, Wong FT (2019) Characterization of Cas proteins for CRISPR-Cas editing in streptomycetes. Biotechnol Bioeng 116:2330–2338. https://doi.org/10.1002/bit.27021

Zeng H, Wen S, Xu W, He Z, Zhai G, Liu Y, Deng Z, Sun Y (2015) Highly efficient editing of the actinorhodin polyketide chain length factor gene in Streptomyces coelicolor M145 using CRISPR/Cas9-CodA(sm) combined system. Appl Microbiol Biotechnol 99:10575–10585. https://doi.org/10.1007/s00253-015-6931-4

Zhang MM, Wong FT, Wang Y, Luo S, Lim YH, Heng E, Yeo WL, Cobb RE, Enghiad B, Ang EL, Zhao H (2017a) CRISPR–Cas9 strategy for activation of silent Streptomyces biosynthetic gene clusters. Nat Chem Biol 13:607–609. https://doi.org/10.1038/nchembio.2341

Zhang X, Wang J, Cheng Q, Zheng X, Zhao G, Wang J (2017b) Multiplex gene regulation by CRISPR-ddCpf1. Cell Discov 3:17018. https://doi.org/10.1038/celldisc.2017.18

Zhang C, Sun C, Huang H, Gui C, Wang L, Li Q, Ju J (2018) Biosynthetic Baeyer–Villiger chemistry enables access to two anthracene scaffolds from a single gene cluster in deep-sea-derived *Streptomyces olivaceus* SCSIO T05. J Nat Prod 81:1570–1577. https://doi.org/10.1021/acs.jnatprod.8b00077

Zhao Y, Li L, Zheng G, Jiang W, Deng Z, Wang Z, Lu Y (2018) CRISPR/dCas9-Mediated Multiplex Gene Repression in *Streptomyces*. Biotechnol J 13:1800121. https://doi.org/10.1002/biot.201800121

Zhao Y, Li G, Chen Y, Lu Y (2020a) Challenges and advances in genome editing technologies in streptomyces. Biomol Ther 10:734. https://doi.org/10.3390/biom10050734

Zhao Y, Tian J, Zheng G, Chen J, Sun C, Yang Z, Zimin AA, Jiang W, Deng Z, Wang Z, Lu Y (2020b) Multiplex genome editing using a dCas9-cytidine deaminase fusion in Streptomyces. Sci China Life Sci 63(7):1053–1062. https://doi.org/10.1007/s11427-019-1559-y

Zheng Y, Su T, Qi Q (2019) Microbial CRISPRi and CRISPRa systems for metabolic engineering. Biotechnol Bioproc E 24:579–591. https://doi.org/10.1007/s12257-019-0107-5

Chapter 11
Uncultured Actinobacteria and Reverse Engineering and Artificial Intelligence Role in Future

Bhagwan Narayan Rekadwad, Wen-Jun Li, Juan M. Gonzalez, Khalid Parwez, Punchappady Devasya Rekha, and Arun A. Bhagwath

Abstract Actinobacteria have been known to the world since the mid-eighteenth century, when Ferdinand Cohn portrayed them as ropy critters. In the nineteenth century, these unidentified morphons became famous for their broad array of applications and for causing maladies. Over the last five decades, actinomycetes have been classified and their names have been validly published as representative taxa. Actinobacteria have preserved their overall genetic integrity and kept newer genes that code for bioactive substances, as evidenced by their capacity to produce a myriad of bioactive compounds. As of today, we have hardly cultured 1% of existing microorganisms including Actinobacteria. Some *Streptomyces* strains were exploited for enhanced production of new bioactive compounds. In times of the bioinformatics era, insufficient and developing laboratory methods for enumeration of potential species are acting as a barrier to obtaining the maximum product in downstream processing. This issue may be addressed by using the reserve-engineering approach. Published reports indicate that new or high production can be achieved through either strain improvement or the remaining microbial dark matter consists of growing hidden gems for an awaited potential application. Reverse engineering of microorganisms, on the other hand, improves the efficacy of a specific target gene to obtain high product value, which can be seen in the expression of the tylosin biosynthetic gene cluster. Hidden gems among the phylum Actinobacteria need to be discovered and reverse engineered for antibiotics and

B. N. Rekadwad (✉) · K. Parwez · P. D. Rekha · A. A. Bhagwath
Yenepoya Research Centre, Yenepoya (Deemed to be University), University Road, Deralakatte, Mangalore, Karnataka, India

W.-J. Li
State Key Laboratory of Biocontrol, Guangdong Provincial Key Laboratory of Plant Resources and Southern Marine Science and Engineering Guangdong Laboratory (Zhuhai), School of Life Sciences, Sun Yat-Sen University, Guangzhou, China

J. M. Gonzalez
Microbial Diversity and Microbiology of Extreme Environments Research Group, Agencia Estatal Consejo Superior de Investigaciones Científicas, IRNAS-CSIC, Seville, Spain

L. Karthik (ed.), *Actinobacteria*, https://doi.org/10.1007/978-981-16-5835-8_11

novel drugs. Therefore, advancement in research machine learning and artificial intelligence seems to be incredible tools. Hence, uncultivated Actinobacteria can be used for the benefit of mankind by combining reverse engineering with machine learning (ML) and artificial intelligence (AI).

Keywords Streptomyces · Microbial dark matter · The phylum Actinobacteria · Reverse engineering · Antibiotics

Abbreviations

AI	Artificial Intelligence
AR	Antimicrobial resistance
CNN	Convolution of Neural Network
CRISPR	Clustered regularly interspaced short palindromic repeats
CRISPR-BEST	CRISPR-base editing systems
CSS	Colony Spectral Signature
Cu	Copper
DNA	Deoxyribonucleic acid
dsDNA	DNA double-strand
Fe	Ferrous (Iron)
GC	Guanine and Cytosine
Gln	Glutamine synthetase
HIS	Hyper Spectral Imaging technique
IJSEM	International Journal of Systematics and Evolutionary Microbiology
MALDI	Matrix-Assisted Laser Desorption Ionization (MALDI),
ML	Machine Leaning
MS	Mass Spectrometry
RNA	Ribonucleic acid
rRNA	ribosomal RNA

11.1 Introduction

11.1.1 Boom of Actinobacterial Taxonomy

Ferdinand Cohn had portrayed "Actinomycetes" in 1875. In coalescence first from nasolacrimal duct, he noticed a capillaceous morphon. Streptothrix foersteri has been the name he decided to give to a ropy organism (Waksman 1959). Subsequently, C.O. Harz enumerated Actinomyces bovis from the bovinae lumpy jaw later in the triennium (Harz 1977–1978). Greek refers Actinomycetes as "ray fungus" (Lechevalier and Lechevalier 1981). Representative taxa of phylum *Actinobacteria* are kenspeckle for their vast application for many decades. In regard to classification and taxonomy of *Actinomycetes*, the first meeting was held in 1966 at the Gause Institute of New Antibiotics Russian Academy of Medical Sciences, Russia (Küster 1967). The taxonomy of *Actinobacteria* was poorly justified and facing conflict in naming representatives as type material till the early 80s. Since the mid-80s, almost there is no type of material of the phylum *Actinobacteria* validly reported in IJSEM. Finally, a recommendation for a new hierarchic categorization system based on the 16S rRNA gene sequence has been proposed as a result of the microbial taxonomist's effort (Stackebrandt et al. 1997). Validly published representative types of Actinobacteria in IJSEM (LPSN, https://lpsn.dsmz.de/) opened a gold mine for actinobacterial taxonomy, which began in the year 2000 and has lasted for two millennia. The current scenario is that many species published in other journals specified by Bacteriological Code need validation provided that new additions will be high in future (Fig. 11.1). Most of these discovered actinobacterial representatives are potential producers of bioactive compounds and will have a promising application in near future.

Members of the uncultivated Actinobacterial Candidatus *Aquicultoria*, Candidatus *Humimicrobiia*, and Candidatus *Geothermincolia* have encoded the Wood–Ljungdahl pathway, in which enzymes including hydrogenases and the carbon monoxide dehydrogenase/acetyl-CoA synthase complex were inherited vertically from their last common ancestor, according to published studies (Jiao et al. 2021). As a result, genome and metagenome data are used to construct an artificial, novel, dynamic organism that can replicate like a natural cell and reproduce artificially coded compounds. Nowadays, a reverse genetics, or reverse genome engineering, approach will greatly aid in encoding genes necessary for the production of novel antibiotics and bioactive compounds in artificial organisms via normal cell division (Pelletier et al. 2021). While CRISPR-Cas9 tools have made genetic manipulation of *Actinomycetes* much easier, there are still concerns—How do dsDNA breaks and astray results cause genome instability? To address these concerns, a break-free ds DNA and single nucleotide with high fidelity-base editing technique for *Streptomyces* spp. has been developed, with clear applications in evaluating editing properties and genome-wide astray effects. The CRISPR-BEST toolkit allows for Csy4-based multiplexing to simultaneously target multiple genes of interest. As a result, CRISPR-BEST is a significant advancement over established genetic manipulation methods for engineering *Streptomyces* spp., remarkably for those strains whose genome cannot be edited using conventional CRISPR-Cas9 (Tong et al. 2019). Existing cultivable antimicrobial gene coding diversity of Actinobacterial members could be potential reason behind their applicability in various field.

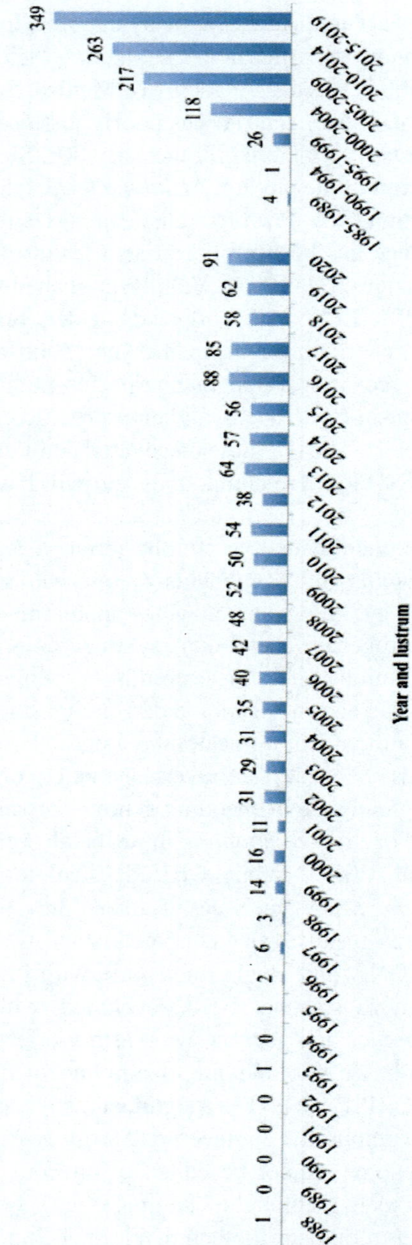

Fig. 11.1 Number of validly published types of material and Lustrum of Actinobacteria type listed by IJSEM The initial genome sequence of *Streptomyces* was published in 2002. *Streptomyces coelicolor*, a model actinomycete, had this genome sequence. *S. coelicolor* has gene clusters for secondary metabolites, although four metabolites were produced under conventional laboratory

11.2 Actinobacterial Reverse Engineering

Reverse engineering of microorganisms allows us to study particular target gene for improvement of efficacy of organisms and achieve high production of particular gene product (Buetti-Dinh et al. 2020). Reverse engineering is critical in the discovery of novel qualities, the enhancement of existing product value, and the promotion of efficient expression methods (Luo et al. 2010). For instance, expression of tylosin biosynthetic gene cluster (85 kb in size) has considerable improved rate of antibiotic synthesis and reduced time of production (Lum et al. 2004). Actinobacteria are kenspeckle group of high G + C containing the dominant consanguineous phyla (Bellassi et al. 2020) that makes their genome unique for investigations. Analyses of the 16S and 23S rRNA genes, the occurrence of conserved indels of some proteins, and characteristic gene rearrangements all help the phylum "Actinobacteria" (Ludwig et al. 2012). Notably, *glnA* serves as an important biomarker for phylogenetic analysis of Actinobacteria in addition to rRNA. Moreover, multiple copies (*glnA, glnA2, glnA3,* and *glnA4*) have been identified in Mycobacteria. Analyses suggest that glnA1 and glnA2 genes were inherited from primitive ancestor of Actinobacteria, whereas glnA4 and glnA3 genes were acquired from during speciation of Actinobacteria. Mycobacteria having *gln* encoded by glnA4 and glnA3 genes experience circumlocution in reductive evolution, whereas those encoded by *glnA1* and *glnA2* are more conserved (Hayward et al. 2009). Pervasive existence in various habitats makes them the most common microflora in soil, freshwater, and marine waters. Being common in the total microbiome, pave diverse functions in the environment include decomposition of cellulosic and chitin containing organic matter (Santos-Júnior et al. 2010), plays a pivotal role in carbon sequestration (Ahmed et al. 2019) and toxic metal – Fe (II) & Cu (II) sequestration (Mohr et al. 2021). Therefore, Actinobacteria have key roles in restoration of soil health and supply micronutrients (Novara et al. 2020).

Fig. 11.1 (continued) conditions, according to this published report. Several genome sequences from the genus *Streptomyces* are currently available. However according to genome mining assessments, fewer than 10% of antibiotic producers' genetic potential is actually being used, signifying that there appears to be an infinite number of genetic reservoirs waiting to be discovered. Furthermore, metagenomic data suggest presence of potential antibiotic producing Actinobacteria in nature just waiting to be discovered and studied. As a result, roughly eight decades after Selman Waksman first described the genus *Streptomyces*, the first antibiotic actinomycin derived from an actinomycete, the genus *Streptomyces* continues to be a goldmine for the discovery of new antibiotics (Mast and Stegmann 2019). Technological advancements in genetic analyses have the ability to hoard genomic DNAs from everything on Earth. New techniques can scoop total DNA and DNA signatures from various organisms and environment. The composition of extracted DNA from any sample mimics their nature in a given time. Scientists are constantly attempting to preserve useful uncultivated or cultivated biotechnologically essential microorganisms. Actinobacteria are famous for their antibiotics producing ability. Apart from developing antibiotics and bioactives, Actinobacteria also code for genes that carry out biochemical reactions

Albeit, Actinobacteria are responsible for enzymatic activities in soil and rhizosphere soil. Apart from this, there are many important places other than freshwater and marine niches where actinobacteria play pivotal roles that include natural places such as volcanic cave, hypersaline soil, desert, and limestone and live organisms, viz. actinmycorrhizal plants, earthworm castings, sponges, and insect guts (Selim et al. 2021). Interestingly, a peculiar odor that comes from dry soil after rain is called as Petrichor (Neff 2018). This smell is an organic compound secreted by some Actinobacteria known as geosmin. Earthy smelling odorant geosmin encoded SCO6073 gene in *Streptomyces coelicolor* strain A3(2) (Jiang et al. 2007). A few members of the phylum *Actinobacteria* include *Corynebacterium* spp. (Yatera and Mukae 2020), *Mycobacterium* spp. (Meijers et al. 2020; To et al. 2020), *Nocardia* spp. (Martínez-Barricarte 2020; Vautrin et al. 2021), and *Rhodococcus* spp. (*Savory* et al. 2020). Despite most important antibiotics and drugs (e.g., Avermectin) producing abilities of *Streptomyces* spp., some strains were reported as pathogenic to certain groups of organisms (Ismail et al. 2020; Evangelista-Martínez et al. 2020).

Streptomyces species, for instance *Streptomyces avermitilis,* possesses a drug called as avermectin. Furthermore, *S. avermitilis* produces a series of 16-carbon pentacyclic macrolactone type-I polyketides with excellent antihelminthic activity against a variety of worm and arthropod species. The production avermectin have reported to be lesser than expected value. Therefore, effective reverse biological engineering was inducted in avermectin coding genes from parent strain 3–115. The avermectin biosynthesis genes were upregulated in the high-producing strain, particularly the pathway-specific regulatory gene *aveR*. The upstream promoter region of *aveR* is expected to recognize *hrdB*. Error-prone PCR was used to create a mutant library of the hrdB gene, which was then selected using high-throughput screening. To prove improved production using reverse biological engineering, a mutant strain A56 was investigated. A56 was grown in a 180-m^3 fermenter to test the production efficiency of mutant strains on large scale, and it proved to be quite productive. As per Fig. 11.2, A56 produced more avermectin than the parent strain, and the output of avermectin *B1a* in A56 was 53% higher than in the parent strain 3–115. This suggested that A56 retained the wild strain's industrial characteristics and could be well suited to large-scale production (Zhuo et al. 2010). This proves that expression of polyketide pathway in *Streptomyces* is harmonizing approach for production of valuable polyketide. Deletion of entire endogenous actinorhodin cluster offers an opportunity for expression of foreign polyketide pathways that improves overproduction and makes *Streptomyces* strains as efficient cell factory (Nah et al. 2013).

Similarly, growing antimicrobial resistance (AR) in microorganisms raised a need of new antibiotics and bioactive compound that helps to tackle AR. Research has been prioritized in many countries. Scientists are extensive searching novel antibiotics to fulfill future need and reduce antimicrobial resistance using new generation tools including reverse engineering, machine learning, and artificial intelligence in antibiotics research. As of today, approximately 60% antibiotics were obtained through either direct culture or genetic engineering from the genus *Streptomyces* spp. (Arefa et al. 2021). Therefore, raised demand will be accomplished by reverse

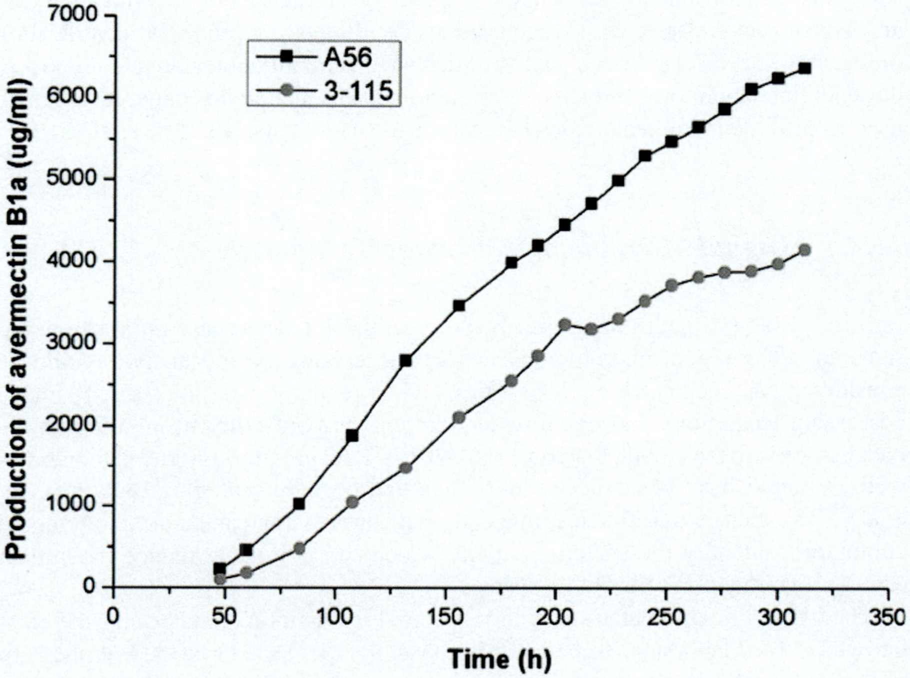

Fig. 11.2 Prospective of avermectin B1a production by wild strain 3–115 and mutant strain A56 (Note: This image has been cited with written permission from PNAS to use original figure for publication) (Zhuo et al. 2010)

engineering of the anaerobic (Barka et al. 2015) and novel uncultivated strains of Actinobacteria (Merino et al. 2020; Jiao et al. 2021). Therefore, gene replacement strategies as mentioned previously becomes an invincible approach for strain improvement and construction of new cells for synthetic biology.

11.3 Machine Learning and Artificial Intelligence Approach

The traditional method of identification of Actinobacteria through culture is a more cost and time intense and exhausting process where chances of human or technical errors cannot be ruled out. The modern technologies like ML, MS, MALDI, and image processing software shall be the alternative to traditional methods of identification of bacteria. These modern technologies are less expensive, less time consuming, and highly accurate. Metabolomics is considered as the modern technology to identification of bacteria. MALDI-MS is the metabolomics tool, which is generally used for identification of the bacterial strains and their species (Pérez-Llarena and Bou 2016). The MALDI-MS is a probe-based technique, which requires very sophisticated instruments and sample preparation expertise for identification and analysis of bacterial species. These machines use fluorescent probe for identification of bacterial species. However, the instruments are generally not available in most of the scientific laboratory, and another limitation of this technique is that, it cannot be used as differentiation techniques for unknown bacterial species (Lee et al. 2010).

11.3.1 Digital Microbiological Image Technology

Bacterial Colony images analysis has the potential for bacterial Identification and screening. The digital microbiological imaging technology and analysis could be considered as dependable diagnostics. The modern computer-aided image processing technology and colony counting software are efficient in counting the bacterial cells in the colonies (Ferrari et al. 2017). Recently, food-borne pathogens as well as Bacilli have been successfully identified by light-scattering tools (Chiang et al. 2015). Hence, machine learning and image processing can be used to train the computer to identify the bacterial pattern, which can ultimately reduce the human error in identification of bacterial species.

The ML-based automated system for bacterial recognition could be considered at par with skilled laboratory technician. Because, the ML based system with the help of the biological images provided by life science research labs could produce more dependable model for recognition of bacteria (Arrigoni et al. 2017). Various bacterial species identification are being done using various techniques, e.g., Hyper Spectral Imaging (HIS) technique and Convolution of Neural Network (CNN) technique. These techniques use Multilayer Perceptron (MLP) automation system for identification/differentiation of bacterial species. These methods work in three steps: (i) Segmentation, (ii) Feature extraction, and (iii) Classification (Srisukkham et al. 2017). The segmentation is the process of cropping the bacterial colony image obtained from the culture plates and creating the Colony Spectral Signature (CSS). CNN can be applied to classify the genera and species of bacteria with 97% accuracy (Zieliński et al. 2017). The Convolution of Neural Network (CNN) technique works in three stages, where initially the image is uploaded and then machines process the input and finally CNN is used to classify the images (Bahrami and Sajedi 2019). To summarize, we can assume that by applying those modern technologies of deep learning, we can achieve the accuracy of a human expert in identification of bacterial species.

11.4 Future Perspectives

Existing collection of cultivated taxa of the phylum Actinobacteria and metagenomic analysis of microbial dark matter shows an evidence that members of the phylum Actinobacteria have potential genes and codes novel bioactive compounds.

Unavailability of appropriate production system or unsettled laboratory protocols become a barrier to get maximum product value from these species. This issue may be addressed by using reserve-engineering approach. Indeed, reverse-engineering need improved ways and analysis to achieve desired product value. Therefore, the enactment of Machine Learning cannot be ruled out in Microbiology and especially in the field of identification and differentiation of bacterial species. Instant live image analysis and AI are all set to replace the existing human intellectual evaluation of identification of bacterial species by observing bacterial colonies and biochemical reactions. Macro-morphological analysis is an important technique among all prevalent technology for differentiation of bacterial species. However, this technology is limited by the phyla of those bacteria that do not produce a typical macro-morphological pattern. The order Actinobacteria produces highly distinct colonies on culture plates; hence the analysis of these bacteria using AI seems incredibly unimaginative. However, the evolution of image processing algorithms will help us in identification and differentiation of Actinobacteria species using AI in future.

11.5 Conclusion

The AI-based tool for microorganisms including Actinobacteria can help to encode complex pathways that may inherit vertically from their last common ancestor or acquired through horizontal gene transfer. Thereby, Actinobacteria are maintaining a continuous flow of genes incorporated in either way. New technologies such as reverse genetics or reverse genome engineering will have great aid to elucidate changes in phenotypes those re-coded for genes (antibiotics and bioactive compounds) artificially. Moreover, advanced genetic engineering tools offer a substantial opportunity to manipulate actinobacteria for biotechnological exploitation.

References

Ahmed AAQ, Odelade KA, Babalola OO (2019) Microbial inoculants for improving carbon sequestration in agroecosystems to mitigate climate change. In: Leal Filho W (ed) Handbook of climate change resilience. Springer, Cham. https://doi.org/10.1007/978-3-319-71025-9_119-1

Arefa N, Sarker AK, Rahman MA (2021) Resistance-guided isolation and characterization of antibiotic-producing bacteria from river sediments. BMC Microbiol 21:116. https://doi.org/10.1186/s12866-021-02175-5

Arrigoni S, Turra G, Signoroni S (2017) Hyperspectral image analysis for rapid and accurate discrimination of bacterial infections: a benchmark study. Comput Biol Med 88:60–71. https://doi.org/10.1016/j.compbiomed.2017.06.018

Bahrami M, Sajedi H (2019) Image concept detection in imbalanced datasets with ensemble of convolutional neural networks. Intell Data Anal 23:1131–1144. https://doi.org/10.3233/IDA-184327

Barka EA, Vatsa P, Sanchez L, Gaveau-Vaillant N, Jacquard C, Meier-Kolthoff JP, Klenk HP, Clément C, Ouhdouch Y, van Wezel GP (2015) Taxonomy, physiology, and natural products of Actinobacteria. Microbiol Mol Biol Rev 80:1–43. https://doi.org/10.1128/MMBR.00019-15

Bellassi P, Cappa F, Fontana A, Morelli L (2020) Phenotypic and genotypic investigation of two representative strains of *Microbacterium* species isolated from micro-filtered milk: growth capacity and spoilage-potential assessment. Front Microbiol 11:554178. https://doi.org/10.3389/fmicb.2020.554178

Buetti-Dinh A, Herold M, Christel S, Hajjami ME, Delogu F, Ilie O, Bellenberg S, Wilmes P, Poetsch A, Sand W, Vera M, Pivkin IV, Friedman R, Dopson M (2020) Reverse engineering directed gene regulatory networks from transcriptomics and proteomics data of biomining bacterial communities with approximate Bayesian computation and steady-state signalling simulations. BMC Bioinformatics 21:23. https://doi.org/10.1186/s12859-019-3337-9

Chiang P, Tseng M, He Z, Li C (2015) Automated counting of bacterial colonies by image analysis. J Microbiol Methods 108:74–82. https://doi.org/10.1016/j.mimet.2014.11.009

Evangelista-Martínez Z, Contreras-Leal EA, Corona-Pedraza LF, Gastélum-Martínez E (2020) Biocontrol potential of *Streptomyces* sp CACIS-15CA against phytopathogenic fungi causing postharvest fruit diseases. Egypt J Biol Pest Co 30:117. https://doi.org/10.1186/s41938-020-00319-9

Ferrari A, Lombardi S, Signoroni A (2017) Bacterial colony counting with convolutional neural networks in digital microbiology imaging. Pat Recognit 61:629–640. https://doi.org/10.1016/j.patcog.2016.07.016

Harz CO (1977–1978) Actinomyces bovis, ein neuer Schimmel in den Geweben des Rindes. Deutsche Zeitschrift für Thiermedizin 5:125–140

Hayward D, Helden PD, Wild IJF (2009) Glutamine synthetase sequence evolution in the mycobacteria and their use as molecular markers for Actinobacteria speciation. BMC Evol Biol 9:48. https://doi.org/10.1186/1471-2148-9-48

Ismail S, Jiang B, Nasimi Z, Inam-ul-Haq M, Yamamoto N, Danso Ofori A, Khan N, Arshad M, Abbas K, Zheng A (2020) Investigation of *Streptomyces scabies* causing potato scab by various detection techniques, its pathogenicity and determination of host-disease resistance in potato germplasm. Pathogens 9(9):760. https://doi.org/10.3390/pathogens9090760

Jiang J, He X, Cane DE (2007) Biosynthesis of the earthy odorant geosmin by a bifunctional *Streptomyces coelicolor* enzyme. Nat Chem Biol 3:711–715. https://doi.org/10.1038/nchembio.2007.29

Jiao JY, Fu L, Hua ZS, Liu L, Salam N, Liu PF, Lv AP, Wu G, Xian WD, Zhu Q, Zhou EM, Fang BZ, Oren A, Hedlund BP, Jiang HC, Knight R, Cheng L, Li WJ (2021) Insight into the function and evolution of the Wood–Ljungdahl pathway in Actinobacteria. ISME J. https://doi.org/10.1038/s41396-021-00935-9

Küster E (1967) International Committee on Nomenclature of Bacteria Subcommittee on Taxonomy of *Actinomycetes*. Int J Syst Bacteriol 17:145–156

Lechevalier HA, Lechevalier MP (1981) The prokaryotes: a handbook of habitat, isolation and identification of bacteria, vol II. Springer-Verlag, Berlin

Lee DY, Bowen BP, Northen TR (2010) Mass spectrometry-based metabolomics, analysis of metabolite-protein interactions, and imaging. BioTechniques 49:557–565

Ludwig W, Euzéby J, Schumann P, Busse HJ, Trujillo ME, Kämpfer P, Whitman WB (2012) Road map of the phylum Actinobacteria. In: Goodfellow M et al (eds) Bergey's Manual® of systematic bacteriology. Springer, New York. https://doi.org/10.1007/978-0-387-68233-4_1

Lum AM, Huang J, Hutchinson R, Kao CM (2004) Reverse engineering of industrial pharmaceutical-producing actinomycete strains using DNA microarrays. Metab Eng 6:186–196. https://doi.org/10.1016/j.ymben.2003.12.001

Luo J, Wang J, Ma TM, Sun Z (2010) Reverse engineering is critical in the discovery of novel qualities, the enhancement of existing product value, and the promotion of efficient expression methods. PLoS ONE. https://doi.org/10.1371/journal.pone.0009182

Martínez-Barricarte R (2020) Isolated *Nocardiosis*, an unrecognized primary immunodeficiency? Front Immunol 11:590239. https://doi.org/10.3389/fimmu.2020.590239

Mast Y, Stegmann E (2019) Actinomycetes: the antibiotics producers. Antibiotics 8:105. https://doi.org/10.3390/antibiotics8030105

Meijers AS, Troost R, Ummels R, Maaskant J, Speer A, Nejentsev S, Bitter W, Kuijl CP (2020) Efficient genome editing in pathogenic mycobacteria using *Streptococcus thermophilus* CRISPR1-Cas9. Tuberculosis 124:101983. https://doi.org/10.1016/j.tube.2020.101983

Merino N, Kawai M, Boyd ES, Colman DR, McGlynn SE, Nealson KH, Kurokawa K, Hongoh Y (2020) Single-cell genomics of novel Actinobacteria with the Wood–Ljungdahl pathway discovered in a serpentinizing sytem. Front Microbiol 11:1031. https://doi.org/10.3389/fmicb.2020.01031

Mohr JF, Baldeweg F, Deicke M, Morales-Reyes CF, Hoffmeister D, Wichard T (2021) Frankobactin metallophores produced by nitrogen-fixing *Frankia* actinobacteria function in toxic metal sequestration. J Nat Prod 84:1216–1225. https://doi.org/10.1021/acs.jnatprod.0c01291

Nah JH, Kim HJ, Lee HN, Lee MJ, Choi SS, Kim ES (2013) Identification and biotechnological application of novel regulatory genes involved in streptomyces polyketide overproduction through reverse engineering strategy. Biomed Res Int. https://doi.org/10.1155/2013/549737

Neff EP (2018) Stop and smell the geosmin. Lab Anim 47:270. https://doi.org/10.1038/s41684-018-0161-1

Novara A, Catania V, Tolone M, Gristina L, Laudicina VA, Quatrini P (2020) Cover crop impact on soil organic carbon, nitrogen dynamics and microbial diversity in a mediterranean semiarid vineyard. Sustainability 12:3256. https://doi.org/10.3390/su12083256

Pelletier JF, Sun L, Wise KS, Assad-Garcia N, Karas BJ, Deerinck TJ, Ellisman MH, Mershin A, Gershenfeld N, Chuang RY, Glass JI, Strychalski EA (2021) Genetic requirements for cell division in a genomically minimal cell. Cell 184:2430–2440. https://doi.org/10.1016/jcell202103008

Pérez-Llarena FJ, Bou G (2016) Proteomics as a tool for studying bacterial virulence and antimicrobial resistance. Front Microbiol 7:410. https://doi.org/10.3389/fmicb.2016.00410

Santos-Júnior CD, Pan S, Zhao XM, Coelho LP (2010) Macrel: antimicrobial peptide screening in genomes and metagenomes. PeerJ 8:e10555. https://doi.org/10.7717/peerj.10555

Savory EA, Weisberg AJ, Stevens DM, Creason AL, Fuller SL, Pearce EM, Chang JH (2020) Phytopathogenic *Rhodococcus* have diverse plasmids with few conserved virulence functions. Front Microbiol 11:1022. https://doi.org/10.3389/fmicb.2020.01022

Selim MSM, Abdelhamid SA, Mohamed SS (2021) Secondary metabolites and biodiversity of actinomycetes. J Genet Eng Biotechnol 19:72. https://doi.org/10.1186/s43141-021-00156-9

Srisukkham W, Zhang L, Neoh CN, Todryk S, Lim CP (2017) Intelligent leukaemia diagnosis with bare-bones PSO based feature optimization. Appl Soft Comput 56:405–419. https://doi.org/10.1016/j.asoc.2017.03.024

Stackebrandt E, Rainey FA, Ward-Rainey NL (1997) Proposal for a new hierarchic classification system, Actinobacteria classis nov. Int J Syst Evol Microbiol. https://doi.org/10.1099/00207713-47-2-479

To K, Cao R, Yegiazaryan A, Owens J, Venketaraman V (2020) General overview of nontuberculous mycobacteria opportunistic pathogens: *Mycobacterium avium* and *Mycobacterium abscessus*. J Clin Med 9:2541. https://doi.org/10.3390/jcm9082541

Tong Y, Whitford CM, Robertsen HL, Blin K, Jørgensen TS, Klitgaard AK, Gren T, Jiang X, Weber T, Lee SY (2019) Highly efficient DSB-free base editing for streptomycetes with CRISPR-BEST. Proc Natl Acad Sci 116:20366–20375. https://doi.org/10.1073/pnas.1913493116

Vautrin F, Pujic P, Paquet C, Bergeron E, Mouniée D, Marchal T, Salord H, Bonnet JM, Cournoyer B, Winiarski T, Louzier V, Rodriguez-Nava V (2021) Microbial risk assessment of *Nocardia cyriacigeorgica* in polluted environments, case of urban rainfall water. Comput Struct Biotechnol J 19:384–400. https://doi.org/10.1016/j.csbj.2020.12.017

Waksman SA (1959) The Actinomycetes—nature, occurrence and activities, vol I. Williams and Wilkins Company, Baltimore, MD

Yatera K, Mukae H (2020) Drastic change in the lung microbiome induced by mechanical ventilation. Respir Investig 58:425–426. https://doi.org/10.1016/j.resinv.2020.07.005

Zhuo Y, Zhang W, Chen D, Gao H, Tao J, Liu M, Gou Z, Zhou X, Ye BC, Zhang Q, Zhang S, Zhang LX (2010) Reverse biological engineering of hrdB to enhance the production of avermectins in an industrial strain of *Streptomyces avermitilis*. Proc Natl Acad Sci U S A 107:11250–11254. https://doi.org/10.1073/pnas.1006085107

Zieliński B, Plichta A, Miszta K, Spurek P, Brzychczy-Włoch M, Ochońska D (2017) Deep learning approach to bacterial colony classification. PLoS One 12:1–14. https://doi.org/10.1371/journal.pone.0184554

Chapter 12
Cultivation and Diversity of Marine Actinomycetes: Molecular Approaches and Bioinformatics Tools

Dalip Singh Rathore, Amit Kumar Sharma, Ankita Dobariya, Hashti Ramavat, and Satya P. Singh

Abstract Marine habitats are still largely unexplored for their microbial diversity and ecosystem analysis. Among the marine microbial population, the actinomycetes are significant component. The actinomycetes are gram-positive bacteria with high G+C contents, displaying filamentous cell morphology and chalky white, powdery texture of the colonies. Isolation and cultivation of the actinomycetes is a difficult and tricky part of their study. Cultivation approaches can be bifurcated into the traditional and high-throughput techniques. Designing suitable media mimicking the in situ native conditions is highly significant to trap the majority of the actinomycetes diversity of the marine habitat. Molecular techniques like ARDRA, DGGE, TGGE and habitat or group specific signature approaches are used to investigate the taxonomy and diversity of the actinomycetes. Nowadays, newer approaches, such as metagenomics and metabolomics, are explored to investigate and analyze community-level diversity and metabolic behavior of the marine actinomycetes. These microorganisms play vital role in the ecosystem and project potential avenues for various applications.

Keywords Marine actinomycetes · Cultivation approaches · High-throughput cultivation · Isolation chips · Microbial traps · Metagenomics and metabolomics

12.1 Introduction

The marine habitats encompass the largest area on the planet harboring largely unexplored microbial diversity. Oceans and sea water constitute majority of the marine habitats and serves as the cradle of the microbial diversity (Fenical et al. 1999; Keller and Zengler 2004; Rappe and Giovannoni 2003; Rathore et al. 2019). Microorganisms of the marine habitats are haloalkaliphilic in nature due to their

---

D. S. Rathore · A. K. Sharma · A. Dobariya · H. Ramavat · S. P. Singh (✉)
UGC-CAS Department of Biosciences, Saurashtra University, Rajkot, Gujarat, India

L. Karthik (ed.), *Actinobacteria*, https://doi.org/10.1007/978-981-16-5835-8_12

abilities to survive and grow under the dual extremities of alkaline pH and high salinity (Bruns et al. 2003; Connon and Giovannoni, 2002; Schut et al. 1993). Marine habitats are minimally explored; however, during the last few decades, many studies are reported in the literature. Majority of such studies have focused on the bacterial diversity leading to the description and discovery of number of new halophilic microbial species and genera (Kurtböke 2012; Lazzarini et al. 2000; Tiwari and Gupta 2012). Actinomycetes of marine origin possess promising pharmacologically important compounds (Gartner et al. 2011; Hughes et al. 2009; Jensen et al. 2007). Marine actinomycetes are tedious to isolate and cultivate as it depends on the enrichments, sample pretreatment, designing, and selection of growth media (Jensen et al. 1991, 2005; Rathore et al. 2019; Sheikh et al. 2019).

Marine actinomycetes are reported from diverse habitats like the marine sediments, soil, coastlines, deep sea region, in association with the marine fauna, soda lakes, brines, and other saline habitats (Maldonado et al. 2005b; Mincer et al. 2002; Pathom-aree et al. 2006; Stach et al. 2003; Weyland 1969). Actinomycetes of these habitats possess efficient survival mechanisms and ability to grow at the alkaline pH and high salt concentrations. Thus, majority of the marine actinomycetes are physiologically "haloalkaliphilic" in nature. Several genera like *Streptomyces, Nocardiopsis, Micromonospora, Frankia, Pseudonocardia, Prauseria, Salinispora, Aeromicrobium, Salinibacterium, Streptosporangium, Actinomadura, Streptomonospora, Saccharopolyspora, Rhodococcus, Nonomuraea, Gordonia, Actinopolyspora,* and *Marinophilus* are reported from the saline habitats (Gohel and Singh 2018a, b; Gontang et al. 2007; Jensen et al. 2005; Kennedy et al. 2009; Maldonado et al. 2005a; Mincer et al. 2005; Sharma et al. 2021; Sheikh et al. 2019; Rathore et al. 2021a, b). Halophilic actinomycetes are described as moderate halophiles (requiring 15% NaCl) and extreme halophiles (requiring 30% NaCl). The survival mechanisms of the moderate and extreme haloalkaliphilic actinomycetes into the marine environment include "high salts-in" and "organic salts in" to enable the cellular machinery to function properly (Maldonado et al. 2005a, 2009; Zhang et al. 2006). Sodium, potassium, and chlorides help the prokaryotic cell in the maintenance of the osmotic pressure (Selvin et al. 2004).

12.2 Cultivation Strategies and Diversity of the Marine Actinomycetes

Since only limited knowledge is available on the isolation and cultivation of the haloalkaliphilic actinomycetes, it is quite difficult to trap the diversity of such microorganisms of the marine environment (Epstein 2013; Hameş-Kocabaş and Ataç 2012; Kurm et al. 2019; Nichols 2007; Pham and Kim 2012; Ward and Bora 2006). Approximately, 99.9% of microorganisms are yet to be cultivated (Akbari et al. 2014; Kikani et al. 2017; Purohit and Singh 2013; Raiyani and Singh 2020). Staley and Konopka (1985) proposed a term "great plate count anomaly" which

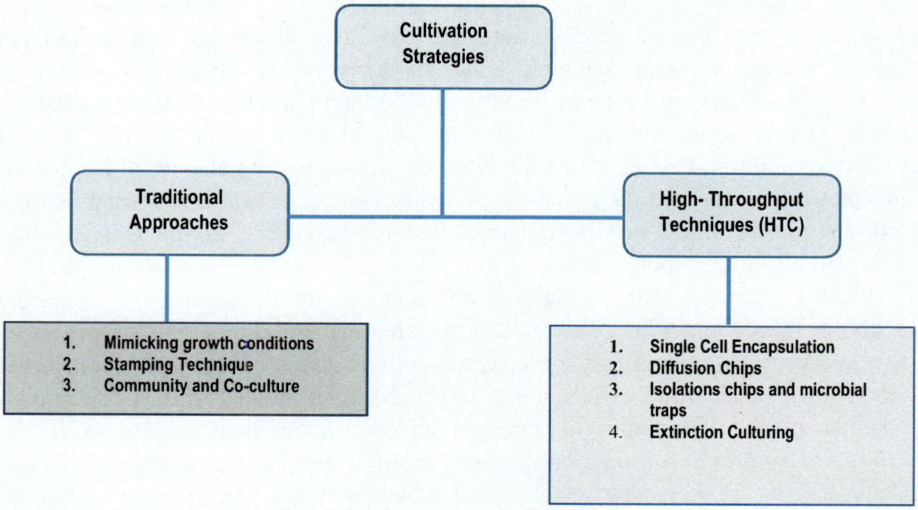

Fig. 12.1 Different cultivation techniques for the marine actinomycetes

states "the vast discrepancy between the number of cells into an environmental sample and the colonies formed on the traditional media."

In recent past, attempts have been made to ease the restrictions of non-cultivability through various approaches and tools (Gohel and Singh 2018a, b; Maldonado et al. 2005b; Mincer et al. 2002; Pathom-aree et al. 2006; Rathore et al. 2019; Sheikh et al. 2019; Sharma et al. 2020, 2021; Stach et al. 2003; Weyland 1969). Synergistic approaches have proved highly significant in the cultivation of marine actinomycetes rather than depending on a single method. In the cultivation of marine actinomycetes, the pretreatment of the samples is important. Pretreatments of the samples include heat treatment, serial dilution, calcium carbonate treatment, enrichment, shaking with glass beads, phenol treatment, and filtration techniques. Cultivation techniques for the isolation of actinomycetes can be categorized into two parts (Fig. 12.1):

1. Traditional Techniques
2. High-Throughput Cultivation (HTC)

12.2.1 Traditional Techniques

The traditional approaches of the cultivation do not require any advanced tools and techniques and thus can be implemented with limited resources.

The simple cultivation technique includes the designing of the growth media in various combinations of nutrients in view of the environmental conditions of the particular habitat (Gandhimathi et al. 2008; Kennedy et al. 2009). Earlier, it was observed that cultivation of microorganisms from sea water requires comparatively lower concentrations of nutrients into the growth medium and incorporation of certain selected nutrients captured better diversity of the organisms (Magarvey et al. 2004). Excess of nutrients into the growth medium inhibited or restricted the appearance of actinomycetes, a condition referred as "substrate killing." Use of undefined and blind media resulted into monotonous growth behavior of the actinomycetes with much lesser diversity. A better choice is to initially establish the nutrient profile of the particular habitat before proceeding to the isolation and cultivation of the actinomycetes.

Phenol treatment of the samples is an efficient way to culture the rare marine actinomycetes. Phenol being a biocide kills other bacteria, fungi, and *Streptomyces* species, therefore supporting the propagation of rare genera of marine actinomycetes (Hayakawa et al. 1991a; Nonomura 1988). Samples treated with 1.5% phenol resulted in the cultivation of rare genera like *Actinomadura, Microbispora, Micromonospora, Nocardia*, and *Nonomuraea* from the sea water (Hayakawa et al. 1991b, 2004; Istianto et al. 2012; Khamna et al. 2009; Qiu et al. 2008). Incorporation of gentamicin antibiotics into the growth medium supports the growth of rare genera of marine actinomycetes (Okami 1988; Qiu et al. 2008).

Treatment of the marine sediment samples with the calcium carbonate also favored the growth of rare marine actinomycetes genera (Alferova and Terekhova 1988; Natsume et al. 1989). While the exact reason behind this support is not yet clear, calcium ions may influence the pH, affecting the propagation of alkaliphilic actinomycetes (Otoguro et al. 2001; Tiwari and Gupta 2012). Another school of thought assumes the induction of mycelia formation by calcium on the solid substratum favoring the growth of rare actinomycetes (Subramani and Aalbersberg 2013).

Previously, the treatment of marine sediment samples with microwaves was reported to enhance growth of actinomycetes (Bulina et al. 1997; Wang et al. 2013; Xue et al. 2010; Yang et al. 2008). Microwaves irradiate the fungal and other undesired microorganisms and thus facilitating the growth of the rare genera of marine actinomycetes (Miguélez et al. 1993; Niyomvong et al. 2012).

Centrifugation of samples also increases the chance of rare actinomycetes genera to grow. This method selectively segregates the non-motile microorganism like fungi and *Streptomycetes* species (Hayakawa et al. 2000; Hong et al. 2009; Qin et al. 2009). Similarly, applications of chemo-attractants and chlorinated compounds for the isolation of the actinomycetes have proved significant to enhance the growth rare actinomycetes (Hayakawa 2008). The motile spores of *Actinoplanes, Dactylosporangium*, and *Catenuloplanes* are efficiently attracted by the xylose, chloride, bromide, and vanillin (Hong et al. 2009).

In the stamping technique, the marine sediment samples are dried in a laminar airflow and thereafter grounded into powder. The grounded powder is used to inoculate the growth media with the help of nylon swab where each of the fibers acts as individual inoculants (Gontang et al. 2007; Jensen et al. 2005; Mincer et al. 2002). Moreover, use of this swab in continuation manner provides the dilution effect where the chances of propagation of individual cells increases.

In yet another approach of co-culturing, the probability of the isolation of rare actinomycetes from marine environment is enhanced. D'Onofrio et al. (2010) established that cultivation of two microorganism in close proximity supports the growth of the individual organism. Uncultured bacterium *Maribacter polysiphoniae* was cultivated in the presence of helper strain *Micrococcus luteus*. Similarly, *Bacillus megaterium* supported the growth of other uncultivable bacteria (D'Onofrio et al. 2010; Kaeberlein et al. 2002; Nichols et al. 2008). Interspecies communication in microorganisms is well known and they communicate with each other by quorum sensing. Another phenomenon responsible for this mutual growth is co-metabolism. Some species are not able to grow on precursor metabolites and they require intermediate compounds for their growth (Stewart 2012). This phenomenon is usual in the bioremediation using consortia where compounds degraded by one species are utilized by the successive species. However, co-culture is successful in the minimal media where the concentrations of nutrients are comparatively low (McInerney et al. 2008).

12.2.2 High-Throughput Cultivation Techniques

Advanced molecular techniques and microscopic examination of samples have suggested that a large portion of the microbial load is still uncultivable by the conventional techniques. Conventional techniques like enrichments and pure culture approaches often yield fast-growing organisms. High-throughput cultivation is an advanced and novel technique where bacteria in samples are separated and incubated individually. In this approach, the bacteria are individually propagated so they are least affected by the fast-growing or weedy bacteria. Moreover, low concentrations of nutrients mimic the natural environmental habitat and ultimately enhancing the chances of rare actinomycetes. High-Throughput Cultivation (HTC) can be categorized into following segments.

12.2.2.1 Extinction Culturing

The extinction culturing method is a powerful tool for culturing the rare marine actinomycetes. In this method, the marine sediment samples are diluted in such a way that only 1–10 cells remain in individual microcells (Button et al. 1993; Henson et al. 2020; Tiwari and Gupta 2012). Thereafter, the cultures are analyzed by flow cytometry. Undiluted samples in large quantity can be used (Benítez et al. 2021; Bruns et al. 2003). The major drawback of the method is fast-growing bacteria suppress the growth of slow grower and therefore the proper dilution is recommended for the isolation of rare species (Rappé and Giovannoni 2003; Stingl et al. 2007).

12.2.2.2 Encapsulation of Single Cells

In this method, the samples are distributed in small quantities and environmental cells are encapsulated into gel microdroplets for large-scale actinomycetes cultivation under low nutrient conditions. It follows the detection of microcolonies using flow cytometry (Czechowska et al. 2008; Hu et al. 2021; Lewis et al. 2020; Pensold and Zimmer-Bensch 2020). In this process, the cells from environmental samples are concentrated using gradient centrifugation. Initially, the cells are mixed with the melted agar followed by the mixing it with oil and microdroplets developed ranging between 20 and 70 μm (Hu et al. 2020; Lewis et al. 2020; Wang et al. 2021; Zengler et al. 2002). The cells concentration is very critical before development of microdroplets (Hengoju et al., 2020; Keller and Zengler, 2004; Toledo et al. 2006; Wang et al. 2021; Zhang et al. 2017). These microdroplets are then incubated with nutrient medium after proper washing and filtration. In this way, the uncultivable microorganisms can be cultivated using proper growth medium. The microcapsule can be examined by fluorescence-activated cell sorting (FACS). The slow grower organisms can be distinguished from fast grower in such a way that the fast grower will propagate very promptly and burst the capsule (D'Onofrio et al. 2010; Hu et al. 2021; Pensold and Zimmer-Bensch 2020; Zengler et al. 2002). In this method, the microorganisms involved in cell communication can also be propagated as the nutrient are free to move from one capsule to another.

12.2.2.3 Diffusion Chambers, Isolation Chips, and Microbial Traps

The motive of cultivable approaches is to provide almost natural conditions to the microorganisms into in vitro conditions. It is very difficult to provide almost natural conditions due to the limitation of laboratory conditions. Such techniques employ the diffusion chambers, isolation chips, and microbial traps to meet the natural conditions. In diffusion chamber technique, the environmental samples are serially diluted with the sea water, blended with agar, and incubated in diffusion chambers for the propagation (Chaudhary et al. 2019; Kaeberlein et al. 2002; Marcolefas et al. 2019). The diffusion chambers are then incubated in naturally simulated environmental conditions and possibly in the natural habitat from where the samples were collected (Jung et al. 2018).

Microbial traps are a type of diffusion chambers designed to specifically grow the filamentous microorganisms such as actinomycetes. In this method, two different pore sized agar layers are made where thick larger and smaller pore permits the growth of actinomycetes in natural conditions (Kaeberlein et al. 2002; Luo et al. 2020; Zhang et al. 2005). Once the filamentous actinomycetes appeared in micro colonies, the slice of that particular agar is put on another growth media for the propagation of the individual colonies. This method allows every actinomycetes to grow in the natural habitat. The media and growth conditions are created as per the requirement of the habitat (Gavrish et al. 2008; Lewis et al. 2020; Lodhi et al. 2020).

Isolation chips (ichips) are more sophisticated and precisely made diffusion chambers. Generally, the ichips are made up of thousands of small sized pores where the individual cells can stick and thereafter pores are covered with semipermeable membranes where the nutrients penetrate but the cells are held in particular pores, allowing the individual microorganisms to develop into microcolonies (Alessi et al. 2018; Berdy et al. 2017; Lewis et al. 2020; Liu et al. 2021). The initial sample load is very crucial in the successful operation of ichips. It is advised to possibly dilute the samples to a smallest cell counts and then apply on the sophisticated isolation chips to get slow grower and rare species. Once the uncultivable species appear on the chips, it can be grown on the conventional petri dishes after repeatedly transfer (Cantrell et al. 2017; Li et al. 2020; Nichols et al. 2010).

12.3 Phylogenetic and Phenotypic Analysis of the Marine Actinomycetes

Cultivability provides a comprehensive account about the phenotypic characteristics of the microorganisms (Demain and Sanchez 2009; Newman and Cragg 2007). Genetic traits, phylogenetic relatedness, and molecular identification are represented in the phylogenetic tree, while the phenotypic characteristics are either described in text format or Phenogram. Phenotypic characteristics represent the expressive behavior of the marine actinomycetes while phylogenetic trends are useful in the identification of the microorganisms. Phenograms represent phenotypic characteristics of the organisms in the form of clusters for the assessment of the diversification.

The phylogenetic characteristics of the marine actinomycetes can be adjudged on the basis of the 16SrRNA gene sequences, Denature Gradient Gel Electrophoresis (DGGE), Temperature Gradient Gel Electrophoresis (TGGE), Random Fragment Length Polymorphisms (RFLP), Amplified Ribosomal DNA Restriction Analysis (ARDRA), and other genetic techniques (Bredholdt et al. 2007; Gohel and Singh 2018a, b; Mutka et al. 2006). The analysis of the microbial diversity by ARDRA and DGGE displays significant heterogeneity among the identical species.

Metabolic investigations of the marine actinomycetes include pH profile, salt tolerance, sugar fermentation capabilities, various substrates utilization, antibiotic susceptibility, antagonistic activities, morphological features, cultural characteristics, secretion of the extracellular enzymes, cell wall sugars, and polar lipids analysis (Cross 1981; Gohel and Singh 2018a, b; Rathore et al. 2019, 2020, 2021b; Sharma et al. 2020, 2021; Sheikh et al. 2019; Vasavada et al. 2006). Marine actinomycetes are least studied with respect to the phenotypic characteristics. The disadvantage of the non-cultivable approach is inability of the cultivation of species (Keller and Zengler, 2004; Rappé and Giovannoni 2003). In most of the cases, the sequenced species are not identified due to the lack of reference data availability. Thus, cultivability provides the reference baseline for the identification of the marine microorganisms. There are instances of genetically similar species being significantly different in their cultivable behavior. Same species can display different morphology on different nutrient media. Media variation leads to the changing patterns in pigmentation, intensity in the aroma, colony appearance, and mycelial structures. The extent of "halophily" is highly regulated by the cellular machinery and adaptation strategies. The cultivability of a particular microorganism also relies on the thermodynamic and kinetic feasibility (Berdy 2005; Harrison et al. 2013; Oren 2011; Rathore et al. 2021a, b; Schuchmann and Müller 2014). Majority of the microorganisms present in the sea water fail to grow due to lack of suitable conditions and absence of the association with the hosts or some other organisms (Demain and Sanchez 2009; Oren 2011; Rathore et al. 2021a ; Sheikh et al. 2018). The identical species of *Nocardiopsis alba* differ in their sugar fermentation capabilities, hydrogen sulfide production capabilities, citrate utilization pattern, and casein hydrolysis properties (Gohel and Singh 2018a, b; Rathore et al. 2021b; Thakrar and Singh 2019; Thakrar et al. 2018).

The List of Prokaryotic Names withstanding in nomenclature (LPSN) suggests about the addition of many species of actinomycetes in recent past (Parte 2014). Several genera are incorporated for their better descriptions and demarcation (Parte 2018; Parte et al. 2020). Further, marine actinomycetes require more inorganic salts in comparison with the organic compounds, a trait developed probably due to the composition of sea water. Phenogram and phylogram based studies of the marine actinomycetes provide excellent approach for the diversity analysis. Phenogram are constructed using UPGMA method in PAST software by converting the phenotypic data into binary matrix (Bhatt et al. 2018). "Phenons" are defined by applying threshold similarity limits and thereafter clusters are defined and compared with their phylograms (Rathore et al. 2021b). A significant difference between the phylogenetic and phenotypic characteristics has been observed in genetically identical actinomycetes. The species of the same phylogenetic clusters grouped in different clusters in the Phenogram as they possess different expressive behavior. *Nocardiopsis* and *Streptomyces* genera are reported as dominant in the Arabian Sea (Gohel and Singh 2018a, b; Rathore and Singh 2021; Rathore et al. 2021b; Sharma et al. 2020, 2021).

12.4 Metagenomics and Metabolomics Approaches for the Analysis of Microbial Diversity

12.4.1 Metagenomics Approach

Microorganisms in marine habitats reflect variability in their diversity, cellular and physiological adaptations, molecular strategies, and genetic heterogeneity. The microorganism from such habitats are significant source of enzymes, antibiotics, and other secondary metabolites. However, only 0.001–0.1% of the microorganisms from marine habitat have been studied and explored (Alma'abadi et al. 2015; Beale et al. 2017; Kikani and Singh 2021; Raiyani and Singh 2020; Rathore et al. 2019; Rathore and Singh 2021; Raval et al. 2014; Sheikh et al. 2019; Turnbaugh and Gordon 2008; Weiland-Bräuer et al. 2017). Metagenomics is an approach to study the uncultivable microbial diversity by direct extraction of environmental DNA. It provides a new platform for the exploration of novel biocatalysts and genes having potential biotechnological applications (Alma'abadi et al. 2015, Chauhan and Gohel 2020; Chauhan et al. 2021; Heidelberg et al. 2010; Purohit and Singh 2009; Raiyani et al. 2020). DNA sequencing tools and polymerase chain reaction (PCR) have opened new opportunities to understand the uncultivable communities, their phylogenetic diversity, metabolic traits and function in the marine biogeochemical cycle (Abid et al. 2018; Akbari et al. 2014; Beale et al. 2017; Singh and Pelaez 2008). Metagenomic DNA can be expressed by cloning in a heterologous host such as *Escherichia coli* and *Streptomyces lividans. Escherichia coli* is used to express partial biosynthetic pathway, whereas *Streptomyces lividans* is used to complete biosynthetic pathway (Charlop-Powers et al. 2014; Singh and Pelaez 2008).

Metagenomics acquired significance due to advancement in sequencing and thus our abilities for the sequence and function-based analysis of the microbial communities (Fig. 12.2). The first approach deals with the profiling of cDNA, identification of potential target sequences, and analysis of the diversity. Whereas the second approach includes the construction of metagenomic library and identification of active genes by following their expression (Charlop-Powers et al. 2014; Felczykowska et al. 2012; Kikani et al. 2017; Mahapatra et al. 2019). Two prominent methods used in sequence-based approach include PCR and hybridization-based techniques, both dependent on conserved DNA sequences of the databases. Due to some limitation of the function-based and sequence-based approaches, the next-generation sequencing is widely used for metagenome sequencing (Barone et al. 2014; Douglas et al. 2020; Felczykowska et al., 2012; Kumar Awasthi et al. 2020; Nguyen et al. 2020). The metagenomic information obtained from the sequencing system further needs to be processed by appropriate bioinformatics tools (Barone et al. 2014; McCombie et al. 2019; Gu et al. 2021; Siddhapura et al. 2010; Slatko et al. 2018).

Metagenomics played a crucial role in discovering many novel extremozymes such as esterase, proteases, laccase, mercuric reductase, protease, lipase, glycosidase, and hydrolases (Dangar et al. 2017; Edet et al. 2017; Jeon et al. 2009; Purohit and Singh 2013; Sharma et al. 2020, 2021; Thumar and Singh 2009). Cold adapted enzymes have excellent activity at lower and moderate temperature than their mesophilic and thermophilic counterparts. By constructing and using the metagenomic library followed by its screening, a novel esterase Est97 was isolated from the Arctic intertidal zone (Jeon et al. 2009). This enzyme displays 60% relative activity at 20 °C and lower thermostability, indicating its potential utilization in cold environment (Fu et al. 2013). Further, marine habitat is an excellent source of halophile, thermophile, hyperthermophile, and barophiles (Barone et al. 2014; Rathore et al. 2020, 2021a, b; Sharma et al. 2020, 2021; Thumar and Singh 2009). Long back, metagenomics approach was employed to analyze and characterize the Sargasso Sea ocean microbial community by using complex bioinformatics tools

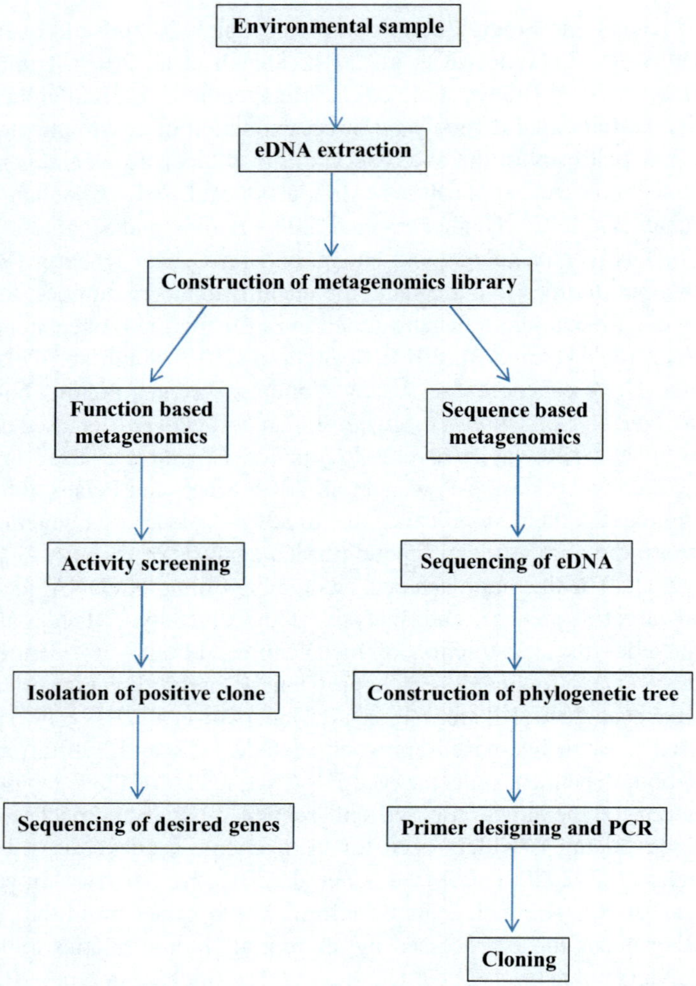

Fig. 12.2 Metagenomics approach for the microbial diversity study

(Abid et al. 2018). The study revealed 1800 different genomes, 48 unknown bacterial phylotypes, and 1.2 million unknown genes. Based on shotgun sequence analysis, they identified novel proteorhodopsin like genes from an outside group of proteobacteria (Abid et al. 2018; Chu et al. 2008). Marine microorganisms are huge sources of various compounds, and according to one estimate, 70% originating from actinobacteria. During the recent years, metagenomic techniques have been employed for the genome sequencing of marine actinomycetes and exploration of their metabolic potential. The Next-Generation Sequencing provides two distinct sides for the study of various biomolecules from actinomycetes: unexploited biosynthetic pathways of the uncultivable actinomycetes and uncharacterized/unknown biosynthetic pathways of the cultivable actinomycetes (Mahapatra et al. 2019; Niu 2018). More recently, the bacterial diversity based on the metagenomics approach of

Kachhigadh (Shivrajpur), Dwarka and Alang-Sosiya ship breaking yard, Bhavnagar of the Coastal Gujarat (India) has been investigated (Raiyani and Singh 2020). The study revealed a total 27 different classes from Kachchigadh and 65 from the Alang sea water, with 1.45% being the actinobacteria. Next-Generation Sequencing and bioinformatics tools are essential for the success of metagenomics studies and developing new platforms to investigate marine actinobacterial communities (Mahapatra et al. 2019).

12.4.2 Metabolomics Approach

Metabolomics is a different approach from other omics such as metagenomics, metatranscriptomics, and metaproteomics (Abid et al. 2018). This approach is applied for the detection of primary and secondary metabolites in living systems (Ali and Farag 2020; Llewellyn et al. 2015). Lipids, carbohydrates, and amino acids are primary metabolites and play pivotal roles in the cellular structures, growth, and basic metabolisms. While, on the other hand, the secondary metabolites, such as pigments, antimicrobial compounds, and other molecules, help the organisms for the survival and sustenance in their specific ecological niche (Lee et al. 2004; Llewellyn et al. 2015). Due to variable chemical nature and different physical and chemical characteristics of the metabolites, there is no specific analytical technique to determine and analyze all metabolites. Some of the communally employed tools and techniques are nuclear magnetic resonance (NMR) spectroscopy and liquid chromatography (LC) or gas chromatography (GC) allied with mass spectroscopy (MS) for profiling of the broad array of metabolites (Abid et al. 2018; Bose et al. 2015; Goulitquer et al. 2012). ^{1}H NMR is used in quantitative analysis for profiling the metabolites from various habitats. While LC-MS is helpful in identification of the components with better resolution and higher sensitivity. The main advantage of this technique is no requirement of the derivatization of the detectable molecules (Ali and Farag 2020; Gowda and Raftery 2014; Joghee and Jayaraman 2014; Macintyre et al. 2014).

Metabolomics (Fig. 12.3) has provided new opportunities to discover natural products, explore the chemical diversity, and identify biological and molecular mechanism for the production of the metabolites (Abid et al. 2018; Reverter et al. 2020). Marine biodiversity is a rich source of novel biological compounds such as antibacterial, antifungal, antiviral, antitumor, and other therapeutic agents. The actinomycetes in particular are known to produce wide range of bioactive compounds and till date approximately 7500 bioactive compounds have been extracted from *Streptomyces* spp. (Forner et al. 2013; Ong et al. 2019; Reverter et al. 2020). The marine *streptomyces* strains have been reported for their antitumor activity against cancer cell line with reduced cytotoxicity against normal cell line (Sudha and Masilamani 2012). Genus *Salinispora* is a marine actinomycetes, cultivated from the tropical and subtropical marine sediments. These actinobacterial species produce important pharmaceutical secondary metabolites such as anticancer and

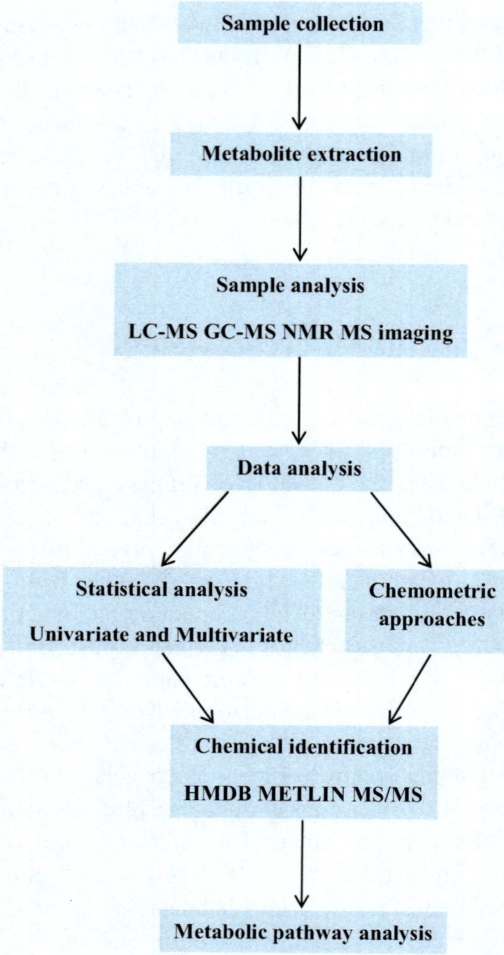

Fig. 12.3 Metabolic approach for the microbial diversity study

antibacterial components. *Salinispora* has been reported as the first obligate marine actinomycetes based on their growth condition. This genus is unable to grow in the absence of salt. However, few years ago, one new species has been found which can survive at a lower concentration of salt (Bose et al. 2015).

Metagenomics and metatranscriptomics studies of Tara Oceans project was unable to completely identify the activities of marine microorganisms and contribution of their biomolecules (Sogin et al. 2019). Along with statistical tools, metabolomics can discover the relationship between metabolites, organisms, physiology, and environment (Llewellyn et al. 2015). On the basis of GC-MS and statistical methods, 727 metabolites were identified in an Australian marine habitat (Beale et al. 2017). Among these metabolites, 38 were common in all study sites, and 17 being statistically variable. LC-MS identifies various metabolites produced by marine actinomycetes strain *Salinispora arenicola* isolated from Great Barrier Reef sponges (Bose et al. 2015). Based on the metabolomics approach, the production of rifamycins from an actinomycetes strain was investigated and the variation in the production of secondary metabolites as function of salt concentrations monitored. The metabolomic studies of the marine habitats, however, are limited by the interference in the MS process by high salt concentrations in the samples, which decreases the reproducibility of the extraction process. The polysaccharides and pigments of the seaweeds are removed by the organic solvents during the extraction process (Goulitquer et al. 2012). The metabolomics of marine habitats needs further investigations for better understanding of the metabolic interaction, signaling pathways, and the chemical complexity (Ghosh et al. 2017).

12.5 Marine Actinobacteria: Diversity, Molecular Signature/s, and Bioinformatics Tools

The Ocean represents an ecosystem with many unique forms of *Actinomycetes* (Sharma et al. 2020, 2021). The culture-dependent and culture-independent diversity of the marine actinomycetes, *Salinispora*, *Nocardiopsis*, *Streptomyces,* and others are reported (Datta et al. 2017; Dangar 2018; Gohel et al. 2018; Mincer et al. 2005; Raiyani and Singh 2020; Sharma and Singh 2016; Sharma 2017; Sharma et al. 2021) from various habitats. *Actinomycetes* possess a relatively larger genome size as compared to bacteria. The molecular signatures and genomes of the phylum Actinobacteria has been earlier described (Gao and Gupta 2012; Ventura et al. 2007). The variation in the diversity of halophiles is directly or indirectly associated with the genomic profiles. The halo/alkaliphilic organisms display characteristic profiles distinct from those of the non-halophilic organisms, such as the specific genomic signatures for salt adaptation (Paul et al. 2008; Raval et al. 2018).

There are few reports of Actinobacteria based on complete genome sequences that can tolerate different salt concentrations in the range of 5–20% and variable alkaline pHs. Some of the representative actinomycetes include *Saccharomonospora marina, Salinispora tropica, Nocardiopsis salina, Nocardiopsis alkaliphila, Saccharomonospora marina*, and *Nocardiopsis halophila* (AL-Tai and Ruan 1994; Wen-Jun et al. 2004; Hozzein et al. 2004; Klenk et al. 2012; Li et al. 2020; Udwary et al. 2007).

Horizontal and lateral gene transfers are common features in bacterial genomes making it dynamic (Brito 2021; Moon et al. 2019; Lawrence Jeffrey and Hendrickson 2005; Ochman et al. 2000). The phylum *Actinobacteria* represents one of the biggest phyla comprising vast species from free-living to complex environments. The size of the genomes varies from 5 to 9 MB (Chandra and Chater 2014; Ventura et al. 2007). Rather small genomes have been reported for several *Actinobacterial* species dwelling in harsh environments (Tiwari and Gupta 2013; Shivlata and Tulasi 2015). A database on the molecular identification of the phylum *Actinobacteria* was earlier reported (www.actionobase.in; Sharma et al. 2012). This phylum is highly diverse and is generally identified based on the branching patterns of the 16S rRNA genes (Gohel et al. 2018). As per an earlier report, the phyla *Actinobacteria* have specific characteristics of homologous insertion of 100 nucleotides in between the helices of 54 and 55 of the 23S ribosomal gene (Roller et al. 1992, 1994). Looking into the recent literature status, there is imminent need for investigating genomic patterns and specific molecular markers for the major clades of actinobacteria. The molecular markers and probes specific to these microorganisms will serve the purpose of identifying novel taxa/genus/species.

Genomic fingerprinting has gained importance in the investigation of the closely related species (Kikani et al. 2015, 2017). Further, different molecular markers, mainly in the internal transcribed spacer region (Ferris et al. 2003), denaturing gradient gel electrophoresis (DGGE) (Gohel and Singh 2018a, b) and amplified ribosomal DNA restriction analysis (ARDRA) (Sharma et al. 2021) are significant tools in investigating the actinomycetes. Various other effective tools include fluorescence in situ hybridization (FISH), (AFLP) amplified-fragment length polymorphism analysis, multi-locus enzyme electrophoresis (MLEEC), rep-PCR genomic fingerprinting, -restriction fragment length polymorphism (T-RFLP), and 16S rRNA sequencing-based phylogenetic analysis (Friedrich et al. 2001; Fulthorpe et al. 1998; Gohel et al. 2015, 2018; Margot et al. 2002; Lee et al. 2003; Rademaker et al. 2000; Whitaker et al. 2003).

Metagenomics, 16S rRNA-RFLP, and other culture-independent techniques were recently developed and adapted to study the microbial diversity including the actinobacterial communities that exist independently or in association with *spongs* and *Songe* sp. (Raiyani and Singh 2020; Xin et al. 2008; Zhang et al. 2008). The phylogenetic affiliation was analyzed on the basis of the patterns generated by *HhaI* digestion of the of 16S r RNA gene sequences of the culturable actinomycetes associated with five marine sponges. Similarly, *Actinobacterial* diversity of the western coast of Arabian Seawater and solar salterns in Tuticorin, India, was analyzed using 16S rDNA-ARDRA to distinguish various taxonomic groups (Gohel et al. 2015; Gohel and Singh 2018a, b; Jose and Jebakumar 2015; Sharma et al. 2021; Singh et al. 2013).

The evolutionary and phylogenetic relationship of the members of *Actinomycetes* depends on the distinction among the species of the same genera of this phylum. Minor reorganization and reclassification of the phylum *Actinomycetes* taxonomy was published by Zhi et al. (2009), based on the signature sequences (16S rRNA) among the ranks of the family and subclass of the *Actinobacteria*. This phylum comprises 219 genera in 50 families (Ludwig et al. 2012; Baltz et al. 2010; Zhi et al. 2009). Many strains of the Algerian arid soils displayed similarities of <99%, clearly indicating the existence of new species. Based on the 16S rDNA sequence analysis, *Streptomyces* is the dominating genus in the Indian salterns. Besides *Streptomyces* and *Nonomuraea* genus, *Nocardia*, *Nocardiopsis*, and *Saccharopolyspora* are also reported from the saline habitats based on the phylogenetic analysis (Gohel and Singh 2018a, b; Gohel et al. 2015, 2017, 2018; Jose and Jebakumar 2015; Sharma et al. 2021).

Acknowledgments The work cited in this review from the SPS laboratory at the Saurashtra University was supported under the UGC-CAS Program, DST-FIST, DBT-Multi-Intuitional Project, MoES (Government of India) Net Working Project, and the Saurashtra University. SPS acknowledges DST-SERB International Travel Fellowships to present his work in Hamburg (Germany), Cape Town (South Africa), and Kyoto (Japan). AKS is grateful for the UGC Basic Science Research (BSR) Fellowship and International Travel Grants from DST-SERB, DBT, and Centre for International Co-operation in Science (CICS) to present his work in Canada. A.D. acknowledges the award of SHoDH Fellowship by the Gujarat Government. SPS also acknowledges award of UGC BSR Faculty Fellowship. DR acknowledges JRF/SRF under the MoES-Net Working Project, New Delhi.

References

Abid F, Zahid MA, Abedin ZU, Nizami SB, Abid MJ, Kazmi SZH, Khan SU, Hasan H, Ali M, Gul A (2018) Omics approaches in marine biotechnology: the treasure of ocean for human betterments. In: Debamalya B, Vasco A (eds) Omics technologies and bio-engineering. Academic Press

Akbari VG, Pandya RD, Singh SP (2014) Extraction of the metagenomic DNA and assessment of the bacterial diversity from the petroleum-polluted sites. Environ Monit Assess 186(10): 6351–6362. https://doi.org/10.1007/s10661-014-3859-6

Alessi AM, Redeker KR, Chong JP (2018) A practical introduction to microbial molecular ecology through the use of isolation chips. Ecol Evol 8(24):12286–12298. https://doi.org/10.1002/ece3.4748

Alferova IV, Terekhova LP (1988) Use of the method of enriching of soil samples with calcium carbonate for isolation of actinomyces. Antibiot Chemother 33(12):888–890

Ali SE, Farag MA (2020) Expanding metabolomics applications to address issues in marine ecology and natural products chemistry. Encycl Mar Biotechnol 3:1827–1842. https://doi.org/10.1002/9781119143802.ch78

Alma'abadi AD, Gojobori T, Mineta K (2015) Marine metagenome as a resource for novel enzymes. GPB 13(5):290–295. https://doi.org/10.1016/j.gpb.2015.10.001

Al-Tai AM and Ruan JS (1994) *Nocardiopsis halophila* sp. nov. a new halophilic actinomycete isolated from soil. Int J Syst Evol 44(3): 474-478. doi: https://doi.org/10.1099/00207713-44-3-474

Baltz RH, Demain AL, Davies JE (2010) Manual of industrial microbiology and biotechnology. American Society for Microbiology Press

Barone R, De Santi C, Palma Esposito F, Tedesco P, Galati F, Visone M, Di Scala A, De Pascale D (2014) Marine metagenomics a valuable tool for enzymes and bioactive compounds discovery. Front Mar Sci 1:38. https://doi.org/10.3389/fmars.2014.00038

Beale DJ, Karpe AV, Ahmed W, Cook S, Morrison PD, Staley C, Sadowsky MJ, Palombo EA (2017) A community multi-omics approach towards the assessment of surface water quality in an urban river system. IJERPH 14(3):303. https://doi.org/10.3390/ijerph14030303

Benítez X, Gonzalez EG, García J, Zúñiga P, de la Calle F, Cuevas C (2021) Detection of a pederin-like compound using a dilution-to-extinction-based platform for the isolation of marine bacteria in drug discovery strategies. Microb Biotechnol 14(1):241–250. https://doi.org/10.1111/1751-7915.13679

Berdy J (2005) Bioactive microbial metabolites. J Antibiot 58(1):1–26. https://doi.org/10.1038/ja.2005.1

Berdy B, Spoering AL, Ling LL, Epstein SS (2017) In situ cultivation of previously uncultivable microorganisms using the ichip. Nat Protoc 12(10):2232. https://doi.org/10.1038/nprot.2017.074

Bhatt HB, Gohel SD, Singh SP (2018) Phylogeny, novel bacterial lineage and enzymatic potential of haloalkaliphilic bacteria from the saline coastal desert of Little Rann of Kutch Gujarat India. 3 Biotech 8(1):1–12. https://doi.org/10.1007/s13205-017-1075-0

Bose U, Hewavitharana AK, Ng YK, Shaw PN, Fuerst JA, Hodson MP (2015) LC-MS-Based metabolomics study of marine bacterial secondary metabolite and antibiotic production in *Salinispora arenicola*. Mar Drugs 13(1):249–266. https://doi.org/10.3390/md13010249

Bredholdt H, Galatenko OA, Engelhardt K, Fjærvik E, Terekhova LP, Zotchev SB (2007) Rare actinomycete bacteria from the shallow water sediments of the Trondheim fjord Norway: isolation diversity and biological activity. Environ Microbiol 9(11):2756–2764. https://doi.org/10.1111/j.1462-2920.2007.01387.x

Brito IL (2021) Examining horizontal gene transfer in microbial communities. Nat Rev Microbiol:1–12. https://doi.org/10.1038/s41579-021-00534-7

Bruns A, Hoffelner H, Overmann J (2003) A novel approach for high throughput cultivation assays and the isolation of planktonic bacteria. FEMS Microbiol Ecol 45(2):161–171. https://doi.org/10.1016/S0168-6496(03)00133-8

Bulina TI, Alferova IV, Terekhova LP (1997) A novel approach to isolation of actinomycetes involving irradiation of soil samples with microwaves. Microbiology (New York, NY) 66(2): 231–234

Button DK, Schut F, Quang P, Martin R, Robertson BR (1993) Viability and isolation of marine bacteria by dilution culture: theory procedures and initial results. Appl Environ Microbiol 59(3): 881–891. https://doi.org/10.1128/aem.59.3.881-891.1993

Cantrell T, Sutton SA, Tan G, Duca ZA, Balayan V, Rajesh N, Rennie V, Haneda S, Matsuura K, Thiel V and Stockton AM (2017) In situ culturing with isolation-chip technology in hydrogeothermal springs. In: AGU fall meeting abstracts, vol. 2017, pp. V51C-0369. doi: https://doi.org/10.1016/j.pss.2014.11.006

Chandra G, Chater KF (2014) Developmental biology of *Streptomyces* from the perspective of 100 actinobacterial genome sequences. FEMS Microbiol Rev 38(3):345–379. https://doi.org/10.1111/1574-6976.12047

Charlop-Powers Z, Milshteyn A, Brady SF (2014) Metagenomic small molecule discovery methods. Curr Opin Microbiol 19:70–75. https://doi.org/10.1016/j.mib.2014.05.021

Chaudhary DK, Khulan A, Kim J (2019) Development of a novel cultivation technique for uncultured soil bacteria. Sci Rep 9(1):1–11. https://doi.org/10.1038/s41598-019-43182-x

Chauhan JV, Gohel SD (2020) Molecular diversity and pharmaceutical applications of free-living and rhizospheric marine actinobacteria. In: Nathani et al (eds) Marine Niche: applications in pharmaceutical sciences. Springer, Singapore

Chauhan JV, Mathukiya RP, Singh SP, Gohel SD (2021) Two steps purification, biochemical characterization, thermodynamics and structure elucidation of thermostable alkaline serine protease from *Nocardiopsis alba* strain OM-5. Int J Biol Macromol 169:39–50. https://doi.org/10.1016/j.ijbiomac.2020.12.061

Chu X, He H, Guo C, Sun B (2008) Identification of two novel esterases from a marine metagenomic library derived from South China Sea. Appl Microbiol Biotechnol 80(4): 615–625. https://doi.org/10.1007/s00253-008-1566-3

Connon SA, Giovannoni SJ (2002) High-throughput methods for culturing microorganisms in very-low-nutrient media yield diverse new marine isolates. Appl Environ Microbiol 68(8): 3878–3885. https://doi.org/10.1128/AEM.68.8.3878-3885.2002

Cross T (1981) Aquatic actinomycetes: a critical survey of the occurrence growth and role of actinomycetes in aquatic habitats. J Appl Microbiol 50(3):397–423. https://doi.org/10.1111/j.1365-2672.1981.tb04245.x

Czechowska K, Johnson DR, van der Meer JR (2008) Use of flow cytometric methods for single-cell analysis in environmental microbiology. Curr Opin Microbiol 11(3):205–212. https://doi.org/10.1016/j.mib.2008.04.006

Dangar K (2018) Diversity of cultivable and Non cultivable actinomycetes from the saline habitats along the coastal Gujarat. Saurashtra University

Dangar KG, Raiyani NM, Pandya RD, Singh SP (2017) Uncultivated lineages and host–microbe interaction in saline environment. In: Singh et al (eds) Understanding host-microbiome interactions: an omics approach. Springer, Singapore

Datta AA, Sharma AK, Kundu R, Singh SP (2017) Diversity and enzymatic profile of bacterial flora in the gut of an estuarine fish *Mugil jerdoni*. Indian J Mar Sci 46(06) http://nopr.niscair.res.in/handle/123456789/42013

Demain AL, Sanchez S (2009) Microbial drug discovery: 80 years of progress. J Antibiot 62(1): 5–16. https://doi.org/10.1038/ja.2008.16

D'Onofrio A, Crawford JM, Stewart EJ, Witt K, Gavrish E, Epstein S, Clard J, Lewis K (2010) Siderophores from neighboring organisms promote the growth of uncultured bacteria. Chem Biol 17(3):254–264. https://doi.org/10.1016/j.chembiol.2010.02.010

Douglas GM, Maffei VJ, Zaneveld JR, Yurgel SN, Brown JR, Taylor CM, Huttenhower C, Langille MG (2020) PICRUSt2 for prediction of metagenome functions. Nat Biotechnol 38(6):685–688. https://doi.org/10.1038/s41587-020-0548-6

Edet UO, Antai SP, Brooks AA, Asitok AD, Enya O, Japhet FH (2017) An overview of cultural molecular and metagenomic techniques in description of microbial diversity. Adv Appl Microbiol:1–19. https://doi.org/10.9734/JAMB/2017/37951

Epstein SS (2013) The phenomenon of microbial uncultivability. Curr Opin Microbiol 16(5): 636–642. https://doi.org/10.1016/j.mib.2013.08.003

Felczykowska A, Bloch SK, Nejman-Faleńczyk B, Barańska S (2012) Metagenomic approach in the investigation of new bioactive compounds in the marine environment. Acta Biochim Pol 59(4). https://doi.org/10.18388/abp.2012_2084

Fenical W, Baden D, Burg M, de Goyet CV, Grimes JD, Katz M, Marcus NH, Pomponi S, Rhines P, Tester P, Vena J (1999) Marine derived pharmaceuticals and related bioactive compounds. In: Fenical W (ed) From monsoons to microbes: understanding the ocean's role in human health. National Academies Press, Washington DC

Ferris MJ, Kühl M, Wieland A, Ward DM (2003) Cyanobacterial ecotypes in different optical microenvironments of a 68 C hot spring mat community revealed by 16S-23S rRNA internal transcribed spacer region variation. Appl Environ Microbiol 69(5):2893–2898. https://doi.org/10.1128/AEM.69.5.2893-2898.2003

Forner D, Berrué F, Correa H, Duncan K, Kerr RG (2013) Chemical dereplication of marine actinomycetes by liquid chromatography–high resolution mass spectrometry profiling and statistical analysis. Anal Chim Acta 805:70–79. https://doi.org/10.1016/j.aca.2013.10.029

Friedrich AB, Fischer I, Proksch P, Hacker J, Hentschel U (2001) Temporal variation of the microbial community associated with the Mediterranean sponge *Aplysina aerophoba*. FEMS Microbiol Ecol 38(2-3):105–113. https://doi.org/10.1111/j.1574-6941.2001.tb00888.x

Fu J, Leiros HKS, de Pascale D, Johnson KA, Blencke HM, Landfald B (2013) Functional and structural studies of a novel cold-adapted esterase from an Arctic intertidal metagenomic library. Appl Microbiol Biotechnol 97(9):3965–3978. https://doi.org/10.1007/s00253-012-4276-9

Fulthorpe RR, Rhodes AN, Tiedje JM (1998) High levels of endemicity of 3-chlorobenzoate-degrading soil bacteria. Appl Environ Microbiol 64(5):1620–1627. https://doi.org/10.1128/AEM.64.5.1620-1627.1998

Gandhimathi R, Arunkumar M, Selvin J, Thangavelu T, Sivaramakrishnan S, Kiran GS, Shanmughapriya S, Natarajaseenivasan K (2008) Antimicrobial potential of sponge associated marine actinomycetes. J Mycol Med 18(1):16–22. https://doi.org/10.1016/j.mycmed.2007.11.001

Gao B, Gupta RS (2012) Phylogenetic framework and molecular signatures for the main clades of the phylum Actinobacteria. MMBR 76(1):66–112. https://doi.org/10.1128/MMBR.05011-11

Gärtner A, Ohlendorf B, Schulz D, Zinecker H, Wiese J, Imhoff JF (2011) Levantilides A and B 20-membered macrolides from a Micromonospora strain isolated from the mediterranean deep sea sediment. Mar Drugs 9(1):98–108. https://doi.org/10.3390/md9010098

Gavrish E, Bollmann A, Epstein S, Lewis K (2008) A trap for in situ cultivation of filamentous actinobacteria. J Microbiol Methods 72(3):257–262. https://doi.org/10.1016/j.mimet.2007.12.009

Ghosh S, Tiwari R, Hemamalini R, Khare SK (2017) Multi-omic approaches for mapping interactions among marine microbiomes. In: Singh et al (eds) Understanding host-microbiome interactions-an omics approach. Springer, Singapore

Gohel SD, Singh SP (2018a) A Molecular phylogeny and diversity of the salt-tolerant alkaliphilic actinobacteria inhabiting coastal Gujarat India. Geomicrobiol J 35(9):775–789. https://doi.org/10.1080/01490451.2018.1471107

Gohel SD, Singh SP (2018b) Thermodynamics of a Ca^{2+} dependent highly thermostable and detergent compatible purified alkaline serine protease from *Nocardiopsis xinjiangensis* strain OM-6. Int J Biol Macromol 113:565–574. https://doi.org/10.1016/j.ijbiomac.2018.02.157

Gohel SD, Sharma AK, Dangar KG, Thakrar FJ, Singh SP (2015) Antimicrobial and biocatalytic potential of haloalkaliphilic actinobacteria. In: Maheshwari D, Saraf M (eds) Halophiles. Springer, Cham

Gohel SD, Sharma AK, Thakrar FJ and Singh SP (2017) Endophytic actinobacteria and their interactions with plant host systems. In: Singh et al. (eds) Understanding host-microbiome interactions: an omics approach, Springer, Singapore

Gohel SD, Sharma AK, Dangar KG, Thakrar FJ, Singh SP (2018) Biology and applications of halophilic and haloalkaliphilic actinobacteria. In: Durvasala V, Rao S (eds) Extremophiles. CRC Press

Gontang EA, Fenical W, Jensen PR (2007) Phylogenetic diversity of gram-positive bacteria cultured from marine sediments. Appl Environ Microbiol 73(10):3272–3282. https://doi.org/10.1128/AEM.02811-06

Goulitquer S, Potin P, Tonon T (2012) Mass spectrometry-based metabolomics to elucidate functions in marine organisms and ecosystems. Mar Drugs 10(4):849–880. https://doi.org/10.3390/md10040849

Gowda GN, Raftery D (2014) Advances in NMR-based metabolomics. Compr Anal Chem 63:187–211. https://doi.org/10.1016/B978-0-444-62651-6.00008-8

Gu W, Deng X, Lee M, Sucu YD, Arevalo S, Stryke D, Federman S, Gopez A, Reyes K, Zorn K, Sample H (2021) Rapid pathogen detection by metagenomic next-generation sequencing of infected body fluids. Nat Med 27(1):115–124. https://doi.org/10.1038/s41591-020-1105-z

Hameş-Kocabaş EE, Ataç UZEL (2012) Isolation strategies of marine-derived actinomycetes from sponge and sediment samples. J Microbiol Methods 88(3):342–347. https://doi.org/10.1016/j.mimet.2012.01.010

Harrison JP, Gheeraert N, Tsigelnitskiy D, Cockell CS (2013) The limits for life under multiple extremes. Trends Microbiol 21(4):204–212. https://doi.org/10.1016/j.tim.2013.01.006

Hayakawa M (2008) Studies on the isolation and distribution of rare actinomycetes in soil. Actinomycetologica 22(1):12–19. https://doi.org/10.3209/saj.SAJ220103

Hayakawa M, Tamura T, Iino H, Nonomura H (1991a) Pollen-baiting and drying method for the highly selective isolation of *Actinoplanes spp.* from soil. J Ferment Bioeng 72(6):433–438. https://doi.org/10.1016/0922-338X(91)90050-Q

Hayakawa M, Sadakata T, Kajiura T, Nonomura H (1991b) New methods for the highly selective isolation of *Micromonospora and Microbispora* from soil. J Ferment Bioeng 72:320–326. https://doi.org/10.1016/0922-338X(91)90080-Z

Hayakawa M, Otoguro M, Takeuchi T, Yamazaki T, Iimura Y (2000) Application of a method incorporating differential centrifugation for selective isolation of motile actinomycetes in soil and plant litter. Antonie van Leeuwenhoek 78(2):171–185. https://doi.org/10.1023/a:1026579426265

Hayakawa M, Yoshida Y, Iimura Y (2004) Selective isolation of bioactive soil actinomycetes belonging to the *Streptomyces violaceusniger* phenotypic cluster. J Appl Microbiol 96(5):973–981. https://doi.org/10.1111/j.1365-2672.2004.02230.x

Heidelberg KB, Gilbert JA, Joint I (2010) Marine genomics: at the interface of marine microbial ecology and biodiscovery. Microb Biotechnol 3(5):531–543. https://doi.org/10.1111/j.1751-7915.2010.00193.x

Hengoju S, Tovar M, Man DKW, Buchheim S, Rosenbaum MA (2020) Droplet microfluidics for microbial biotechnology. Adv Biochem Eng Biotechnol. https://doi.org/10.1007/10_2020_140

Henson MW, Lanclos VC, Pitre DM, Weckhorst JL, Lucchesi AM, Cheng C, Temperton B, Thrash JC (2020) Expanding the diversity of bacterioplankton isolates and modeling isolation efficacy with large-scale dilution-to-extinction cultivation. Appl Environ Microbiol 86(17). https://doi.org/10.1128/AEM.00943-20

Hong K, Gao AH, Xie QY, Gao HG, Zhuang L, Lin HP, Yu HP, Li J, Yao XS, Goodfellow M, Ruan JS (2009) Actinomycetes for marine drug discovery isolated from mangrove soils and plants in China. Mar Drugs 7(1):24–44. https://doi.org/10.3390/md7010024

Hozzein Wael N, Wen-Jun Li, Mohammed Ibrahim A. Ali, Ola Hammouda, Ahmed S. Mousa, Li-Hua Xu, Cheng-Lin Jiang (2004) *Nocardiopsis alkaliphila* sp. nov., a novel alkaliphilic actinomycete isolated from desert soil in Egypt. Int J Syst Evol Microbiol 54 (1): 247-252. doi: https://doi.org/10.1099/ijs.0.02832-0

Hu B, Xu B, Yun J, Wang J, Xie B, Li C, Yu Y, Lan Y, Zhu Y, Dai X, Huang Y (2020) High-throughput single-cell cultivation reveals the underexplored rare biosphere in deep-sea sediments along the Southwest Indian Ridge. Lab Chip 20(2):363–372. https://doi.org/10.1039/C9LC00761J

Hu B, Xu P, Ma L, Chen D, Wang J, Dai X, Huang L, Du W (2021) One cell at a time: droplet-based microbial cultivation, screening and sequencing. J Mar Sci Technol:1–20. https://doi.org/10.1007/s42995-020-00082-8

Hughes CC, MacMillan JB, Gaudêncio SP, Jensen PR, Fenical W (2009) The ammosamides: structures of cell cycle modulators from a marine-derived *Streptomyces species*. Angew Chem Int Ed 48(4):725–727. https://doi.org/10.1002/anie.200804890

Istianto Y, Koesoemowidodo RSA, Saputra H, Watanabe Y, Pranamuda H, Marwoto B (2012) Application of phenol pretreatment for the isolation of rare actinomycetes from Indonesian soil. Microbiol Indones 6:42–47. https://doi.org/10.5454/mi.6.1.7

Jensen PR, Ryan D, William F (1991) Distribution of actinomycetes in near-shore tropical marine sediments. Appl Environ Microbiol 57(4):1102–1108. https://doi.org/10.1128/aem.57.4.1102-1108.1991

Jensen PR, Gontang E, Mafnas C, Mincer TJ, Fenical W (2005) Culturable marine actinomycete diversity from tropical Pacific Ocean sediments. Environ Microbiol 7(7):1039–1048. https://doi.org/10.1111/j.1462-2920.2005.00785.x

Jensen PR, Williams PG, Oh DC, Zeigler L, Fenical W (2007) Species-specific secondary metabolite production in marine actinomycetes of the genus *Salinispora*. Appl Environ Microbiol 73(4):1146–1152. https://doi.org/10.1128/FAEM.01891-06

Jeon JH, Kim JT, Kim YJ, Kim HK, Lee HS, Kang SG, Kim SJ, Lee JH (2009) Cloning and characterization of a new cold-active lipase from a deep-sea sediment metagenome. Appl Microbiol Biotechnol 81(5):865–874. https://doi.org/10.1007/s00253-008-1656-2

Joghee NN, Jayaraman G (2014) Metabolomic characterization of halophilic bacterial isolates reveals strains synthesizing rare diaminoacids under salt stress. Biochimie 102:102–111. https://doi.org/10.1016/j.biochi.2014.02.015

Jose PA, Jebakumar SRD (2015) Taxonomic and antimicrobial profiles of a rare actinomycete isolated from an inland solar saltern (India). Indian J Mar Sci 44:3. http://nopr.niscair.res.in/handle/123456789/34702

Jung D, Seo EY, Owen JS, Aoi Y, Yong S, Lavrentyeva EV, Ahn TS (2018) Application of the filter plate microbial trap (FPMT), for cultivating thermophilic bacteria from thermal springs in Barguzin area, eastern Baikal, Russia. Biosci Biotechnol Biochem 82(9):1624–1632. https://doi.org/10.1080/09168451.2018.1482194

Kaeberlein T, Lewis K, Epstein SS (2002) Isolating "uncultivable" microorganisms in pure culture in a simulated natural environment. Science 296(5570):1127–1129. https://doi.org/10.1126/science.1070633

Keller M, Zengler K (2004) Tapping into microbial diversity. Nat Rev Microbiol 2(2):141–150. https://doi.org/10.1038/nrmicro819

Kennedy J, Baker P, Piper C, Cotter PD, Walsh M, Mooij MJ, Bourke MB, Rea MC, O'Connor PM, Ross RP, Hill C (2009) Isolation and analysis of bacteria with antimicrobial activities from the marine sponge *Haliclona simulans* collected from Irish waters. Mar Biotechnol 11(3): 384–396. https://doi.org/10.1007/s10126-008-9154-1

Khamna S, Yokota A, Lumyong S (2009) Actinomycetes isolated from medicinal plant rhizosphere soils: diversity and screening of antifungal compounds indole-3-acetic acid and siderophore production. World J Microbiol Biotechnol 25(4):649–655. https://doi.org/10.1007/s11274-008-9933-x

Kikani BA, Singh SP (2021) Amylases from thermophilic bacteria: structure and function relationship. Crit Rev Biotechnol 1–17. https://doi.org/10.1080/07388551.2021.1940089

Kikani BA, Sharma AK, Singh SP (2015) Culture dependent diversity and phylogeny of thermophilic bacilli from a natural hot spring reservoir in the Gir Forest Gujarat (India). Microbiology 84(5):687–700. https://doi.org/10.1134/S0026261715050124

Kikani BA, Sharma AK, Singh SP (2017) Metagenomic and culture-dependent analysis of the bacterial diversity of a hot spring reservoir as a function of the seasonal variation. Int J Environ Res 11(1):25–38. https://doi.org/10.1007/s41742-017-0003-9

Klenk HP, Lu M, Lucas S, Lapidus A, Copeland A, Pitluck S, Goodwin LA, Han C, Tapia R, Brambilla EM, Pötter G (2012) Genome sequence of the ocean sediment bacterium *Saccharomonospora marina* type strain (XMU15 T). Stand Genom Sci 6(2):265–275. https://doi.org/10.4056/sigs.2655905

Kumar Awasthi M, Ravindran B, Sarsaiya S, Chen H, Wainaina S, Singh E, Liu T, Kumar S, Pandey A, Singh L, Zhang Z (2020) Metagenomics for taxonomy profiling: tools and approaches. Bioengineered 11(1):356–374. https://doi.org/10.1080/21655979.2020.1736238

Kurm V, Van Der Putten WH, Hol WG (2019) Cultivation-success of rare soil bacteria is not influenced by incubation time and growth medium. PloS one 14(1):0210073. https://doi.org/10.1371/journal.pone.0210073

Kurtböke DI (2012) Biodiscovery from rare actinomycetes: an eco-taxonomical perspective. Appl Microbiol Biotechnol 93(5):1843–1852. https://doi.org/10.1007/s00253-012-3898-2

Lawrence Jeffrey G, Hendrickson H (2005) Genome evolution in bacteria: order beneath chaos. Curr Opin Microbiol 8(5):572–578. https://doi.org/10.1016/j.mib.2005.08.005

Lazzarini A, Cavaletti L, Toppo G, Marinelli F (2000) Rare genera of actinomycetes as potential producers of new antibiotics. Antonie van Leeuwenhoek 78(3):399–405. https://doi.org/10.1023/A:1010287600557

Lee EY, Lee HK, Lee YK, Sim CJ, Lee JH (2003) Diversity of symbiotic archaeal communities in marine sponges from Korea. Biomol Eng 20(4-6):299–304. https://doi.org/10.1016/S1389-0344(03)00034-0

Lee C, Wakeham S, Arnosti C (2004) Particulate organic matter in the sea: the composition conundrum AMBIO. J Hum Environ 33(8):565–575. https://doi.org/10.1579/0044-7447-33.8.565

Lewis WH, Tahon G, Geesink P, Sousa DZ, Ettema TJ (2020) Innovations to culturing the uncultured microbial majority. Nat Rev Microbiol:1–16. https://doi.org/10.1038/s41579-020-00458-8

Li L, Wang C, Nie Y, Yao B, Hu H (2020) Nanofabrication enabled lab-on-a-chip technology for the manipulation and detection of bacteria. Trac-Trend Anal Chem:115905. https://doi.org/10.1016/j.trac.2020.115905

Liu X, Wang M, Nie Y, Wu XL (2021) Isolation chip increases culturable bacterial diversity and reduces cultivation bias. Curr Microbiol 78(5):2025–2032. https://doi.org/10.1007/s00284-021-02474-0

Llewellyn CA, Sommer U, Dupont CL, Allen AE, Viant MR (2015) Using community metabolomics as a new approach to discriminate marine microbial particulate organic matter in the western English Channel. Prog Oceanogr 137:421–433. https://doi.org/10.1016/j.pocean.2015.04.022

Lodhi AF, Zhang Y, Adil M, Deng Y (2020) A novel culturing chip (cChip) can facilitate culturing of unculturable bacteria from aquatic environment.

Ludwig W, Euzéby J, Schumann P, Busse HJ, Trujillo ME, Kämpfer P, Whitman WB (2012) Road map of the phylum actinobacteria. In: Bergey's manual® of systematic bacteriology. Springer, New York NY

Luo R, Pashapour S, Staufer O, Platzman I, Spatz JP (2020) Polymer-based porous microcapsules as bacterial traps. Adv Funct Mater 30(17):1908855. https://doi.org/10.1002/adfm.201908855

Macintyre L, Zhang T, Viegelmann C, Martinez IJ, Cheng C, Dowdells C, Abdelmohsen UR, Gernert C, Hentschel U, Edrada-Ebel R (2014) Metabolomic tools for secondary metabolite discovery from marine microbial symbionts. Mar Drugs 12(6):3416–3448. https://doi.org/10.3390/md12063416

Magarvey NA, Keller JM, Bernan V, Dworkin M, Sherman DH (2004) Isolation and characterization of novel marine-derived actinomycete taxa rich in bioactive metabolites. Appl Environ Microbiol 70(12):7520–7529. https://doi.org/10.1128/AEM.70.12.7520-7529.2004

Mahapatra GP, Raman S, Nayak S, Gouda S, Das G, Patra JK (2019) Metagenomics approaches in discovery and development of new bioactive compounds from marine actinomycetes. Curr Microbiol:1–12. https://doi.org/10.1007/s00284-019-01698-5

Maldonado LA, Fenical W, Jensen PR, Kauffman CA, Mincer TJ, Ward AC, Bull AT and Goodfellow M (2005a) *Salinispora arenicola gen. nov., sp. nov.* and *Salinispora tropica sp. nov.,* obligate marine actinomycetes belonging to the family *Micromonosporaceae*. Int J Syst Evol Microbiol 55(5): 1759-1766. doi: https://doi.org/10.1099/ijs.0.63625-0

Maldonado LA, Stach JE, Pathom-aree W, Ward AC, Bull AT, Goodfellow M (2005b) Diversity of cultivable actinobacteria in geographically widespread marine sediments. Antonie Van Leeuwenhoek 87(1):11–18. https://doi.org/10.1007/s10482-004-6525-0

Maldonado LA, Fragoso-Yáñez D, Pérez-García A, Rosellón-Druker J, Quintana ET (2009) Actinobacterial diversity from marine sediments collected in Mexico. Antonie Van Leeuwenhoek 95(2):111–120. https://doi.org/10.1007/s10482-008-9294-3

Marcolefas E, Altshuler I, Whyte LG (2019) Advanced microbial cultivation methodologies and their applicability in cryoenvironments. In: Yergeau E (ed) Advanced techniques for studying microorganisms in extreme environments. De Gruyter

Margot H, Acebal C, Toril E, Amils R, Puentes JF (2002) Consistent association of crenarchaeal Archaea with sponges of the genus Axinella. Mar Biol 140(4):739–745. https://doi.org/10.1007/s00227-001-0740-2

McCombie WR, McPherson JD, Mardis ER (2019) Next-generation sequencing technologies. Cold Spring Harb Perspect Med 9(11):036798. https://doi.org/10.1101/cshperspect.a036798

McInerney MJ, Struchtemeyer CG, Sieber J, Mouttaki H, Stams AJ, Schink B, Rohlin L, Gunsalus RP (2008) Physiology, ecology, phylogeny, and genomics of microorganisms capable of syntrophic metabolism. Ann N Y Acad Sci 1125(1):58–72. https://doi.org/10.1196/annals.1419.005

Miguélez EM, Martín C, Hardisson C, Manzanal MB (1993) Synchronous germination of Streptomyces antibioticus spores: tool for the analysis of hyphal growth in liquid cultures. FEMS Microbiol Lett 109(2-3):123–129. https://doi.org/10.1111/j.1574-6968.1993.tb06156.x

Mincer TJ, Jensen PR, Kauffman CA, Fenical W (2002) Widespread and persistent populations of a major new marine actinomycete taxon in ocean sediments. Appl Environ Microbiol 68(10): 5005–5011. https://doi.org/10.1128/AEM.68.10.5005-5011.2002

Mincer TJ, Fenical W, Jensen PR (2005) Culture-dependent and culture-independent diversity within the obligate marine actinomycete genus *Salinispora*. Appl Environ Microbiol 71(11): 7019–7028. https://doi.org/10.1128/AEM.71.11.7019-7028.2005

Moon S, Kim YJ, Han S, Hwang MY, Shin DM, Park MY, Lu Y, Yoon K, Jang HM, Kim YK, Park TJ (2019) The Korea biobank array: design and identification of coding variants associated with blood biochemical traits. Sci Rep 9(1):1–11. https://doi.org/10.1038/s41598-018-37832-9

Mutka SC, Carney JR, Liu Y, Kennedy J (2006) Heterologous production of epothilone C and D in *Escherichia coli*. Biochemistry 45(4):1321–1330. https://doi.org/10.1021/bi052075r

Natsume M, Yasui K, Marumo S (1989) Calcium ion regulates aerial mycelium formation in actinomycetes. J Antibiot 42(3):440–447. https://doi.org/10.7164/antibiotics.42.440

Newman DJ, Cragg GM (2007) Natural products as sources of new drugs over the last 25 years. J Nat Prod 70(3):461–477. https://doi.org/10.1021/np068054v

Nguyen KT, Nguyen NL, Tung NV, Nguyen HH, Milhim M, Le TTX, Lai THN, Phan TTM, Bernhardt R (2020) A novel thermostable cytochrome P450 from sequence-based metagenomics of Binh Chau hot spring as a promising catalyst for testosterone conversion. Catalysts 10(9):1083. https://doi.org/10.3390/catal10091083

Nichols D (2007) Cultivation gives context to the microbial ecologist. FEMS Microbiol Ecol 60(3): 351–357. https://doi.org/10.1111/j.1574-6941.2007.00332.x

Nichols D, Lewis K, Orjala J, Mo S, Ortenberg R, O'connor P, Zhao C, Vouros P, Kaeberlein T, Epstein SS (2008) Short peptide induces an "uncultivable" microorganism to grow in vitro. Appl Environ Microbiol 74(15):4889–4897. https://doi.org/10.1128/AEM.00393-08

Nichols D, Cahoon N, Trakhtenberg EM, Pham L, Mehta A, Belanger A, Kanigan T, Lewis K, Epstein SS (2010) Use of ichip for high-throughput in situ cultivation of "uncultivable" microbial species. Appl Environ Microbiol 76(8):2445–2450. https://doi.org/10.1128/AEM.01754-09

Niu G (2018) Genomics-driven natural product discovery in actinomycetes. Trends Biotechnol 36(3):238–241. https://doi.org/10.1016/j.tibtech.2017.10.009

Niyomvong N, Pathom-aree W, Thamchaipenet A, Duangmal K (2012) Actinomycetes from tropical limestone caves. Chiang Mai J Sci 39(3):373–388

Nonomura H (1988) Isolation, taxonomy and ecology of soil actinomycetes. Actinomycetologia 3: 45–54. https://doi.org/10.3209/saj.3_45

Ochman H, Lawrence JG, Groisman EA (2000) Lateral gene transfer and the nature of bacterial innovation. Nature 405(6784):299–304. https://doi.org/10.1038/35012500

Okami Y (1988) Search and discovery of new antibiotics. Actinomycetes Biotechnol:33–67. https://doi.org/10.1016/b978-0-12-289673-6.50007-5

Ong JFM, Goh HC, Lim SC, Pang LM, Chin JSF, Tan KS, Liang ZX, Yang L, Glukhov E, Gerwick WH, Tan LT (2019) Integrated genomic and metabolomic approach to the discovery of potential anti-quorum sensing natural products from microbes associated with marine samples from Singapore. Mar Drugs 17(1):72. https://doi.org/10.3390/md17010072

Oren A (2011) Thermodynamic limits to microbial life at high salt concentrations. Environ Microbiol 13(8):1908–1923. https://doi.org/10.1111/j.1462-2920.2010.02365.x

Otoguro M, Hayakawa M, Yamazaki T, Iimura Y (2001) An integrated method for the enrichment and selective isolation of *Actinokineospora spp.* in soil and plant litter. J Appl Microbiol 91(1): 118–130. https://doi.org/10.1046/j.1365-2672.2001.01372.x

Parte AC (2014) LPSN—list of prokaryotic names with standing in nomenclature. Nucleic Acids Res 42(D1):D613–D616. https://doi.org/10.1093/nar/gkt1111

Parte AC (2018) LPSN–list of prokaryotic names with standing in nomenclature (bacterio. net), 20 years on. Int J Syst Evol Microbiol 68(6):1825–1829. https://doi.org/10.1099/ijsem.0.002786

Parte AC, Carbasse JS, Meier-Kolthoff JP, Reimer LC, Göker M (2020) List of prokaryotic names with standing in nomenclature (LPSN) moves to the DSMZ. Int J Syst Evol Microbiol 70(11): 5607. https://doi.org/10.1099/ijsem.0.004332

Pathom-Aree W, Stach JE, Ward AC, Horikoshi K, Bull AT, Goodfellow M (2006) Diversity of actinomycetes isolated from Challenger Deep sediment (10,898 m) from the Mariana Trench. Extremophiles 10(3):181–189. https://doi.org/10.1007/s00792-005-0482-z

Paul S, Bag SK, Das S, Harvill ET, Dutta C (2008) Molecular signature of hypersaline adaptation: insights from genome and proteome composition of halophilic prokaryotes. Genome Biol 9(4): 1–19. https://doi.org/10.1186/gb-2008-9-4-r70

Pensold D, Zimmer-Bensch G (2020) Methods for single-cell isolation and preparation. In: Yu et al (eds) Single-cell sequencing and methylation. Springer, Singapore

Pham VH, Kim J (2012) Cultivation of unculturable soil bacteria. Trends Biotechnol 30(9): 475–484. https://doi.org/10.1016/j.tibtech.2012.05.007

Purohit MK, Singh SP (2009) Assessment of various methods for extraction of metagenomic DNA from saline habitats of coastal Gujarat (India) to explore molecular diversity. Lett Appl Microbiol 49(3):338–344. https://doi.org/10.1111/j.1472-765X.2009.02663.x

Purohit MK, Singh SP (2013) A metagenomic alkaline protease from saline habitat: cloning, over-expression and functional attributes. Int J Biol Macromol 53:138–143. https://doi.org/10.1016/j.ijbiomac.2012.10.032

Qin S, Li J, Chen HH, Zhao GZ, Zhu WY, Jiang CL, Xu LH, Li WJ (2009) Isolation, diversity, and antimicrobial activity of rare actinobacteria from medicinal plants of tropical rain forests in Xishuangbanna China. Appl Environ Microbiol 75(19):6176–6186. https://doi.org/10.1128/AEM.01034-09

Qiu D, Ruan J, Huang Y (2008) Selective isolation and rapid identification of members of the genus *Micromonospora*. Appl Environ Microbiol 74(17):5593–5597. https://doi.org/10.1128/AEM.00303-08

Rademaker JL, Hoste B, Louws FJ, Kersters K, Swings J, Vauterin L, Vauterin P, de Bruijn FJ (2000) Comparison of AFLP and rep-PCR genomic fingerprinting with DNA-DNA homology studies: *Xanthomonas* as a model system. Int J Syst Evol Microbiol 50(2):665–677. https://doi.org/10.1099/00207713-50-2-665

Raiyani NM, Singh SP (2020) Taxonomic and functional profiling of the microbial communities of Arabian Sea: a metagenomics approach. Genomics 112(6):4361–4369. https://doi.org/10.1016/j.ygeno.2020.07.024

Raiyani NM, Georrge JJ, Herma TH, Singh SP (2020) Designing and evaluation of metagenomics 16S rRNA gene primers. In: Proceedings of the National Conference on Innovations in Biological Sciences (NCIBS). Available at SSRN 3567789. https://doi.org/10.2139/ssrn.3567789

Rappé MS, Giovannoni SJ (2003) The uncultured microbial majority. Annu Rev Microbiol 57(1): 369–394. https://doi.org/10.1146/annurev.micro.57.030502.090759

Rathore DS, Singh SP (2021) Kinetics of growth and co-production of amylase and protease in novel marine actinomycete Streptomyces lopnurensis KaM5. Folia Microbiol 66:303–316. https://doi.org/10.1007/s12223-020-00843-z

Rathore DS, Sheikh M, Gohel S, Singh SP (2019) Isolation strategies abundance and characteristics of the marine actinomycetes of Kachhighadi Gujarat India. Mar Biol Ass India 61(1):72. https://doi.org/10.6024/jmbai.2019.61.1.2028-11

Rathore DS, Malaviya K, Dobariya A, Singh SP (2020) Optimization of the production of an amylase from a marine actinomycetes *Nocardiopsis dassonvillei strain* KaS11. In: Proceedings of the National Conference on Innovations in Biological Sciences (NCIBS). Available at SSRN 3567722. https://doi.org/10.2139/ssrn.3567722

Rathore DS, Sheikh M, Singh SP (2021a) Marine actinobacteria: new horizons in bioremediation. In: Prasad et al (eds) Recent developments in microbial technologies. Springer, Singapore

Rathore DS, Sheikh MA, Gohel SD, Singh SP (2021b) Genetic and Phenotypic Heterogeneity of the *Nocardiopsis alba* strains of Seawater. Curr Microbiol 78(4):1377–1387. https://doi.org/10.1007/s00284-021-02420-0

Raval VH, Pillai S, Rawal CM, Singh SP (2014) Biochemical and structural characterization of a detergent-stable serine alkaline protease from seawater haloalkaliphilic bacteria. Process Biochem 49(6):955–962. https://doi.org/10.1016/j.procbio.2014.03.014

Raval VH, Bhatt HB, Singh SP (2018) Chapter adaptation strategies in halophilic bacteria. In: Durvasula RV, Rao DVS (eds) Extremophiles: from biology to biotechnology. CRC Press

Reverter M, Rohde S, Parchemin C, Tapissier-Bontemps N, Schupp PJ (2020) Metabolomics and marine biotechnology: coupling metabolite profiling and organism biology for the discovery of new compounds. Front Mar Sci 7:1062. https://doi.org/10.3389/fmars.2020.613471

Roller C, Ludwig W, Schleifer KH (1992) Gram-positive bacteria with a high DNA G+ C content are characterized by a common insertion within their 23S rRNA genes. Microbiology 138(6): 1167–1175. https://doi.org/10.1099/00221287-138-6-1167

Roller C, Wagner M, Amann R, Ludwig W, Schleifer KH (1994) In situ probing of Gram-positive bacteria with high DNA G+ C content using 23S rRNA-targeted oligonucleotides. Microbiology 140(10):2849–2858. https://doi.org/10.1099/00221287-140-10-2849

Schuchmann K, Müller V (2014) Autotrophy at the thermodynamic limit of life: a model for energy conservation in acetogenic bacteria. Nat Rev Microbiol 12(12):809–821. https://doi.org/10.1038/nrmicro3365

Schut F, de Vries EJ, Gottschal JC, Robertson BR, Harder W, Prins RA, Button DK (1993) Isolation of typical marine bacteria by dilution culture: growth maintenance and characteristics of isolates under laboratory conditions. Appl Environ Microbiol 59(7):2150–2160. https://doi.org/10.1128/aem.59.7.2150-2160

Selvin J, Joseph S, Asha KRT, Manjusha WA, Sangeetha VS, Jayaseema DM, Antony MC, Denslin Vinitha AJ (2004) Antibacterial potential of antagonistic *Streptomyces sp.* isolated from marine sponge *Dendrilla nigra*. FEMS Microbiol Ecol 50(2):117–122. https://doi.org/10.1016/j.femsec.2004.06.007

Sharma AK (2017) Stability and characteristics of extracellular proteases from the actinomycetes isolated from sea water along the Gujarat coast. Saurashtra University

Sharma AK, Singh SP (2016) Effect of amino acids on the repression of alkaline protease synthesis in haloalkaliphilic *Nocardiopsis dassonvillei*. Biotechnol Rep 12:40–51. https://doi.org/10.1016/j.btre.2016.10.004

Sharma AK, Gohel S, Singh SP (2012) Actinobase: database on molecular diversity phylogeny and biocatalytic potential of salt tolerant alkaliphilic actinomycetes. Bioinformation 8(11):535. https://doi.org/10.6026/97320630008535

Sharma AK, Kikani BA, Singh SP (2020) Biochemical thermodynamic and structural characteristics of a biotechnologically compatible alkaline protease from a haloalkaliphilic Nocardiopsis dassonvillei OK-18. Int J Biol Macromol 153:680–696. https://doi.org/10.1016/j.ijbiomac.2020.03.006

Sharma AK, Kikani BA, Singh SP (2021) Diversity and phylogeny of actinomycetes of Arabian Sea along the Gujarat Coast. Geomicrobiol J 38(4):347–364. https://doi.org/10.1080/01490451.2020.1860165

Sheikh MA, Rathore DS, Gohel SD, Singh SP (2018) Marine actinobacteria associated with the invertebrate hosts: a rich source of bioactive compounds: a review. Cell Tissue Res 18(1)

Sheikh M, Rathore DS, Gohel S and Singh SP (2019) Cultivation and characteristics of the Marine Actinobacteria from the Sea water of Alang Bhavnagar. http://nopr.niscair.res.in/handle/123456789/52802

Shivlata L, Tulasi S (2015) Thermophilic and alkaliphilic actinobacteria: biology and potential applications. Front Microbiol 6:1014. https://doi.org/10.3389/fmicb.2015.01014

Siddhapura PK, Vanparia S, Purohit MK, Singh SP (2010) Comparative studies on the extraction of metagenomic DNA from the saline habitats of Coastal Gujarat and Sambhar Lake Rajasthan (India) in prospect of molecular diversity and search for novel biocatalysts. Int J Biol Macromol 47(3):375–379. https://doi.org/10.1016/j.ijbiomac.2010.06.004

Singh SB, Pelaez F (2008) Biodiversity chemical diversity and drug discovery. Natural compounds as drugs volume I. Prog Drug Res:141–174. https://doi.org/10.1007/978-3-7643-8117-2_4

Singh SP, Thumar JT, Gohel SD, Kikani B, Shukla R, Sharma A, Dangar K (2013) Actinomycetes from marine habitats and their enzymatic potential. In: Marine enzymes for biocatalysis. Woodhead Publishing, pp 191–214. https://doi.org/10.1533/9781908818355.2.191

Slatko BE, Gardner AF, Ausubel FM (2018) Overview of next-generation sequencing technologies. Curr Protoc Mol Biol 122(1):59. https://doi.org/10.1002/cpmb.59

Sogin EM, Puskas E, Dubilier N, Liebeke M (2019) Marine metabolomics: a method for nontargeted measurement of metabolites in seawater by gas chromatography–mass spectrometry. Msystems 4(6). https://doi.org/10.1128/mSystems.00638-19

Stach JE, Maldonado LA, Ward A, Goodfellow M, Bull AT (2003) New primers for the class Actinobacteria: application to marine and terrestrial environments. Environ Microbiol 5(10): 828–841. https://doi.org/10.1046/j.1462-2920.2003.00483.x

Staley JT, Konopka A (1985) Measurement of in situ activities of nonphotosynthetic microorganisms in aquatic and terrestrial habitats. Annu Rev Microbiol 39(1):321–346. https://doi.org/10.1146/annurev.mi.39.100185.001541

Stewart EJ (2012) Growing unculturable bacteria. J Bacteriol 194(16):4151–4160. https://doi.org/10.1128/JB.00345-12

Stingl U, Tripp HJ, Giovannoni SJ (2007) Improvements of high-throughput culturing yielded novel SAR11 strains and other abundant marine bacteria from the Oregon coast and the Bermuda Atlantic Time Series study site. ISME J 1(4):361–371. https://doi.org/10.1038/ismej.2007.49

Subramani R, Aalbersberg W (2013) Culturable rare Actinomycetes: diversity isolation and marine natural product discovery. Appl Microbiol Biotechnol 97(21):9291–9321. https://doi.org/10.1007/s00253-013-5229-7

Sudha S, Masilamani SM (2012) Characterization of cytotoxic compound from marine sediment derived actinomycete *Streptomyces avidinii strain* SU4. Asian Pac J Trop Biomed 2(10): 770–773. https://doi.org/10.1016/S2221-1691(12)60227-5

Thakrar FJ, Singh SP (2019) Catalytic thermodynamic and structural properties of an immobilized and highly thermostable alkaline protease from a haloalkaliphilic actinobacteria *Nocardiopsis alba* Tata-5. Bioresour Technol 278:150–158. https://doi.org/10.1016/j.biortech.2019.01.058

Thakrar FJ, Kikani BA, Sharma AK, Singh SP (2018) Stability of alkaline proteases from haloalkaliphilic actinobacteria probed by circular dichroism spectroscopy. Appl Biochem Microbiol 54(6):591–602. https://doi.org/10.1134/S0003683818100022

Thumar JT, Singh SP (2009) Organic solvent tolerance of an alkaline protease from salt-tolerant alkaliphilic *Streptomyces clavuligerus strain* Mit-1. J Ind Microbiol Biotechnol 36(2):211. https://doi.org/10.1007/s10295-008-0487-6

Tiwari K, Gupta RK (2012) Rare actinomycetes: a potential storehouse for novel antibiotics. Crit Rev Biotechnol 32(2):108–132. https://doi.org/10.3109/07388551.2011.562482

Tiwari K, Gupta RK (2013) Diversity and isolation of rare actinomycetes: an overview. Crit Rev Microbiol 39(3):256–294. https://doi.org/10.3109/1040841X.2012.709819

Toledo GE, Green WA, Gonzalez RA, Christoffersen LE, Podar MI, Chang HW, Hemscheidt TH, Trapido-Rosenthal HG, Short JM, Bidigare RR, Mathur EJ (2006) High throughput cultivation for isolation of novel marine microorganisms. Oceanography 19(2):120–124. https://doi.org/10.5670/oceanog.2006.75

Turnbaugh PJ, Gordon JI (2008) An invitation to the marriage of metagenomics and metabolomics. Cell 134(5):708–713. https://doi.org/10.1016/j.cell.2008.08.025

Udwary DW, Zeigler L, Asolkar RN, Singan V, Lapidus A, Fenical W, Jensen PR, Moore BS (2007) Genome sequencing reveals complex secondary metabolome in the marine actinomycete *Salinispora tropica*. Proc Natl Acad Sci 104(25):10376–10381. https://doi.org/10.1073/pnas.0700962104

Vasavada SH, Thumar JT, Singh SP (2006) Secretion of a potent antibiotic by salt-tolerant and alkaliphilic actinomycete *Streptomyces sannanensis strain* RJT-1. Curr Sci:1393–1397. https://doi.org/10.13140/2.1.2616.3847

Ventura M, Canchaya C, Tauch A, Chandra G, Fitzgerald GF, Chater KF, van Sinderen D (2007) Genomics of actinobacteria: tracing the evolutionary history of an ancient phylum. Microbiol Mol Biol Rev 71(3):495–548. https://doi.org/10.1128/MMBR.00005-07

Wang DS, Xue QH, Zhu WJ, Zhao J, Duan JL, Shen GH (2013) Microwave irradiation is a useful tool for improving isolation of actinomycetes from soil. Microbiology 82(1):102–110. https://doi.org/10.1134/S0026261712060161

Wang F, Li M, Huang L, Zhang XH (2021) Cultivation of uncultured marine microorganisms. Mar Life Sci Technol 3:117–120. https://doi.org/10.1007/s42995-021-00093-z

Ward AC, Bora N (2006) Diversity and biogeography of marine actinobacteria. Curr Opin Microbiol 9(3):279–286. https://doi.org/10.1016/j.mib.2006.04.004

Weiland-Bräuer N, Langfeldt D, Schmitz RA (2017) Construction and screening of marine metagenomic large insert libraries. In: Metagenomic. Humana Press, New York, NY, pp 23–42. https://doi.org/10.1007/978-1-4939-6691-2_3

Wen-Jun L, Park D-J, Tang S-K, Dong W, Lee J-C, Xu L-H, Kim C-J, Jiang C-L (2004) Nocardiopsis salina sp. nov. a novel halophilic actinomycete isolated from saline soil in China. Int J Syst Evol Microbiol 54(5):1805–1809. https://doi.org/10.1099/ijs.0.63127-0

Weyland H (1969) Actinomycetes in North Sea and Atlantic ocean sediments. Nature 223(5208): 858–858. https://doi.org/10.1038/223858a0

Whitaker RJ, Grogan DW, Taylor JW (2003) Geographic barriers isolate endemic populations of hyperthermophilic archaea. Science 301(5635):976–978. https://doi.org/10.1126/science.1086909

Xin Y, Huang J, Deng M, Zhang W (2008) Culture-independent nested PCR method reveals high diversity of actinobacteria associated with the marine sponges *Hymeniacidon perleve* and *Sponge sp*. Antonie Van Leeuwenhoek 94(4):533–542. https://doi.org/10.1007/s10482-008-9270-y

Xue Q, Dua CM, Wang LN, Lin YB (2010) The influence of microwave irradiation to the isolation effect of soil Actinomycetes. Chin J Microbiol 3:19–24. https://doi.org/10.1134/S0026261712060161

Yang B, Xue QH, Chen ZQ, Zhou YQ, Zhang XL, Xu YJ, Guo ZY (2008) Effects of microwave irradiation on isolation of soil actinomycetes. J Appl Ecol 19(5):1091–1098

Zengler K, Toledo G, Rappé M, Elkins J, Mathur EJ, Short JM, Keller M (2002) Cultivating the uncultured. PNAS 99(24):15681–15686. https://doi.org/10.1073/pnas.252630999

Zhang L, An R, Wang J, Sun N, Zhang S, Hu J, Kuai J (2005) Exploring novel bioactive compounds from marine microbes. Curr Opin Microbiol 3:276–281. https://doi.org/10.1016/j.mib.2005.04.008

Zhang H, Lee YK, Zhang W, Lee HK (2006) Culturable actinobacteria from the marine sponge *Hymeniacidon perleve*: isolation and phylogenetic diversity by 16S rRNA gene-RFLP analysis. Antonie Van Leeuwenhoek 90(2):159–169. https://doi.org/10.1007/s10482-006-9070-1

Zhang H, Zhang W, Jin Y, Jin M, Yu X (2008) A comparative study on the phylogenetic diversity of culturable actinobacteria isolated from five marine sponge species. Antonie van Leeuwenhoek 93(3):241–248. https://doi.org/10.1007/s10482-007-9196-9

Zhang Q, Wang T, Zhou Q, Zhang P, Gong Y, Gou H, Xu J, Ma B (2017) Development of a facile droplet-based single-cell isolation platform for cultivation and genomic analysis in microorganisms. Sci Rep 7(1):1–11. https://doi.org/10.1038/srep41192

Zhi XY, Li WJ, Stackebrandt E (2009) An update of the structure and 16S rRNA gene sequence-based definition of higher ranks of the class Actinobacteria with the proposal of two new suborders and four new families and emended descriptions of the existing higher taxa. Int J Syst Evol Microbiol 59(3):589–608. https://doi.org/10.1099/ijs.0.65780-0

Chapter 13
Antimicrobial Potential and Metabolite Profiling of Marine Actinobacteria

Jignasha Thumar and Satya P. Singh

Abstract Over 90% volume of the Earth's crust is covered by oceans. Many natural product-based drug discovery programs are being run and funded by developed countries. Marine organisms harbor incredibly diverse natural products with novel pharmaceutical applications. Among all the marine microorganisms, actinomycetes remain the most popular because of their capacity to produce a wide range of secondary metabolites that can be developed into drugs for treatment of wide range of diseases in human, agriculture, and veterinary sectors. Further, these compounds also hold the potential in treatment of life-threatened infections in humans. Numerous antibacterial, antifungal, cytotoxic, neurotoxic, antiviral, and antitumor compounds against new targets including AIDS, anti-inflammation, aging process, and immunosuppression have been characterized from marine actinomycetes. *Streptomyces* is the most prominent genus studied so far in this regard. However, many rare actinomycete genera have also been reported to produce a diverse array of antimicrobial compounds including polyenes, peptides, macrolides, aminoglycosides, polyether, etc. This chapter highlights the metabolite profiling of marine actinomycetes with respect to current status on drug discovery programs. It further stresses on the emergence of discovery of new antimicrobial metabolites, as the replacement of already existing ones, due to serious problem of antibiotic resistance among the human pathogens.

Keywords Marine actinomycetes · Metabolite profiling · Antibiotic resistance · Antimicrobial metabolites · Drug discovery

J. Thumar (✉)
Department of Microbiology, Government Science College, Gandhinagar, Gujarat, India

S. P. Singh
Department of Biosciences, Saurashtra University, Rajkot, Gujarat, India

L. Karthik (ed.), *Actinobacteria*, https://doi.org/10.1007/978-981-16-5835-8_13

Abbreviations

AGS	Human gastric adenocarcinoma cells
DKP	Diketopiperazine
ECD	Electron capture detector
FDA	Food and drug administration
GC-MS	Gas chromatography mass spectrometry
HepG-2	Human liver cancer cell lines
HPLC	High performance liquid chromatography
HRESIMS	High resolution electrospray ionization mass spectrometry
HRTOFMS	High resolution time-of-flight mass spectrometry
IC_{50}	Half-maximal inhibitory concentration
KB	Keratin-forming tumor cell lines
LC-MS	Liquid chromatography mass spectrometry
LU-1	Lung cancer cell lines
MAC	*Mycobacterium avium* complex
MCF-7	Breast cancer cell line
MiaPaca-2	1-Pancreatic carcinoma cell lines
MIC	Minimum inhibitory concentration
MRSA	Methicillin-resistant *Staphylococcus aureus*
NMR	Nuclear magnetic resonance
TLC	Thin layer chromatography
VRE	Vancomycin-resistant *Enterococci*
WHO	World Health Organization

13.1 Introduction

Emergence of antibiotic resistance in pathogens has become an alarming problem over the globe. In addition, the decline in the discovery and development of new antibiotics has created havoc in the health sector (Genilloud 2017; Durand et al. 2019). The development of multiple drug resistance in the pathogenic strains reduced susceptibility to antimicrobial compounds and modification of the target drugs has led to an increase in deaths caused by the infectious diseases worldwide. These pathogenic bacteria possess a number of virulent factors, some encoded in plasmids, bacteriophages, and the bacterial chromosomes. Such organisms can also colonize in a biofilm protecting the cells against therapeutic antibacterial agents (Brander et al. 2005; Lino and Degracious 2006). According to the list on the fetal human pathogens, released by World Health Organization (WHO) in 2017, there are a total of 12 bacterial families having multiple drug resistance (WHO 2017). O'Neill (2016) reported that approximately, 7,00,000 deaths occur every year due to multi-drug-resistant pathogens, and this may increase to ten million per year by 2050, if the current trend continues. Organisms may develop multiple drug resistance by various mechanisms; such as presence of antibiotic degrading enzymes, antibiotic altering enzymes, and gene transfer processes like conjugation, transformation, and transduction. Therefore, it necessitates the search of naturally occurring novel antimicrobial compounds to curve the increasing menace of the infection (Vasavada et al. 2006; Thumar et al. 2010).

13.2 Antibiotics: Past and Present

Nature is the great treasure of millions of prokaryotes and eukaryotes which includes approximately 0.5 million plant species, 10^{11}–10^{12} microbial species and 1.5 million fungi. Unfortunately, only a small fraction out of it (approximately 250,000–300,000) has been documented (Berdy 2012; Locey and Lennon 2016). The microbial metabolites are used as the main bioactive scaffold for the development of the novel antibiotics instead of using the already known synthetic combinational treasure of molecules to develop novel drugs (Challinor and Bode 2015). The period spanning 1950–1960 is considered as "The golden age of antibiotics." During this time phase, the large-scale cultivation of microorganisms and extraction of secondary metabolites for the identification of novel antimicrobial compounds was carried out. Genus *Streptomyces* alone is identified as the huge source of novel antimicrobial compounds including antibacterial, antifungals, antiprotozoal, and antivirals. US Food and Drug administration (FDA) gave approval to approximately 1211 small molecule drugs during 1981–2014, among which approximately 65% accounted for natural chemicals/compounds (Newman and Cragg 2016; Noman Van 2016).

13.2.1 Antibiotics from Actinomycetes: Research and Developments

During the last 76 years of research on the actinomycetes for novel bioactive metabolites for human welfare, more than 5000 bioactive compounds were explored and investigated. During this period, the actinomycetes research advanced in various dimensions, from isolation and screening techniques to molecular approaches including post-genomic research for metabolites (Demain and Sanchez 2009; Subramani and Aalbersberg 2012). According to a report by Subramani and Sipkema (2019), during 2007–2017, approximately 177 new species of marine actinomycetes were isolated from geographically rare habitats and belonged to 33 families including three novel families and 29 new genera. The single genus *Streptomyces* produces more than 80% of all actinomycetes origin antibiotics (Subramani and Aalbersberg 2013). Ten major classes of antibiotics are produced by actinomycetes including oligomycin-type macrolids, polyene macrolids, daunomycin-type anthracyclines, non-actin type cyclopolylactones, aminoglycosides, streptothricin, nigericin-type polyethers, cyclopolylactones, quinoxaline-peptides, and actinomycins (Berdy 2012).

13.2.2 *Marine Actinomycetes: The Source of Novel Antimicrobial Compounds*

It is believed that till date we could explore only a small portion of marine microbes. Because of limited accessibility and lack of proper leads, many unique biomolecules from different marine microbial communities are waiting to be discovered. The major pharmaceutical companies are at the verge of losing interest from natural products of microbial origin and focusing on alternative discovery approaches, such as combinational chemistry (Koehn and Carter 2005). This paradigm shift is because of the over-exploitation of the microbial resources and continued rediscovery of compounds that are already in use. However, natural product research has renewed the interest because of significant rise in the demand of novel compounds to treat drug-resistant microbial infections (Li and Vedaras 2009). This is mainly due to the low returns from alternative discovery platforms. It included the exploration of microbial wealth from poorly and less attended habitats, a concept based on the assumption that organisms evolve new bioactive metabolites in order to adapt to the unusual/extreme environments (Letzel et al. 2013). In the light of this knowledge, marine actinomycetes have recently focused attention with emphasis on their biocatalytic potential and pharmaceutically important secondary metabolites (Sharma et al. 2020; Rathore et al. 2021).

Actinomycetes are a group of industrially important microorganisms because of their capability to produce a range of commercially viable products in various sectors; including agriculture, healthcare, veterinary, food, and nutrition (Sisi et al. 2020; Thakrar and Singh 2019; Thumar and Singh 2009). As per the records until October, 2016, the domain Bacteria includes 30 currently recognized phyla, the Actinobacteria being one of the largest phyla with 6 families, about 18 orders, almost 63 families and more than 370 genera (Subramani and Sipkema 2019). Despite a critical role in biogeochemical cycles, the actinomycetes also produce a variety of enzymes (Thumar and Singh 2007a, b; Chen et al. 2020) and therapeutic compounds (Sisi et al. 2020). There are approximately 500,000 naturally occurring biological compounds, from which approximately.

70,000 are microbially derived molecules and 29% are solely derived from actinomycetes. Actinomycetes are Gram-positive, high G + C (>55%) bacteria which were earlier misbelieved as an intermediate link between bacteria and fungi. Being saprophytic in nature, they are the dominant group of soil microflora involved in recycling of organic matter. The metabolites obtained from actinomycetes range from enzymes, antitumor agents, immunity-modifiers, enzyme inhibitors, cytotoxic molecules to vitamins, and nutritional material.

Approximately, 70% of the surface of planet Earth is covered by oceans, accounting for nearly 97% of total water and possessing 80% of the life. There are 15 exclusively marine phyla out of total 33 known animal phyla (Margulis and Chapman 2009). The marine habitats vary in their ecological pressure with respect to available nutrients, pressure, light, oxygen, predation, competition for space, etc. In order to survive under such extreme conditions, marine organisms have developed unique survival strategies, such as secretion of potent and novel secondary metabolites (Skropeta and Wei 2014). Various unexplored or underexplored ecosystems are the most promising sources of novel actinomycetes (Dhakal, et al. 2017). Many of these compounded are afforded by marine actinomycetes belonging to deep sea sediments, marine sponges, marine invertebrates, plants, and coral reefs (Zhang et al. 2005; Thomas et al. 2010; Vynne et al. 2011; Blunt et al. 2013; Viegelmann et al. 2014).

13.2.2.1 Bioactive Compounds from Marine Actinomycetes with Novel Pharmaceutical Potential

Research on pharmaceutically active metabolites from marine actinomycetes is emerging as a hot spot since a decade. A significant number of varied and novel molecules have been isolated from marine-derived actinomycetes. A new molecule, 3-(4-hydroxybenzyl) piperazine-2,5-dione was obtained from a marine *Streptomyces* sp. (Sobolevskaya et al. 2007). Molecular structure of the compound was drawn on the basis of NMR and mass spectroscopy. Its cytotoxic activity was checked on sperm and eggs of the sea urchin *Strogylocentrotus intermedius*.

Actinomycetes exhibit a tremendous taxonomic diversity ranging from the most typical genus *Streptomyces* to rare and exotic non-*Streptomyces* genera including *Dietzia, Salinispora, Marinophilus, Rhodococcus, Solwaraspora, Salinibacterium, Williamsia, Verrucosispora, and Aeromicrobium*, and thereby, increasing the possibilities of new potent bioactive metabolites (Valliappan et al. 2014). There are many compounds from marine actinomycetes, which have been selected for the pharmaceutical trial based on their strong potential. For instance, Diazepinomicin—a dibenzodiazepine alkaloid extracted from *Micromonospora* strain, which exhibited significant antitumor activities. Further, it is also nominated for clinical trials in phase II for the treatment of human glioblastoma cancer (Charan et al. 2004; Mason et al. 2012).

Salinispora is a newly described genus of obligate actinomycetes and also a rich source of such compounds (William and Jensen 2006; Williams et al. 2007a). Diverse categories of secondary metabolites such as cyanosporaside A, saliniketal A and B (Williams et al. 2007b) and sporolide A (Buchanan et al. 2005) have been discovered from this actinomycete on the basis of numerous chemical investigations. Recent studies highlighted *Salinispora* and its extraordinary biosynthetic diversity (Jensen et al. 2015). Interestingly, Salinosporamide A, a β-lactone-γ-lactam obtained from *Salinispora tropica* could enter clinical trials soon after its discovery to cure multiple myeloma.

13.3 Metabolite Profiling of Marine Actinobacteria

Majority of the drug discovery programs are oriented around actinobacteria because of their abundant resourcefulness for discovery of numerous lead metabolites. Further, the emergence of unique metabolic pathways provides them an ability to synthesize diverse categories of bioactive metabolites which are rarely available in terrestrial habitats. Marine actinomycetes hold an important position in drug discovery programs in comparison to terrestrial counter parts, mainly because of their unique metabolic pathways and rich molecular library (Yang et al. 2019). Many new biologically active compounds have been isolated from marine actinomycetes from the year 2015 to 2021 as highlighted in Table 13.1.

13.3.1 Antibacterial Activities

Antibacterial substances are significant in the control of infectious diseases which may cause deaths due to drug resistance among the pathogens. The microbial pathogens have developed resistance against various antibacterial compounds. Marine actinobacteria are being used to develop effective newer drugs without any side effects (Table 13.1)

13.3.1.1 Antibacterial Compounds from Marine-Derived Streptomyces sp.

Reports say that out of 100% bioactive metabolites isolated from actinomycetes till date, more than 70% were derived from *Streptomyces* and rest from other rare actinomycete species. Until recently, a range of antibacterial compounds have been reported from marine-derived *Streptomyces* sp. Hassan et al. (2015) identified Salinamide F **(1),** from the culture broth of *Streptomyces* sp., having antibacterial activity against a range of bacterial pathogens including *Enterococcus faecalis, Enterobacter cloacae, Haemophilus influenzae,* and *Neisseria gonorrhoeae*. Chemical analysis of Salinamide F by HRTOFMS revealed its molecular formula $C_{51}H_{71}N_7O_{16}$. Similarly, aranciamycins I and J **(2)** from *Streptomyces sp.* CMB0150 showed moderate-to-severe activity against *Mycobacterium tuberculosis*, Gram-positive *Bacillus subtilis*, and human cancel cell lines with IC_{50} values 0.7–1.7 μM, >1.1 μM and >7.5, respectively (Khalil et al. 2015). *Streptomyces* sp.SNM5 has been reported to produce Hormaomycins B and C **(3)** under altered cultural conditions (Bae et al. 2015a). Very similar to this, rocheicoside A **(5)**—a cytosine type nucleotides discovered from *Streptomyces rochei* 06CM016 demonstrated significant antimicrobial activity (Aksoy et al. 2016). Similarly, Lacret and co-workers (2016) reported napyradiomycin MDN-0170 **(7)** from *Streptomyces zhaozhouensis* CA-271078 with antibacterial (against methicillin-resistant

Table 13.1 Novel bioactive metabolites from marine actinomycetes (From year 2015–2021)

Sr. no.	The organism	Name of the compound	Biological activity	Reference
01	*Streptomyces* sp.	Salinamide F	Antibacterial	Hassan et al. (2015)
02	*Streptomyces* sp. CMB0150	Aranciamycins I and J	Antibacterial	Khalil et al. (2015)
03	*Streptomyces* sp.SNM5	Hormaomycins B and C	Antibacterial	Bae et al. (2015a)
04	*Streptomyces* sp.	Mohangamides A and B	Antifungal	Bae et al. (2015b)
05	*Streptomyces rochei* 06CM016	Rocheicoside A	Antibacterial and antifungal	Aksoy et al. (2016)
06	*Streptomyces zhaozhouensis* CA-185989	Ikarugamycin derivatives	Antifungal	Lacret et al. (2015)
07	*Streptomyces zhaozhouensis* CA-271078	Napyradiomycin MDN-0170	Antibacterial and antifungal	Lacret et al. (2016)
08	*Streptomyces* sp. SCSGAA 0027	Nahuoic acids B-E	Antibacterial	Nong et al. (2016)
09	*Nocardiopsis* sp. SCSIO 10419, SCSIO 04583, SCSIO KS107	α-pyrones (1–8)	Antibacterial	Zhang et al. (2016)
10	*Streptomyces* sp. 182SMLY	Polycyclic anthraquinones	Antibacterial	Liang et al. (2016)
11	*Micromonospora* sp. 5–297	Tetrocarcins N and O	Antibacterial	Tan et al. (2016)
12	*Nocardiopsis* sp. G057	Compounds 1–12	Antibacterial and antifungal	Thi et al. (2016a)
13	*Micromonospora* sp. G019	Quinoline alkaloid and 1,4- dioxane derivative	Antibacterial	Thi et al. (2016b)
14	*Verrucosispora* sp. MS 100047	1-Hydroxy-2, 5-dimethyl benzoate	Antibacterial	Huang et al. (2016)
15	*Streptomyces* sp.*IMB094*	Neo-actinomycins A and B	Antibacterial and antifungal	Wang et al. (2017)
16	*Streptomyces* sp.SUK 25	Diketopiperazine derivatives	Antibacterial and cytotoxic	Alshaibani et al. (2017)
17	*Streptomyces* sp. HZP-2216E	N-arylpyrazinone	Antibacterial and cytotoxic	Zhang et al. (2017a)
18	*Streptomyces* sp. HZP-2216E	Indolizinium alkaloids and Bifilomycins	Antibacterial and cytotoxic	Zhang et al. (2017b)
19	*Streptomyces* sp. EGY1	Sharkquinone	Antitumor	Abdelfattah et al. (2017)
20	*Streptomyces* sp. M-207	Lobophorin K	Antibacterial and cytotoxic	Brana et al. (2017)
21	*Streptomyces chartreusis* NA02069	Streptazolins A and B	Antibacterial	Yang et al. (2017)

(continued)

Table 13.1 (continued)

Sr. no.	The organism	Name of the compound	Biological activity	Reference
22	*Micromonospora* sp. RJA4480	Ansa microlides (1–4)	Antibacterial	Williams et al. (2017)
23	*Micromonospora harpali* SCSIO GJ089	Spirotetronate aglycones	Antibacterial	Gui et al. (2017)
24	*Kribella* sp. MI481-42F6	Kribellosides	Antifungal	Igarashi et al. (2017)
25	*Actinomadura* sp. DSMS-114	Methylbenz[a]anthracene-7, 12-quinone	Antibacterial	Kurata et al. (2017)
26	*Thermoactinomyces vulgaris* ISCAR 2354	Thermoactinoamide A	Antibacterial	Teta et al. (2017)
27	Actinomycete HF-11225	Nivelactum B	Antibacterial	Chen et al. (2018)
28	*Streptomyces pratensis*	New angucycline-type antibiotics	Antibacterial	Akhter et al. (2018)
29	*Streptomyces coeruleorubidus* GRG 4	Bis (2-Ethylhexyl) phthalate (BEP)	Antibacterial and antitumor	Rajivgandhi et al. (2018)
30	*Streptomyces* sp. LHW52447	Actinomycins D1-D4	Antibacterial	Jiao et al. (2018)
31	*Streptomyces cyaneofuscatus* M-169	Anthramycin B	Antibacterial	Rodriguez et al. (2018)
32	*Streptomyces seoulensis* A 01	Streptoceomycin 1	Antibacterial	Zhang et al. (2018a)
33	*Streptomyces sp. ZZ745*	Bagremycins (F-G)	Antibacterial	Zhang et al. (2018b)
34	*Streptomyces xinghaiensis* SCSIO S15077	Tunicamycin E	Antibacterial and antifungal	Zhang et al. (2018c)
35	*Streptomyces* sp. IMB7–145	Niphimycins C-E	Antibacterial and antifungal	Hu et al. (2018)
36	*Nocardiopsis* sp.	Terretonin N-1	Antibacterial	Hamed et al. (2018a)
37	*Streptomyces mutabilis* sp. MII	Borrelidin B	Anticancer	Hamed et al. (2018b)
38	*Micromonospora carbonacea* LS276	Tetrocarcin Q	Antibacterial	Gong et al. (2018)
39	*Streptomyces chartreusis* XMA39	Medermycin, Streptoxepinmycin A-D	Antibacterial and antifungal	Jiang et al. (2018)
40	*Nocardiopsis* sp. CNQ-115	Fluvirucin	Antibacterial	Leutou et al. (2018)
41	*Lechevalieria aerocolonigenes* K 10–0216	Pyrizomicins A and B	Antibacterial and antifungal	Kimura et al. (2018)
42	*Kocuria marina* CMGS2	Kocumarin	Antibacterial	Uzair et al. (2018)

(continued)

Table 13.1 (continued)

Sr. no.	The organism	Name of the compound	Biological activity	Reference
43	*Streptomyces* sp. G212	Novel metabolites	Antibacterial and antifungal	Cao et al. (2019a)
44	*Streptomycetes* sp. G248	Lavandulylated flavanoids	Antibacterial	Cao et al. (2019b)
45	*Streptomycetes* sp. strain 271,078	Napyradiomycins	Antibacterial and cytotoxic	Carretero-Monila et al. (2019)
46	*Streptomycetes albolongus* CA-186053	Medermycin analog MDN-0171	Antibacterial	Lacret et al. (2019)
47	*Streptomycetes puniceus*	Diketopiperazines	Antifungal	Kim et al. (2019)
48	*Streptomyces* sp. ZZ741	Streptoglutirimides	Antifungal, antibacterial, and cytotoxic	Zhang et al. (2019a)
49	*Streptomyces* sp. SCSIO 41	Aranciamycin and Isotirandamycin	Cytotoxic and antibacterial	Cong et al. (2019)
50	*Streptomyces althioticus* MSM3	Desertomycin G	Antitumor and antibacterial	Brana et al. (2019)
51	*Streptomyces* sp. OPMA 1730	Nosiheptides, Griseoviridin, and Etamycin	Antibacterial	Hosoda et al. (2019)
52	*Streptomyces sp.* ZZ820	Streptoprenylindoles A-C	Antibacterial	Yi et al. (2019)
53	*Streptomyces atratus* SCSIOZH16	Atratumycin	Antibacterial	Sun et al. (2019)
54	*Salinispora arenicola* BRA-213	Salinapththoquinones	Antibacterial	Da Silva et al. (2019)
55	*Verucosispora* sp. SCSIO	Kendomycins	Antibacterial	Zhang et al. (2019b)
56	*Streptomyces* sp. G246	Lavandulylated flavanoids	Antibacterial and antifungal	Cao et al. (2020)
57	*Streptomyces* sp.EG1	Mersaquinone	Antibacterial	Kim et al. (2020)
58	*Streptomyces* sp.4506	Lobophorin L and M	Antibacterial	Luo et al. (2020)
59	*Streptomyces* sp.	*n*-hexadecanoic acid, tetradecanoic acid, and pentadecanoic acid	Antifungal	Sangkanu et al. (2021)

Staphylococcus aureus) and antifungal properties (against *Aspergillus niger* and *Candida albicans*). The compound was studied with respect to its structure on the basis of molecular modeling in combination with nOe—nuclear overhauser effect NMR spectroscopy—and coupling constant analysis. *Streptomyces* sp. SCSGAA 0027 yielded nahuoic acids B-E (**8**); a novel nahuoic acid with SETD8 inhibition activity. Compound 1–5 showed antibiofilm activity against *Shewanella onedensis* MR-1 biofilms (Nong et al. 2016).

Neo-actinomycins A and B **(15)** were extracted from *Streptomyces* sp. IMB094 which displayed strong antibacterial activity against VRE (vancomycin-resistant *Enterococci*). Structure elucidation by spectroscopic analysis confirmed the presence of tetracyclic 5H-oxazolo (4,5-b) phenoxazine (Wang et al. 2017). *Streptomyces* sp. SUK 25 produced five active diketopiperazine (DKP) derivatives **(16)** which displayed significant activities against multi-drug-resistant *Staphylococcus aureus* (Alshaibani et al. 2017). Streptazolins A and B **(21)** were isolated, together with already reported streptazolin, from *Streptomyces chartreusis* NA02069, which displayed weak anti-*Bacillus subtilis* activity with MIC value of 64 μM. While compound A inhibited acetylcholinesterase (AchE) activity under in vitro conditions with IC_{50} value 50.6 μM, compound B was not active at all (Yang et al. 2017). Novel angucycline-type antibiotics 1 and 2 **(28)** from *Streptomyces pratensis* NA-ZhouA1 showed antibacterial activities against *Klebsiella pneumoniae*, *Escherichia coli,* and MRSA (methicillin-resistant *Staphylococcus aureus*) (Akhter et al. 2018). Bis (2-ethylhexyl) phthalate (BEP) **(29)** produced by *Streptomyces coeruleorubidus* GRG 4, inhibited CR (colistin resistant) *Klebsiella pneumoniae,* and *Pseudomonas aeruginosa* (Rajivgandhi et al. 2018). Recently, Jiao et al. (2018) reported actinomycins D1–D4 **(30)** from the culture broth of Streptomyces sp. LHW52447. They exhibited strong antibacterial activities against MRSA (MIC- 0.125–0.25 μg/ml).

Anthramycin B **(31)**, a potent anti-tubercular compound against *Mycobacterum tuberculosis* (MIC 0.03 μg/ml) has been isolated from *Streptomyces cyaneofuscatus* M-169. The structure elucidation of the compound revealed the presence of lactone carbonyl on first carbon and oxygenated enol on third carbon. Further, the ability of the organism to produce anthramycin B at very high quantities (17.7 mg/L) was evident during the studies (Rodriguez et al. 2018). A rare macrodilactone named Streptoceomycin 1 **(32)** with anti-microaerophilic bacterial activity has been extracted from *Streptomyces seoulensis* A 01. When characterized to unfold the structural details, it was found to possess a pentacyclic ring along with the ether bridge (Zhang et al. 2018a). Two Bagremycins analogs; F and G **(32)** were obtained from *Streptomyces* sp. ZZ745. Both the compounds were highly active against *Escherichia coli* and showed the MIC values 41.8 (F) and 61.7 (G) μM, respectively (Zhang et al. 2018b). Same way, *Streptomyces xinghaiensis* SCSIOS15077 is reported to produce tunicamycin E by Zhang et al. (2018c). Very high to moderate activities against *Bacillus thuringiensis* W102 and *Bacillus thuringiensis* BT01 were evident based on the MIC values (range: 0.0008–2 μg/ml). Further, four new naphthoquinones named Medermycin **(39)** and Streptoxepinmycin A-D were found in the extracts of *Streptomyces chartreusis* XMA39 (Jiang et al. 2018). These compounds afforded the antibacterial compounds against *E. coli* and MRSA along with antifungal activities against *Candida albicans*.

Cao et al. (2019a) reported novel metabolites **(43)** with antibacterial and antifungal activities from marine-derived *Streptomycetes sp.* G212. Nuclear magnetic resonance (NMR) and other analysis confirmed the presence of three new lavandulylated flavonoids (44) which showed significant inhibitory activities against multi-drug-resistant *Mycobacterium tuberculosis* H37Rv. Recently, Carretero-Monila et al. (2019) reported four new napyradiomycins (1–3, 5) **(45)** from

Streptomycetes sp. strain 271,078 with detailed characterization. While compound 1 had a functionalized prenyl side chains of napyradiomycin—A series, compound 2 and 3 harbored rings of chlorocyclohexane resembling to napyradiomycin B. The authors further identified compound 5 to be a new class of napyradiomycins on the basis of its cyclic ether ring and designated the compound as napyradiomycin D1. All the compounds also displayed remarkable inhibitory activities against *Mycobacterium tuberculosis*, *Staphylococcus aureus,* and cytotoxic activity against human liver cancer cell lines (Hepatoma G2). Lacret and co-workers (2019) isolated a new Medermycin analog MDN-0171 **(46)** from marine-derived *Streptomycetes albolongus* CA-186053 which showed potent activity against MRSA (methicillin-resistant *Staphylococcus aureus*) and *E. coli.* Streptoglutirimides A-J **(48)** with antibacterial (methicillin-resistant *Staphylococcus aureus*; MIC: 08–12 μg/ml), antifungal (*Candida albicans;* MIC: 08–20 μg/ml) and cytotoxic (human glioma U87MG and U251 cells with IC_{50} values 1.5–3.8 μM) activities was reported by Zhang et al. (2019a). They elucidated the structure of these compounds based on their HRESIMS data, ECD calculations, X-ray diffraction experiments, and NMR spectroscopic analysis.

Mycobacterium is a multi-drug-resistant organism and is known to cause serious diseases in humans including *Mycobacterium avium* complex (MAC). Cultivation of *Streptomyces* sp. OPMA 1730 yielded Griseoviridin, Nosiheptides, and Etamycin **(51)**. Interestingly, these compounds showed portent activities against *Mycobacterium avium* and *M. intracellulare* with MIC in the range of 0.024–1.5 μg/ml (Hosoda et al. 2019). Streptoprenylindoles A-C **(52)** was isolated from *Streptomyces* sp. ZZ820, which reflected the antibacterial activity against MRSA (Yi et al. 2019). Recently, Sun et al. (2019) reported atratumycin **(53)** from *Streptomyces atratus* SCSIOZH16 with broad spectrum antibacterial activity. The organic extract of sponge-derived *Streptomyces* sp. G246 yielded two new lavandulylated flavonoids **(56)**. These metabolites had a broad spectrum antibacterial activity against a range of Gram-positive (*Bacillus subtilis* and *Staphylococcus aureus*) and Gram-negative bacteria (*Enterococcus faecalis*, *Salmonella enterica*, *Pseudomonas aeruginosa*) (Cao et al. 2020). Similarly, Kim and co-workers (2020) reported mersaquinone **(57)** from *Streptomyces sp.* EG1 which displayed antibacterial activity against MRSA (MIC- 3.36 μg/ml). Luo et al. (2020) reported two new spirotetronates **(58)** natural products from marine *Streptomyces* sp.4506 with strong antibacterial activities.

13.3.1.2 Antibacterial Compounds from Marine-Derived NOCARDIOPSIS sp.

Genus *Nocardiopsis* is known for its biotechnologically versatile and ecologically important nature. Many species of *Nocardiopsis* have been reported to belong to hyper saline locations. Diverse antibacterial compounds including terphenyls, alkaloids, polyketides, quinoline alkaloids, amines, proteins, thiopeptides, and phenzines have been studied from this genus. Eight new α-pyrones **(9)** were obtained from

Nocardiopsis sp. SCSIO 10419, SCSIO 04583, and SCSIO KS107. They displayed antibacterial activity against *Bacillus cereus* and *Micrococcus luteus* (Zhang et al. 2016). The structure analysis revealed that the side chain was important to decide the characteristic high wavelength ECD transition. Similarly, *Nocardiopsis* sp. G057 afforded the secretion of 12 compounds each with different chemical properties (**12**). While antibacterial activity of compound 1 was evident against *E. coli* (MIC 16 μg/ml), compound 2 and 3 displayed the activity against both, Gram-positive and Gram-negative bacteria and the yeast *candida albicans,* respectively (Thi et al. 2016a).

Terpenes have emerged as an interesting group of bioactive metabolites these days, may be because of their diverse skeletal compositions. Soil-state fermentation of *Nocardiopsis* sp. yielded a highly oxygenated terretonin N-1 **(36)**—a unique tetracyclic 6-hydroxymeroterpenoid. While its antibacterial activity against Gram-positive *Staphylococcus warneri* was very significant, very low activity was detected against Gram-negative *E.coli* (7 mm) (Hamed et al. 2018a). Recently, Fluvirucin B6 **(40)**—a 14-membered macrolactum was extracted from *Nocardiopsis sp.*CNQ-115. Surprisingly, it exhibited weak antibacterial activity against Gram-positive *Bacilli* and no effect at all on Gram-negative bacteria (Leutou et al. 2018).

13.3.1.3 Antibacterial Compounds from Marine-Derived Micromonospora sp.

Genus *Micromonospora* has been established as a vigorous model for the drug discovery module since its discovery before 100 years. It is still emerging as an untapped resource of many drug leads because of its unique chemical diversity. *Micromonospora* sp. 5–297 produced two new tetrocarcins N- and O-glycosidic spirotetronate antibiotics (**11**). Structural analysis revealed that tetrocarcin O is the derivative of tetrocarcin N. Both the compounds were able to inhibit the growth of *Bacillus subtilis* with MIC ranging from 02 μg/ml (tetrocarcin N) to 64 μg/ml (tetrocarcin O). Similarly, *Micromonospora* sp.G019 secreted quinoline alkaloid as well as 1,4-dioxine derivative **(13)**. While quinoline alkaloid showed antibacterial activity against human pathogens including *Enterococcus faecalis*, *Salmonella enterica,* and *Escherichia coli*, the 1, 4-dioxane derivative was effective against *Enterococcus faecalis* and *Candida albicans* (MIC- 32 μg/ml and 64 μg/ml, respectively) (Thi et al. 2016b). Ansa microlides 1–4 **(22)** were obtained from *Micromonospora* sp. RJA4480. These four antibiotics showed very high antibacterial activity against prominent human pathogens including methicillin-resistant *Staphylococcus aureus*, *Escherichia coli*, and *Mycobacterium tuberculosis* having MIC values of 0.0009, 0.0003, and 0.0009 (compound 1); 0.0001, 0.00083, and 0.0009 μg/ml (compound 2); 0.8, 1.8, and 7.0 μg/ml (compound 3); 0.06, 0.40, and 1.80 (compound 4) μg/ml, respectively (Williams et al. 2017).

Two spirotetronate aglycones **(23)**, 22-dehydroxymethyl-kijanolide and 8-hydroxy-22-dehydroxymethyl-kijanolide, were separated from *Micromonospora harpali* SCSIO GJ089. Both the compounds displayed very high activity against *Bacillus subtilis* and *B. thuringiensis* with MIC values ranging from 0.016 to 8.0 μg/

ml (Gui et al. 2017). The fermentation broth of *Micromonospora carbonacea* LS276 yielded a new spirotetrone Tetrocarcin Q **(38).** Bearing a glycosyl group, the compound possessed moderate potency (MIC; 12.5 μM), when tested against *Bacillus subtilis* ATCC 63501. Presence of a unique sugar (2-deoxy-allose) at C-9 position of the compound was reported for the first time from spirotetronate glycosides (Gong et al. 2018).

13.3.1.4 Antibacterial Compounds from Other Marine-Derived Actinomycetes

As stated earlier, there are only a few rare non-*Streptomyces* actinomycete genera have been identified from marine sources in recent past. Bulk cultivation of *Verrucosispora* sp. MS 100047 afforded the production of a new glycerol 1-hydroxy-2, 5-dimethyl benzoate—a salicylic acid derivative **(14)**. It exhibited selective activity against methicillin-resistant *Staphylococcus aureus* (MRSA) with MIC 12.5 μg/ml. In addition; the compound also displayed significant anti-tubercular activity (Huang et al. 2016). Kurata et al. (2017) reported the extraction and structure elucidation of *Actinomadura* sp. DS-MS-1145 derived, 6, dihydrol-1-8, dihydroxy-3-methylbenz(a)anthracene-7, 12-quinone **(25).** The purified compound possessed very strong activity when tested against Gram-positive *Staphylococcus aureus*. However, scare activities were evident against Gram-negative, *E. coli*; yeast, *Candida albicans* and fungi, *Aspergillus brasiliensis*. The molecular formula of the compound was $C_{19}H_{14}O_4$ with the molecular weight 306.0966 (Kurata et al. 2017). Thermoactinoamide A **(26)**—a lipophilic cyclopeptide antibiotic was obtained from thermophilic bacteria—*Thermoactinomyces vulgaris* ISCAR 2354. The cyclic hexapeptide displayed potent activity against *Staphylococcus aureus* with MIC value 35 μM (Teta et al. 2017). Nivelactum B **(27)** was obtained from actinomycete HF-11225, which displayed antibacterial activities against a range of pathogens.

The culture broth of very rare actinomycete *Lechevalieria aerocolonigenes* K 10-0216 yielded Pyrizomicins A and B **(41)**, which exhibited strong activity against a range of pathogenic bacteria. Interestingly, the results of NMR and mass spectroscopy proposed them as the new thiazolyl pyridine compounds (Kimura et al. 2018). A unique ultraviolet (UV) bioactive kocumarin **(42)** was obtained from *Kocuria marina* CMGS2 isolated from a sea weed *Pelvetia canaliculata*. It showed potent activity against pathogenic bacteria including MRSA (range of MIC; 15–20 μg/ml) and fungal isolates (minimum fungal inhibitory concentration; 15–25 μg/ml). The chemical structure elucidation studies confirmed the compound to be 4-[(Z)-2 phenyl ethenyl] benzoic acid (Uzair et al. 2018). Salinaphthoquinones **(54)** with broad spectrum antimicrobial activities were obtained from *Salinispora arenicola* BRA-213 (Da Silva et al. 2019). The solvent extracts of *Verucosispora* sp. SCSIO 07399 yielded three new analogs (B-D) of kendomycin **(55)** with very good antibacterial activities. The compounds were very effective against six Gram-positive bacteria with 0.5–8.0 μg/ml (range) of MIC values (Zhang et al. 2019b).

13.3.2 Antifungal Activities

While numerous antibiotics have been isolated from a range of marine microorganisms, studies to discover potent compounds against fungal pathogen are still at the limit. Marine actinobacteria can be a hidden treasure for the exploration of many antifungal metabolites. As discussed in the Table 13.1 Bae et al. (2015b), reported mohangamides A and B **(4)** from *Streptomyces* sp. which strongly inhibited *Candida albicans* isocitrate lyase. When studied by chromatographic and spectroscopic analysis, the compound showed a novel structure with dilactone-ethered pseudodimeric peptides having 14 different amino acids and two unusual acyl chains. Similarly, Ikarugamycin derivatives **(6)** from *Streptomyces zhaozhouensis* CA-185989 showed remarkable antifungal activities, when tested against *Candida albicans* (MIC; 2–4 µg/ml) and *Aspergillus fumigatus* (MIC; 4–8 µg/ml) (Lacret et al. 2015). Antifungal cocktail included three new tetramic acid macrolactams (polycyclic) with four already identified compounds. Further, the authors claimed that compound-1 from the above mixture was a newly isolated natural compound by them and hence, was given the trivial name isokarugamycin. Capping enzymes are different in terms of the structure and function in yeast, when compared to mammalian system. Cultivation of *Kribbella* sp. MI481-42F6 yielded Kribellosides **(24)**—RNA 5′-triphosphatase inhibitor which belong to the alkyl glyceryl ethers. Kribellosides inhibited *Saccharomyces cerevisiae* and secured the minimum inhibitory concentration in the range of 3.12–100 µg/ml. In addition, it also suppressed the activity of intracellular RNA 5'triohosphatase, named Cet1p from the same organism (Igarashi et al. 2017). Interestingly, tunicamycin E **(34)** with moderate antifungal activities (MIC; 02–1 µg/ml) against fuconazole-resistant *Candida albicans* ATCC96901 has been reported for the first time from *Streptomyces xinghaiensis* SCSIOS15077, isolated from the marine mud sample (Zhang et al. 2018c).

Antifungal activities of five Diketopiperazines **(47)** from marine *Streptomycetes puniceus,* against *Candida albicans,* were explained by Kim et al. (2019). Cyclo (L-Phe-L-Val) was a potent inhibitor with 27 µg/ml half-maximal inhibitory concentration. Streptoglutirimides A-J having antifungal (*Candida albicans;* MIC: 08–20 µg/ml), antibacterial (MRSA; MIC: 08–12 µg/ml), and cytotoxic (against human glioma U87MG and U251 cells with IC_{50} values 1.5–3.8 µM) activities was reported from *Streptomyces sp.* ZZ741 by Zhang et al. (2019a). They elucidated the structure of these compounds based on their HRESIMS data, ECD calculations, X-ray diffraction experiments, and NMR spectroscopic analysis.

Most recently, Sangkanu et al. (2021) extracted and identified n-hexadecanoic acid, tetradecanoic acid, and pentadecanoic acid **(59)** from *Streptomyces* sp. All the compounds were capable enough to inhibit *Talaromyces marneffei*—a thermally dimorphic pathogenic fungus.

13.3.3 Anticancer Activities

Mankind has witnessed many serious health problems such as cancer. Cao et al. (2019a) emphasised that the second most common reason of deaths in human females is breast cancer. While a number of metabolites with anticancer properties are known in recent years, there is need for extensive efforts in this direction. The immense development in the cancer research has geared up the search for anticancer compounds from natural resources. In this direction, many marine actinobacteria are also being studied with respect to their potential to produce antitumor, anticancer, and cytotoxic compounds. The literature suggests that only limited studies have focused on finding bioactive metabolites (Table 13.1) as anticancer agents from marine actinobacteria.

Cultivation of *Streptomyces* sp. 182SMLY produced two new polycyclic anthraquinones **(10)**. Proliferation and progression of glioma—a type of cancer in the glial cells of brain, was suppressed by these compounds (identified as streptoanthraquinone and N-acetyl-N-demethylmayamycin) with IC_{50} values >14–31 and 6.4–5 μM, respectively (Liang et al. 2016). *Nocardiopsis* sp. G057 was identified to produce 12 new compounds **(12).** These compounds displayed strong cytotoxic activity against keratin-forming tumor (KB) cell lines, lung cancer cell lines (LU-1), human liver cancer cell lines (HepG-2), and breast cancer cell line (MCF-7). However, compound 1 and 2 displayed poor effect ($IC_{50;}$ 128 μg/ml) against KB and LU cell lines even at high concentrations (Thi et al. 2016a). *Streptomyces* sp. IMB094-derived neo-actinomycins A and B **(15)** exhibited strong cytotoxic activities against adenocarcinomic human alveolar (A549) and human colon cancer cell lines (HCT116) with IC_{50} values 65.8 and 38.7 nM, respectively (Wang et al. 2017). Five active diketopiperazine (DKP) derivatives **(16)** were obtained from endophytic *Streptomyces sp.* SUK 25 which displayed low toxicity against human hepatoma HepaRG cell line (Alshaibani et al. 2017). Marine green algae *Ulva pertusa* associated *Streptomyces* sp. HZP-2216E secreted N-arylpyrazinone derivative **(17)** which selectively inhibited the cell division of malignant glioma cells. In addition, Streptoarylpyrazinone A was identified as a rare compound existing as a zwitterion from natural sources (Zhang et al. 2017a). Very similar to this, a novel indolizinium alkaloid, named streptopertusacin A, **(18)** was reported in the extracts of *Streptomyces* sp. HZP2216E. Chemical degradation, electronic circular calculations and nOe confirmed it to be a novel compound. Interestingly, it not only inhibited methicillin-resistant *Staphylococcus aureus*, but also affected of human glioma cells with great potency (Zhang et al. 2017b). Marine coral *Lophelia pertusa* – derived *Streptomyces* sp. M-207 afforded to produce Lobophorin K **(20)**. The compound managed to show very strong activity against two human cell lines; 1-pancreatic carcinoma (MiaPaca-2) and 2-breast adenocarcinoma (Brana et al. 2017). The activity of the compounds on human cell lines may establish *Streptomyces* sp. M-207 as the potential candidate for the treatment of highly prevailing breast cancer. Nivelactum B (1), a new macrolactum derivative **(27)** with antifungal activities has been demonstrated from marine-derived actinomycete HF-11225, which showed weak cytotoxic and antifungal activity (Chen et al. 2018). Sponge- associated *Streptomyces* sp. LHW52447 produce four actinomycins D1-D4 **(30)** that possess an oxazole unit into the central phenoxazinone chromohpore. When studied for the cytotoxicity potential, D1-D4 showed the activity against WI-38 human diploid lung fibroblasts (Jiao et al. 2018).

Niphimycins C-E was produced by *Streptomyces* sp. IMB7–145 **(35)**. Hu et al. (2018) proposed their full configuration on the basis of studies on their biosynthetic gene clusters in ketoreductase and enoylreductase domains. The cytotoxicity of niphimycins C, E, and F was evident against cancerous human HeLa cell lines (IC_{50} range: 3.0–9.0 μM). N-acetylborrelidin B **(37)**—a naturally new microlide antibiotic was obtained from *Streptomyces mutabilis* sp. MII which demonstrated a potent cytotoxic effect even in crude extract against carcinoma cell lines of human cervix (KB-3-1) under in vitro conditions (Hamed et al. 2018b). The fermentative cultivation of *Streptomyces sp.* SCSIO 41 afforded aranciamycin and isotirandamycin **(49)**, which displayed in vitro cytotoxic activities against K560 cell lines with IC_{50} values; 22, 1.8, and 12.1 μM, respectively (Cong et al. 2019).

13.3.4 Antitumor Activities

Among various treatment strategies to combat cancer, chemotherapy remains the main and the most efficient treatment. Marine actinomycetes have been recently focused with respect to their metabolic and physiological abilities with the potential to produce antitumor compounds (Table 13.1) (Olano et al. 2009). Abdelfattah et al. (2017) reported a new ana-quinonoid tetracene, Sharkquinone **(19)** from the ethyl acetate extracts of *Streptomyces* sp. EGY1. Quantum chemical calculations and detailed spectral analysis revealed the structure of the compound, which displayed strong ability to overcome necrosis factor-related apoptosis in human gastric adenocarcinoma (AGS) cells. *Streptomyces coeruleorubidus* GRG 4 afforded to produce bis (2-ethylhexyl) phthalate (BEP) **(29)** which displayed very strong activity antitumor activities. It inhibited the proliferation and progression of human lung cancer cells in 24 h of treatment at the concentration of 100 μg/ml along with oxidative damage. Compound was extracted in methanol followed by TLC and HPLC analysis. Presence of carbonyl group was confirmed followed by GC-MS and LC-MS that further confirmed the compound to be BEP (Rajivgandhi et al. 2018).

Desertomycin G **(50)** was obtained from *Streptomyces althioticus* MSM3. It was first time reported to show antitumor activity against colorectal adenocarcinoma cells (DLD-1) and human breast cancer adenocarcinoma (MCK-7) cell lines. Desertomycin G also displayed moderate antibacterial activity against *Clostridium perfringens, Bacteroide fragilis, Haemophilus influenzae,* and *Neisseria meningitidis* (Brana et al. 2019).

13.4 Conclusion

The world is at urgent need of new drugs, especially antibiotics, where the unexplored and underexplored sources remain the natural products. New methodologies, such as genome sequencing in conjunction with molecular genetics, bioinformatics, and understanding of the regulatory and biosynthetic pathways would lead to develop rare molecules for diverse uses including pharmaceuticals. Several analytical approaches such as molecular networking, peptidogenomics and glycogenomics are clubbed with advance mass spectra-based analysis and investigations, making it possible to search strains that eliminate the randomness in the traditionally associated approaches. In the exploration of new resources for the novel bioactive molecules, the marine environment catches more attention because of the tremendous physiological variations among the organisms and also the metabolites of pharmaceutical interest. Expensive studies on the metabolite profiling of marine actinomycetes opened the hidden treasure of the capabilities, these fraction of microorganisms hold, with respect to the production of natural products with antibacterial, antifungal, antiviral, and antitumor properties. They are even diverse with respect to their structural skeletons including polyketides, caprolactones, lynamicins, sterols, terpenoids, cyclic hexapeptides, and nitrogen-containing compounds (e.g., alkaloids and peptides). However, the blending of traditional knowledge and modern analytical will certainly lead to the discovery of many new antimicrobial metabolites to combat the novel infectious agents.

Acknowledgments The work described this review from the SPS laboratory at the Saurashtra University was supported by UGC-CAS program, DST-FIST, DBT-Multi-Institutional Project, MoES (Government of India) Net Working Project, and the Saurashtra University. SPS acknowledges DST-SERB International Travel Fellowships to present his work in Hamburg (Germany), Cape Town (South Africa), and Kyoto (Japan). JT acknowledges financial support from DBT under Bio-Care women scientist research project and DST-SERB International Travel Fellowships to present her work in Cape Town (South Africa). SPS also acknowledge award of UGC BSR Faculty Fellowship. JT acknowledges infrastructure facilities at the Department of Microbiology, Government Science College, Gandhinagar, Gujarat, India.

References

Abdelfattah MS, Elmallah MIY, Mohamed AA, Ishibashi M (2017) Sharkquinone, a new ana-quinonoid tetracene derivative from marine-derived *Streptomyces sp.* EGY1 with TRAIL resistance-overcoming activity. J Nat Med 71(3):564–569. https://doi.org/10.1007/s11418-017-1086-5

Akhter N, Liu YQ, Auckloo BN, Shi YT, Wang KW, Chen JJ, Wu XD, Wu B (2018) Stress–driven discovery of new angucycline–type antibiotics from a marine *Streptomyces pratensis* NA-ZhouS1. Mar Drugs 16:331. https://doi.org/10.3390/md16090331

Aksoy SC, Uzel A, Bedir E (2016) Cytosine–type nucleosides from marine–derived *Streptomyces rochei* 06CM016. J Antibiot 69:51–56. https://doi.org/10.1038/ja.2015.72

Alshaibani MM, Mohamad ZN, Jalil J, Sidik NM, Ahmad SJ, Kamal N, Edrada-Ebel RJ (2017) Isolation, purification, and characterization of five active diketopiperazine derivatives from endophytic *Streptomyces* SUK 25 with antimicrobial and cytotoxic activities. Microbiol Biotechnol 27(11):2074. https://doi.org/10.4014/jmb.1608.08032

Bae M, Chung B, Oh KB, Shin J, Oh DC (2015a) Hormaomycins B and C: new antibiotic cyclic depsipeptides from a marine mud fat–derived *Streptomyces sp.* Mar Drugs 13:5187–5200. https://doi.org/10.3390/md13085187

Bae M, Kim H, Moon K, Nam SJ, Shin J, Oh KB, Oh DC (2015b) Mohangamides A and B, new dilactone-tethered pseudodimeric peptides inhibiting *Candida albicans* isocitrate lyase. Org Lett 17:712–715. https://doi.org/10.1021/ol5037248

Berdy J (2012) Thoughts and facts about antibiotics: where we are now and where we are heading. J Antibiot 65:385–395. https://doi.org/10.1038/ja.2012.27

Blunt JW, Copp BR, Keyzers RA, Munro MH, Prinsep MR (2013) Marine natural products. Nat Prod Rep 30:237–323. https://doi.org/10.1038/nrd4683

Brana AF, Sarmiento-Vizcaíno A, Osset M, Pérez-Victoria I, Martín J, De Pedro N, Díaz C, Vicente F, Reyes F, García LA, Blanco G (2017) Lobophorin K, a new natural product with cytotoxic activity produced by *Streptomyces sp.* M-207 associated with the deep–sea coral *Lophelia pertusa*. Mar Drugs 15:144. https://doi.org/10.3390/md15050144

Brana AF, Sarmiento-Vizcaíno A, Pérez-Victoria I, Martín J, Otero L, Palacios-Gutiérrez JJ, Fernández J, Mohamedi Y, Fontanil T, Salmón M, Cal S, Reyes F, García LA, Blanco G (2019) Desertomycin G, a new antibiotic with activity against *Mycobacterium tuberculosis* and human breast tumor cell lines produced by *Streptomyces althioticus* MSM3, isolated from the Cantabrian Sea Intertidal Macroalgae *Ulva* sp. Mar Drugs 17(2):114. https://doi.org/10.3390/md17020114

Brander C, Zamfir O, Geoffroy S, Laurans G, Ariet G, Thien HV, Gourious S, Picard B, Denamur E (2005) Genetic background of *E. coli* and extended-spectrum beta-lactamase type. Emerg Infect Dis 11:54–81. https://doi.org/10.3201/eid1101.040257

Buchanan GO, Williams PG, Feling RH, Kauffman CA, Jensen PR, William F (2005) Sporolides A and B: structurally unprecedented halogenated macrolides from the marine actinomycete *Salinispora tropica*. Org Lett 7(13):2731–2734. https://doi.org/10.1021/ol050901i

Cao DT, Nguyen TL, Tran VH, Doan-Thi-Mai H, Vu-thi Q, Nguyen MA, Le-Thi H, Chau VM, Pham VC (2019a) Synthesis, structure and antimicrobial activity of novel metabolites from a marine actinomycete in Vietnam's East Sea. Nat Prod Commun 14:121–124. https://doi.org/10.1177/1934578X1901400132

Cao DD, Trinh TTV, Mai HDT, Vu VN, Le HM, Thi QV, Nguyen MA, Duong TT, Tran DT, Chau VM, Ma R, Shetye G, Cho S, Murphy BT, Pham VC (2019b) Antimicrobial lavandulylated flavonoids from a sponge-derived *Streptomyces* sp. G248 in East Vietnam Sea. Mar Drugs 17 (9):529. https://doi.org/10.3390/md17090529

Cao DD, Do TQ, Mai HDT, Thi QV, Nguyen MA, Thi HML, Tran DT, Chau VM, Thung DC, Pham VC (2020) Antimicrobial lavandulylated flavonoids from a sponge-derived actinomycete. Nat Prod Res 34(3):413–420. https://doi.org/10.1080/14786419.2018.1538219

Carretero-Molina D, Ortiz-López FJ, Martín J, Oves-Costales D, Díaz C, de la Cruz M, Cautain B, Vicente F, Genilloud O, Reyes F (2019) New Napyradiomycin analogues from *Streptomyces* sp. Strain CA-271078. Mar Drugs 18(1):22. https://doi.org/10.3390/md18010022

Challinor VL, Bode HB (2015) Bioactive natural products from novel microbial sources. Ann N Y Acad Sci 1354:82–97. https://doi.org/10.1111/nyas.12954

Charan RD, Schlingmann G, Janso J, Bernan V, Feng X (2004) Diazepinomicin, a new antimicrobial alkaloid from a marine *Micromonospora* sp. J Nat Prod 67:1431–1433. https://doi.org/10.1021/np040042r

Chen H, Cai K, Yao R (2018) A new macrolactam derivative from the marine actinomycete HF-11225. J Antibiot 71:477–479. https://doi.org/10.1038/s41429-017-0021-z

Chen XL, Wang Y, Wang P, Zhang YZ (2020) Proteases from the marine bacteria in the genus *Pseudoalteromonas*: diversity, characteristics, ecological roles, and application potentials. Mar Life Sci Technol 2:309–323. https://doi.org/10.1007/s2995-020-00058-8

Cong Z, Huang X, Liu Y, Liu Y, Wang P, Liao S, Wang B, Zhou X, Huang D, Wang J (2019) Cytotoxic anthracycline and antibacterial tirandamycin analogues from a marine-derived *Streptomyces sp.* SCSIO 41399. J Antibiot 72:45–49. https://doi.org/10.1038/s41429-018-0103-6

Da Silva AB, Silveira ER, Wilke DV, Ferreira EG, Costalotufo LV, Torres MCM, Ayala AP, Costa WS, Canuto KM, Araujonobre ARD, Araujo AJ, Filho JDBM, Pessoa ODL (2019) Antibacterial salinaphthoquinones from a strain of the bacterium *Salinispora arenicola* recovered from the marine sediments of St. Peter and St. Paul Archipelago. J Nat Prod 82:1831–1838. https://doi.org/10.1021/acs.jnatprod.9b00062

Demain AL, Sanchez S (2009) Microbial drug discovery: 80 years of progress. J Antibiot 62:5–16. https://doi.org/10.1038/ja.2008

Dhakal D, Pokhrel AR, Shrestha B, Sohng JK (2017) Marine rare actinobacteria: isolation, characterization, and strategies for harnessing bioactive compounds. Front Microbiol 8 (1106):1–8. https://doi.org/10.3389/fmicb.2017.01106

Durand GA, Raoult D, Dubourg G (2019) Antibiotic discovery: history, methods and perspectives. Intl J Antimicrob Agents 53(4):371–382. https://doi.org/10.1016/j.ijantimicag.2018.11.010

Genilloud O (2017) Actinomycetes: still a source of novel antibiotics. Nat Prod Rep 34 (10):1203–1232. https://doi.org/10.1039/c7np00026j

Gong T, Zhen X, Li XL, Chen JJ, Chen TJ, Yang JL, Zhu P (2018) Tetrocarcin Q, a new spirotetronate with a unique glycosyl group from a marine–derived actinomycete *Micromonospora carbonacea* LS276. Mar Drugs 16(2):74. https://doi.org/10.3390/md16020074

Gui C, Zhang SW, Zhu XC, Ding WJ, Huang HB, Gu YC, Duan YW, Ju JH (2017) Antimicrobial spirotetronate metabolites from marine derived *Micromonospora harpali* SCSIO GJ089. J Nat Prod 80:1594–1603. https://doi.org/10.1021/acs.jnatprod.7b00176

Hamed A, Abdel-Razek AS, Frese M, Stammler H, Elhaddad AF, Ibrahim TM, Sewald N, Shaaban M (2018a) Terretonin N: a new meroterpenoid from *Nocardiopsis sp*. Molecules 23:299. https://doi.org/10.3390/molecules23020299

Hamed A, Abdel-Razek AS, Frese M, Wibberg D, Elhaddad AF, Ibrahim TM, Kalinowski J, Sewald N, Shaaban M (2018b) N-Acetylborrelidin B: a new bioactive metabolite from *Streptomyces mutabilis sp.* MII. Z Naturforsch C – J Biosci 73:49–57. https://doi.org/10.1515/znc-2017-0140

Hassan HM, Degen D, Jang KH, Ebright RH, Fenical W (2015) Salinamide F, new depsipeptide antibiotic and inhibitor of bacterial RNA polymerase from a marine-derived *Streptomyces* sp. J Antibiot 68:206–209. https://doi.org/10.1038/ja.2014.122

Hosoda K, Koyama N, Kanamoto A, Tomoda H (2019) Discovery of nosiheptide, griseoviridin, and etamycin as potent anti-mycobacterial agents against *Mycobacterium avium* complex. Molecules 24:1495. https://doi.org/10.3390/molecules24081495

Hu Y, Wang M, Wu C, Tan Y, Li J, Hao X, Duan Y, Guan Y, Shang X, Wang Y, Xiao C, Gan M (2018) Identification and proposed relative and absolute configurations of niphimycins C-E from the marine-derived *Streptomyces sp.* IMB7–145 by genomic analysis. J Nat Prod 81:178–187. https://doi.org/10.1021/acs.jnatprod.7b00859

Huang P, Xie F, Ren B, Wang Q, Wang J, Wang Q, Abodeimageed WM, Liu MM, Han JY, Oyeleye A, Shen JZ, Song FH, Dai HQ, Liu XT, Zhang LX (2016) Anti-MRSA and anti-TB metabolites from marine-derived *Verrucosispora sp.* MS100047. Appl Microbiol Biotechnol 100:7437–7447. https://doi.org/10.1007/s00253-016-7406-y

Igarashi M, Sawa R, Yamasaki M, Hayashi C, Umekita M, Hatano M, FuJiwara T, Mizumoto K, Nomoto A (2017) Kribellosides, novel RNA 5′–triphosphatase inhibitors from the rare actinomycete *Kribbella sp.* MI481-42F6. J Antibiot 70:582–589. https://doi.org/10.1038/ja.2016.161

Jensen PR, Bradley SM, William F (2015) The marine actinomycete genus *Salinispora*: a model organism for secondary metabolite discovery. Nat Prod Rep 32(5):738–751. https://doi.org/10.1039/c4np00167b

Jiang YJ, Zhang DS, Zhang HJ, Li JQ, Ding WJ, Xu CD, Ma ZJ (2018) Medermycin-type naphthoquinones from the marine–derived *Streptomyces sp.* XMA39. J Nat Prod 81:2120–2124. https://doi.org/10.1021/acs.jnatprod.8b00544

Jiao WH, Yuan W, Li ZY, Li J, Li L, Sun JB, Gui YH, Wang J, Ye BP, Lin HW (2018) Anti-MRSA actinomycins D1–D4 from the marine sponge-associated *Streptomyces sp.* LHW52447. Tetrahedron 74:5914–5919. https://doi.org/10.1016/j.tet.2018.08.023

Khalil ZG, Raju R, Piggott AM, Salim AA, Blumenthal A, Capon RJ (2015) Aranciamycins I and J, antimycobacterial anthracyclines from an Australian marine-derived *Streptomyces* sp. J Nat Prod 78:949–952. https://doi.org/10.1021/acs.jnatprod.5b00095

Kim H, Hwang JY, Shin J, Oh KB (2019) Inhibitory effects of diketopiperazines from marine-derived *Streptomyces puniceus* on the isocitrate lyase of *Candida albicans*. Molecules 24 (11):2111. https://doi.org/10.3390/molecules24112111

Kim MC, Cullum R, Hebishy AMS, Mohamed HA, Faraag AHI, Salah NM, Abdelfattah MS, Fenical W (2020) Mersaquinone, a new tetracene derivative from the marine-derived *Streptomyces* sp. EG1 exhibiting activity against methicillin-resistant *Staphylococcus aureus* (MRSA). Antibiot Basel 9(5):252. https://doi.org/10.3390/antibiotics9050252

Kimura T, Inahashi Y, Matsuo H, Suga T, Iwatsuki M, Shiomi K, Takahashi Y, Omura S, Nakashima T (2018) Pyrizomicins A and B: Structure and bioactivity of new thiazolyl pyridines from *Lechevalieria aerocolonigenes* K10–0216. J Antibiot 71:606–606. https://doi.org/10.1038/s41429-018-0038-y

Koehn FE, Carter GT (2005) The evolving role of natural products in drug discovery. Nat Rev Drug Discov 4:206–220. https://doi.org/10.1038/nrd1657

Kurata A, Sugiura M, Kokoda K, Tsujimoto H, Numata T, Kato C, Nakasone K, Kishimoto N (2017) Taxonomy of actinomycetes in the deep–sea Calyptogena communities and characterization of the antibacterial compound produced by *Actinomadura sp.* DS-MS-114. Biotechnol Biotechnol Equip 31:1000–1006. https://doi.org/10.1080/13102818.2017.1342563

Lacret R, Oves-Costales D, Gomez C, Gómez C, Diaz C, La Cruz MD, Perezvictoria I, Vicente F, Genilloud O, Reyes F (2015) New ikarugamycin derivatives with antifungal and antibacterial properties from *Streptomyces zhaozhouensis*. Mar Drugs 13:128–140. https://doi.org/10.3390/md13010128

Lacret R, Perez-Victoria I, Oves-Costales D, La Cruz MD, Domingo E, Martin J, Diaz C, Diaz C, Vicente F, Genilloud O, Reyes F (2016) MDN-0170, a New Napyradiomycin from *Streptomyces sp.* Strain CA-271078. Mar Drugs 14(10):188. https://doi.org/10.3390/md14100188

Lacret R, Oves-Costales D, Pérez-Victoria I, de la Cruz M, Díaz C, Vicente F, Genilloud O, Reyes F (2019) MDN-0171, a new medermycin analogue from *Streptomyces albolongus* CA-186053. Nat Prod Res 33(1):66–73. https://doi.org/10.1080/14786419.2018.1434636

Letzel AC, Pidot SJ, Hertweck C (2013) A genomic approach to the cryptic secondary metabolome of the anaerobic world. J Nat Prod 30:392–428. https://doi.org/10.1039/c2np20103h

Leutou AS, Yang I, Le TC, Hahn D, Lim K, Nam S, Fenical W (2018) Fluvirucin B6, a new macrolactam isolated from a marinederived actinomycete of the genus *Nocardiopsis*. J Antibiot 71:609–611. https://doi.org/10.1038/s41429-018-0033-3

Li JW, Vederas JC (2009) Drug discovery and natural products: end of an era or an endless frontier? Science 325:161–165. https://doi.org/10.1126/science.1168243

Liang Y, Xie X, Chen L, Yan SL, Ye XW, Anjum KA, Huang HC, Lian XY, Zhang ZZ (2016) Bioactive polycyclic quinones from marine *Streptomyces sp.* 182SMLY. Mar Drugs 14:1–11. https://doi.org/10.3390/md14010010

Lino A, Deogracious O (2006) The in vitro antibacterial activity of *Annona senegalensis*, *Securidacca longipendiculata* and *Steganotaenia araliacea*-Ugandan medicinal plants. Afri Health Sci 6:31–35. https://doi.org/10.5555/afhs.2006.6.1.31

Locey KJ, Lennon JT (2016) Scaling laws predict global microbial diversity. Proc Natl Acad Sci U S A 113:5970–5975. https://www.pnas.org/content/113/21/5970

Luo M, Tang L, Dong Y, Huang H, Deng Z, Sun Y (2020) Antibacterial natural products lobophorin L and M from the marine-derived Streptomyces sp. 4506. Nat Prod Res. https://doi.org/10.1080/14786419.2020.1797730

Margulis L, Chapman MJ (2009) Kingdoms and domains. In: An illustrated guide to the phyla of life on earth, 1st edn. Elsevier Science, USA

Mason WP, Belanger K, Nicholas G, Vallières I, Mathieu D, Kavan P, Desjardins A, Omuro A, Reymond A (2012) Phase II study of the Ras-MAPK signaling pathway inhibitor TLN-4601 in patients with glioblastoma at first progression. J Neuro Oncol 107:343–349. https://doi.org/10.1007/s11060-011-0747-6

Newman DJ, Cragg GM (2016) Natural Products as sources of new drugs from 1981 to 2014. J Nat Prod 79:629–661. https://doi.org/10.1021/acs.jnatprod.5b01055

Nong XH, Zhang XY, Xu XY, Wang J, Qi SH (2016) Nahuoic acids B-E, polyhydroxy polyketides from the marine-derived *Streptomyces sp.* SCSGAA 0027. J Nat Prod 79:1. https://doi.org/10.1021/acs.jnatprod.5b00805

Norman Van GA (2016) Drugs, devices, and the FDA: Part 1 An overview of approval processes for drugs. JACC: Basic Trans Sci 1:170–179. https://doi.org/10.1016/j.jacbts.2016.03.009

O'Neill J (2016) Tackling drug-resistant infections globally: final report and recommendations. London (UK): The review on antimicrobial resistance. pp. 4–8. https://amr-review.org/sites/default/files/160518_Final%20paper_with%20cover.pdf

Olano C, Méndez C, Salas JA (2009) Antitumor Compounds from Marine Actinomycetes. Mar Drugs 7:210–248. https://doi.org/10.3390/md7020210

Rajivgandhi G, Muneeswaran T, Maruthupandy M, Muthiah CR, Saravanan K, Ravikumar V, Manoharan N (2018) Antibacterial and anticancer potential of marine endophytic actinomycetes *Streptomyces coeruleorubidus* GRG 4 (KY457708) compound against colistin resistant uropathogens and A549 lung cancer cells. Microb Pathog 125:325–335. https://doi.org/10.1016/j.micpath.2018.09.025

Rathore DR, Sheikh M, Gohel GD, Singh SP (2021) Genetic and phenotypic heterogeneity of the *Nocardiopsis alba* strains of sea water. Curr Microbiol 78:1377–1387. https://doi.org/10.3390/v13060950

Rodriguez V, Martín J, Sarmiento-Vizcaíno A, De la Cruz M, García LA, Blanco G, Fernando F (2018) Anthracimycin B, a potent antibiotic against gram-positive bacteria isolated from cultures of the deep-sea actinomycete *Streptomyces cyaneofuscatus* M–169. Mar Drugs 16: 406. https://doi.org/10.3390/md16110406

Sangkanu S, Rukachaisirikul V, Suriyachadkun C, Phongpaichit S (2021) Antifungal activity of marine-derived actinomycetes against *Talaromyces marneffei*. J Appl Microbiol 130 (5):1508–1522. https://doi.org/10.1111/jam.14877

Sharma AK, Kikani BA, Singh SP (2020) Biochemical, thermodynamic and structural characteristics of a biotechnologically compatible alkaline protease from a haloalkaliphilic, *Nocardiopsis dassonvillei* OK-18. Int J Biol Macromol 153:680–696. https://doi.org/10.1016/j.ijbiomac.2020.03.006

Sisi Q, Min G, Huanhuan L, Chunbo Y, Hongji L, Zheling Z, Peng S (2020) Secondary metabolites from marine *Micromonospora*: chemistry and bioactivities. Chem Biodivers. https://doi.org/10.1002/cbdv.202000024

Skropeta D, Wei L (2014) Recent advances in deep-sea natural products. Nat Prod Rep 31:999–1025. https://doi.org/10.1039/c3np70118b

Sobolevskaya MP, Denisenko VA, Moiseenko AS, Shevchenko LS, Menzorova NI, Sbirtsev YT, Kim N, Kuznetsova TA (2007) Bioactive metabolites of the marine actinobacterium *Streptomyces sp.* KMM 7210. Russ Chem Bull 56:838–840. https://doi.org/10.1007/s11172-007-0126-9

Subramani R, Aalbersberg W (2012) Marine actinomycetes: an ongoing source of novel bioactive metabolites. Microbiol Res 167:571–580. https://doi.org/10.1016/j.micres.2012.06.005

Subramani R, Aalbersberg W (2013) Culturable rare actinomycetes: diversity, isolation and marine natural product discovery. Appl Microbiol Biotechnol 97:9291–9321. https://doi.org/10.1007/s00253-013-5229-7

Subramani R, Sipkema D (2019) Marine rare actinomycetes: a promising source of structurally diverse and unique novel natural products. Mar Drugs 17:249. https://doi.org/10.3390/md17050249

Sun CL, Yang ZJ, Zhang CY, Liu ZY, He JQ, Liu Q, Zhang TY, Ma JY (2019) Genome mining of *Streptomyces atratus* SCSIO ZH16: discovery of atratumycin and identification of its biosynthetic gene cluster. Org Lett 21:1453–1457. https://doi.org/10.1021/acs.orglett.9b00208

Tan Y, Hu YY, Wang Q, Zhou HX, Wang YG, Gan ML (2016) Tetrocarcins N and O, glycosidic spirotetronates from a marine derived *Micromonospora sp.* identifed by PCR-based screening. RSC Adv 6:91773–91778. https://doi.org/10.1039/C6RA17026A

Teta R, Marteinsson VT, Longeon A, Klonowski AM, Groben R, Bourguetkondracki M, Costantion V, Mangori A (2017) Thermoactinoamide A, an antibiotic lipophilic cyclopeptide from the icelandic thermophilic bacterium *Thermoactinomyces vulgaris*. J Nat Prod 80:2530–2535. https://doi.org/10.1021/acs.jnatprod.7b00560

Thakrar FJ, Singh SP (2019) Catalytic, thermodynamic and structural properties of an immobilized and highly thermostable alkaline protease from a haloalkaliphilic actinobacteria, *Nocardiopsis alba* Tata-5. Bioresour Technol 278:150–158. https://doi.org/10.1016/j.biortech.2019.01.058

Thi QV, Tran VH, Mai HD, Le CV, Hong MLE, Murphy BT, Chau VM, Pham VC (2016a) Secondary metabolites from an Actinomycete from Vietnam's East Sea. Nat Prod Commun 11: 401–404. https://doi.org/10.1177/1934578x1601100320

Thi QV, Tran VH, Mai HD, Le CV, Hong MLY, Murphy BT, Chau VM, Pham VC (2016b) Antimicrobial metabolites from a marine – derived Actinomycete in Vietnam's East Sea. Nat Prod Commun 11:49–51. https://doi.org/10.1080/14786419.2018.1468331

Thomas TR, Kavlekar DP, LokaBharathi PA (2010) Marine drugs from sponge-microbe association: a review. Mar Drugs 8:1417–1468. https://doi.org/10.3390/md8041417

Thumar JT, Singh SP (2007a) Two-step purification of a highly thermostable alkaline protease from salt-tolerant alkaliphilic *Streptomyces clavuligerus* strain Mit-1. J Chromatogr B 854:198–203. https://doi.org/10.1016/j.jchromb.2007.04.023

Thumar JT, Singh SP (2007b) Secretion of an alkaline protease from salt-tolerant and alkaliphilic, *Streptomyces clavuligerus* strain Mit-1. Braz J Microbiol 38:1–9. https://doi.org/10.1590/S1517-83822007000400033

Thumar JT, Singh SP (2009) Organic solvent tolerance of an alkaline protease from salt-tolerant alkaliphilic *Streptomyces clavuligerus* strain Mit-1. J Ind Microbiol Biotechnol 36:211–218. https://doi.org/10.1007/s10295-008-0487-6

Thumar JT, Dhulia KS, Singh SP (2010) Isolation and partial purification of an antimicrobial agent from halo-tolerant alkaliphilic *Streptomyces aburaviensis* strain Kut-8. World J Microbiol Biotechnol 26(11):2081–2087. https://doi.org/10.1007/s11274-010-0394-7

Uzair B, Menaa F, Khan BA, Mohammad FV, Ahmad VU, Djeribi R, Menaa B (2018) Isolation, purification, structural elucidation and antimicrobial activities of kocumarin, a novel antibiotic isolated from actinobacterium *Kocuria marina* CMG S2 associated with the brown seaweed *Pelvetia canaliculata*. Microbiol Res 206:186–197. https://doi.org/10.1016/j.micres.2017.10.007

Valliappan K, Sun W, Li ZY (2014) Marine actinobacteria associated with marine organisms and their potentials in producing pharmaceutical natural products. Appl Microbiol Biotechnol 98: 7365–7377. https://doi.org/10.1007/s00253-014-5954-6

Vasavada SH, Thumar JT, Singh SP (2006) Secretion of a potent antibiotic by salt-tolerant and alkaliphilic actinomycete *Streptomyces sannanensis* strain RJT-1. Curr Sci 91(4):1393–1397. https://doi.org/10.13140/2.1.2616.3847

Viegelmann C, Parker J, Ooi T, Clements C, Abbott G, Young L, Kennedy J, Dobson ADW, Edrada-Ebel R (2014) Isolation and identification of antitrypanosomal and antimycobacterial active steroids from the sponge *Haliclona simulans*. Mar Drugs 12:2937–2952. https://doi.org/10.3390/md12052937

Vynne NG, Mansson M, Nielsen KF, Gram L (2011) Bioactivity, chemical profiling, and 16S rRNA-based phylogeny of *Pseudoalteromonas* strains collected on a global research cruise. Mar Biotechnol 13:1062–1073. https://doi.org/10.1007/s10126-011-9369-4

Wang Q, Zhang YX, Wang M, Tan Y, Hu XX, He HW, Xiao CL, You XF, Wang YG, Gan ML (2017) Neo-actinomycins A and B, natural actinomycins bearing the 5 H-oxazolo [4, 5-b] phenoxazine chromophore, from the marine-derived *Streptomyces sp.* IMB094. Sci Rep 7:1–8. https://doi.org/10.1038/s41598-017-03769-8

WHO (2017) Global priority list of antibiotic-resistant bacteria to guide research, discovery, and development of new antibiotics. World Health Organization, Geneva (Switzerland). https://www.who.int/medicines/publications/WHO-PPL-Short_Summary_25Feb

William F, Jensen PR (2006) Developing a new resource for drug discovery: marine actinomycete bacteria. Nat Chem Biol 2(12):666–673. https://doi.org/10.1038/nchembio841

Williams PG, Oh DC, Zeigler L, Fenical W (2007a) Species-specific secondary metabolite production in marine actinomycetes of the genus *Salinispora*. J Appl Env Microbiol 73:1146. https://doi.org/10.1128/AEM.01891-06

Williams PG, Asolkar RN, Kondratyuk T, Pezzuto JM, Jensen PR (2007b) Fenical Saliniketals A and B, bicyclic polyketides from the marine actinomycete *Salinispora arenicola*. J Nat Prod 70: 83–88. https://doi.org/10.1021/np0604580

Williams DE, Dalisay DS, Chen J, Polishchuck EA, Patrick BO, Narula G, Ko M, Avgay Y, Li HX, Magarvey NA, Andersen RJ (2017) Aminorifamycins and sporalactams produced in culture by a *Micromonospora sp.* isolated from a Northeastern-Pacifc marine sediment are potent antibiotics. Org Lett 19:766–769. https://doi.org/10.1021/acs.orglett.6b03619

Yang CL, Wang YS, Liu CL, Zeng YJ, Cheng P, Jiao RH, Bao SX, Huang HQ, Tan RX, Ge HM (2017) Strepchazolins A and B: two new alkaloids from a marine *Streptomyces chartreusis* NA02069. Mar Drugs 15:244. https://doi.org/10.3390/md15080244

Yang CF, Qian R, Xu Y, Yi J, Gu Y, Liu X, Yu H, Jiao B, Lu X, Zhanf W (2019) Marine actinomycetes-derived natural products. Curr Top Med Chem 19(31):2868–2918. https://doi.org/10.2174/1568026619666191114102359

Yi WW, Li Q, Song TF, Chen L, Li XC, Zhang ZZ, Lian XY (2019) Isolation, structure elucidation, and antibacterial evaluation of the metabolites produced by the marine-sourced *Streptomyces sp.* ZZ820. Tetrahedron 75:1186–1193. https://doi.org/10.1016/j.tet.2019.01.025

Zhang L, An R, Wang J, Sun N, Zhang S, Hu J, Kuai J (2005) Exploring novel bioactive compounds from marine microbes. Curr Opin Microbiol 8:276–281. https://doi.org/10.1016/j.mib.2005.04.008

Zhang HB, Saurav K, Yu ZQ, Mandi A, Kurtan T, Li J, Tian XP, Zhang QB, Zhang WJ, Zhang CS (2016) α-Pyrones with diverse hydroxy substitutions from three marine–derived *Nocardiopsis* Strains. J Nat Prod 79:1610–1618. https://doi.org/10.1021/acs.jnatprod.6b00175

Zhang Z, Chen L, Zhang X, Liang Y, Anjum K, Chen L, Lian XY (2017a) Bioactive bafilomycins and a new N-arylpyrazinone derivative from marine-derived *Streptomyces sp.* HZP-2216E. Planta Med (18):1405–1411. https://doi.org/10.1055/s-0043-111897

Zhang X, Chen L, Chai W, Lian XY, Zhang Z (2017b) A unique indolizinium alkaloid streptopertusacin A and bioactive bafilomycins from marine-derived *Streptomyces sp.* HZP-2216E. Phytochemistry 144:119–126. https://doi.org/10.1016/j.phytochem.2017.09.010

Zhang B, Wang KB, Wang W, Bi SF, Mei YN, Deng XZ, Jiao RH, Tan RX, Ge HM (2018a) Discovery, biosynthesis, and heterologous production of streptoseomycin, an anti–microaerophilic bacteria macrodilactone. Org Lett 20:2967–2971. https://doi.org/10.1021/acs.orglett.8b01006

Zhang D, Shu CY, Lian XY, Zhang ZZ (2018b) New antibacterial bagremycins F and G from the marine-derived *Streptomyces sp.* ZZ745. Mar Drugs 16:330. https://doi.org/10.3390/md16090330

Zhang SW, Gui C, Shao MW, Kumar PS, Huang HB, Ju JH (2018c) Antimicrobial tunicamycin derivatives from the deep sea-derived *Streptomyces xinghaiensis* SCSIO S15077. Nat Prod Res 34(11):1499–1504. https://doi.org/10.1080/14786419.2018.1493736

Zhang D, Yi W, Ge H, Zhang Z, Wu B (2019a) Bioactive streptoglutarimides A-J from the marine-derived *Streptomyces* sp. ZZ741. J Nat Prod 82(10):2800–2808. https://doi.org/10.1021/acs.jnatprod.9b00481

Zhang S, Xie Q, Sun C, Tian XP, Gui C, Win XJ, Zhang H, Ju JH (2019b) Cytotoxic kendomycins containing the carbacylic ansa scafold from the marine-derived *Verrucosispora* sp. SCSIO 07399. J Nat Prod 82:3366–3371. https://doi.org/10.1021/acs.jnatprod.9b00654

Chapter 14
Pharmacology of FDA-Approved Medicines from Actinobacteria

Rong Ma and Loganathan Karthik

Abstracts Actinobacteria are widely distributed in nature, such as marine, soil, and endophytic. Secondary metabolites derived from actinobacteria are valuable for drug discovery and development (R & D). Besides anti-infectious effects, anticancer, immunosuppressive effects, and other effects are displayed by this kind of metabolites. *Streptomyces* species are the primary producer of up to 70% of compounds isolated from actinobacteria. For clinical interests and practical purposes, only drugs approved by FDA are reviewed in this chapter. These include tigecycline, everolimus, telithromycin, miglustat, daptomycin, amrubicin, biapenem, ertapenem, pimecrolimus, and Gemtuzumab. In addition to the efficacy of medicines, side effects and toxicity are both critical for clinical use. Furthermore, the absorption, distribution, metabolism, and excretion (ADME) of a drug is essential for efficacy in the human body. The same drug displays different parameters of ADME in different nationalities or different genders, such as biological half time ($t_{1/2}$), apparent volume of distribution (V), and clearance (CL). We hope the writing is helpful for the researcher who is occupied in drug R & D.

Keywords Actinobacteria · FDA-approved medicines · Pharmacology · Pharmacokinetics · Clinical application

R. Ma (✉)
Clinical Research Institute of Integrative Medicine, Yueyang Hospital, Shanghai University of Traditional Chinese Medicine, Shanghai, China

Clinical Research Institute of Integrative Medicine, Shanghai Institute of Traditional Chinese Medicine, Shanghai, China

L. Karthik
Synthetic Biology, R&D Center, Salem Microbes Private Limited, Salem, India

L. Karthik (ed.), *Actinobacteria*, https://doi.org/10.1007/978-981-16-5835-8_14

14.1 Introduction

Unlimited resources of unidentified new bioactive compounds, known as natural products (NP), are sustained by nature, and studies based on these resources are beneficial and necessary in the discovery of novel drugs (Ebada et al. 2008). In developing countries, these NPs are used mainly as a source of medicinal drugs because of their availability in abundance and low cost. Among the microbes which are the universal producers of NPs, actinomycetes stand in top among others having about 45% of the microbial bioactive molecules discovered, i.e., over 10,000 bioactive NPs, of which 7600 were synthesized from Streptomyces and 2500 from other rare species of actinomycetes (Abdel-Razek et al. 2020). For centuries, NPs have been serving as the most important source of drugs, and about half the pharmaceutical drugs which are currently in use are derived from NPs.

Growing interest in the antibiotics derived from actinomycetes has been rattling since the actinomycin's discovery in 1940 (Cassell 2001). It is then followed by the discovery of streptomycin, streptothricin, the tetracyclines, chloramphenicol, neomycin, the erythromycins, oleandomycin, novobiocin, nystatin, and many others. Even today, this trend extends with the discovery of novel antibiotics such as thienamycin, daptomycin, carbapenem analogs, and epirubicin (Sivaramkrishna and Girish 2009). In recent years, antibiotic resistance is a serious threat by the pathogens to the medical sector as it is capable of worsening the cure of a disease. Understanding the molecular and cellular basis of drug resistance by bacteria may lead to a conclusion on why or how they are resistant to antibiotics. To overcome this drug resistance, discovering novel antibiotics that are active against antibiotic-resistant bacteria is mandatory (Raja et al. 2010). In discovering novel antibiotics, actinomycetes have proved to be an important source with the number of drugs being significant effectively released into the market and is still in use clinically. It is a fecund source of new antibiotics such as glycopeptides, macrolides, aminoglycosides, rifamycins, tetracyclines, and β-lactams having an antibacterial activity (Genilloud 2017).

Secondary metabolites from actinobacteria have diverse bio-activities in the health sector such as anti-angiogenesis, antioxidant, antiviral, antibacterial, antifungal, antiparasitic, antimalarial, anti-inflammatory, cytotoxic, cytostatic, antitumor, and anticancer activity (Manivasagan et al. 2014). This chapter reviews the pharmacological effects of clinically essential drugs obtained from actinomycetes, as the compounds derived from actinomycetes are available in abundance.

14.2 Tigecycline

Tigecycline, semi-synthetic chlortetracycline, is extracted from *Streptomyces aureofaciens*, which is the 9-tert-butyl-glycylamido derivative of minocycline (Fig. 14.1) (Raja and Prabakaran 2011). Similar to other tetracyclines, tigecycline

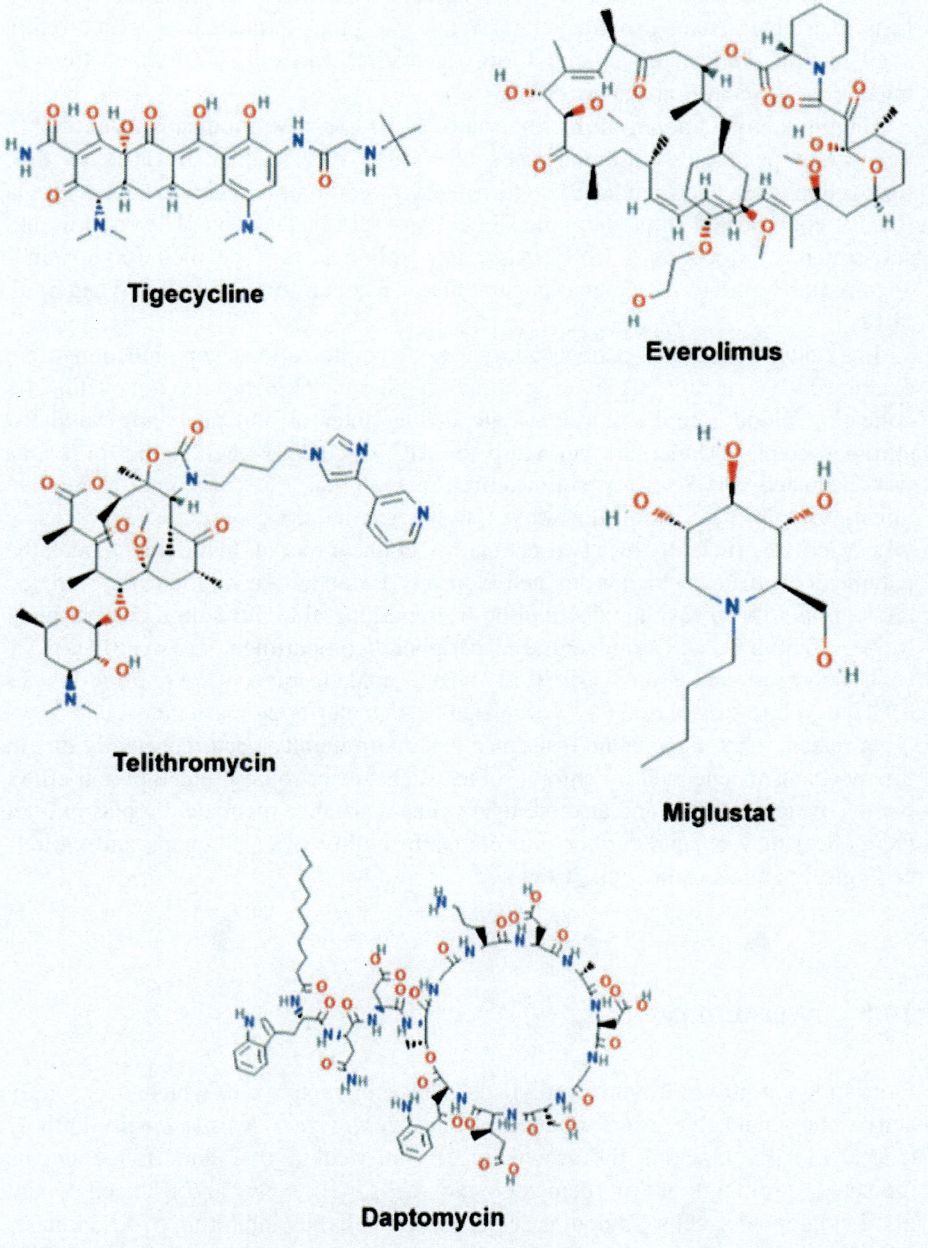

Fig. 14.1 Structure of novel antibiotics produced by actinomycetes (Sources: PubMed Compound Database)

not only has antibacterial activity but also has an increased activity against organisms that are resistant to tetracycline. While the other tetracyclines are currently available for clinical use in oral form, tigecycline can only be utilized through injection (Raja and Prabakaran 2011).

Having a large amount of distribution of 7–10 L/kg (Muralidharan et al. 2005), tigecycline penetrates different types of tissues successfully. Because of this, tigecycline has been authorized for treating community-acquired pneumonia (CAP), complicated intra-abdominal infections (cIAI), and complicated skin and soft-structure infections (cSSSI). Also, tigecycline is not specified for hospital-acquired or ventilator-associated pneumonia or diabetic foot infection (Wang et al. 2017)

In a study, critically Ill patients who showed symptoms for severe infections were examined (Xie et al. 2017) for population pharmacokinetics of tigecycline by collecting blood samples during single dosing interval and analyzing validated chromatography. Under the curvature to MICs (AUC0–24/MIC), the 24-h area was evaluated ($\geq$4.5 for hospital-acquired pneumonia, $\geq$6.96 for intra-abdominal infections, $\geq$17.9 for skin infections). The tigecycline concentrations' time course was described efficiently by a two-compartment linear model. In the final model, the parameter estimates, which is denoted as means $\pm$ standard deviation (SD), were the rate constant for tigecycline distribution from peripheral to the central compartment ($0.29 \pm 0.30\ h^{-1}$) and from central to peripheral compartment ($0.31 \pm 0.16\ h^{-1}$), central compartment volume (72.50 ± 21.18 L), and clearance (CL) (7.50 ± 1.11 L/h). Tigecycline's increased CL was related to a greater body mass index.

In recent years, tigecycline resistance has intermittently occurred mainly due to the mechanisms encoded by chromosomes such as ribosome protection and efflux pumps overexpression, and also the tigecycline resistance mediated by plasmid (tet (X) genes) may expand further into clinically highly risk pathogens and various ecological habitats (Sun et al. 2019).

14.3 Everolimus

Everolimus, a 40-O-(2-hydroxyethyl) derivative of rapamycin, which is an orally active compound, has been produced from *Streptomyces hygroscopicus* initially (Fig. 14.1). By blocking the growth factors interleukin IL-2 and IL-15, and its mediated proliferation of hematopoietic cells (B cells and T cells) and non-hematopoietic cells (smooth vascular muscle cells) by inhibiting p70-S6 kinase, everolimus displays its immunosuppressive effects, thus leading to pulling up G1/S phase of the cell cycle (Raja and Prabakaran 2011).

In the treatment of tumor-like pancreatic ductal adenocarcinoma (Babiker et al. 2019), neuroendocrine tumors (Lee et al. 2018), breast cancer (O'Shaughnessy et al. 2018), tuberous sclerosis complex (Overwater et al. 2019), and solid organ transplantation (van Gelder et al. 2017), everolimus is extensively used. Furthermore, considering pancreatic neuroendocrine tumors, everolimus is considered first-line therapy (Gallo et al. 2017).

14.4 Telithromycin

Derived from *Saccharopolyspora erythraea*, telithromycin is a semi-synthetic derivative of 14-membered macrolide, which holds D-desosamine sugar moiety and also the macrolactone ring (Fig. 14.1). Telithromycin displays an antibacterial effect on pathogens related to the respiratory tract, which are resistant to other macrolides, and through interaction with the 50S bacterial ribosomal subunit peptidyltransferase site, it prevents protein synthesis (Raja and Prabakaran 2011).

Telithromycin is recommended as a treatment for community-acquired pneumonia. Telithromycin was approved in the USA in 2004 and has since been associated with several cases of serious liver damage caused by medications (https://pubmed.ncbi.nlm.nih.gov/31643176/). In a recent in vitro study (Xiong et al. 2021), telithromycin considerably reduced the planktonic cells of both *E. faecium* and *E. faecalis*. In addition, telithromycin and ampicillin combinedly enhanced antibiofilm activity because of their strong ability to prevent biofilms formed by *E. faecalis*. Due to these in vitro antibacterial and antibiofilm activities, it is suggested that telithromycin may be a probable candidate for treating enterococcal infections.

14.5 Miglustat

Miglustat is an equivalent of nojirimycin isolated from *Streptomyces lavendulae* broth filtrate and involves reverse inhibition of glucosylceramide synthase, a ceramide specific glucosyltransferase that catalyzes glucocerebroside formation, in turn decreasing glucosylceramide storage in tissue (Fig. 14.1). For the patients who are unable to take enzyme replacement therapy, miglustat was approved to be used as a therapeutic drug for treating type 1 Gaucher disease. Gaucher disease, linked with pathological glucosylceramide accumulation in the monocyte or macrophage lineage cells, is a progressive lysosomal storage disorder. In the treatment of type 1 Gaucher disease, enzyme replacement therapy has been made available using alglucerase derived from the human placenta (Ceredase) (Raja and Prabakaran 2011).

Type C Niemann-Pick disease (NP-C), a rare, neurodegenerative, autosomal recessive disease, is associated with a diverse variety of progressive neurological manifestations (Pineda et al. 2018). Among both adults and children, miglustat is directed to treat progressive neurological manifestations. Rapid growth had been recorded in the clinal experience with miglustat since its approval in 2009. Using different approaches, miglustat's effects over neurological NP-C manifestations were evaluated with benefits differing from visible clinical improvements to cellular changes in the brain and enhanced survival (Pineda et al. 2018). In a large number of patients, miglustat-based therapy was linked with neurological manifestation stabilization. Considering this miglustat therapy, the tolerability and safety associated with it were unswerving with the previous reports (Patterson et al. 2020).

14.6 Daptomycin

Daptomycin, an antibacterial agent, is a cyclin lipopeptide derived from *Streptomyces roseosporus* and was approved for treating complicated skin and skin structure infections (cSSSIs) (Fig. 14.1). Binding to cell membranes of bacteria, daptomycin affects its membrane potential, hindering DNA synthesis, RNA, and proteins (Raja and Prabakaran 2011).

Daptomycin is vigorous against gram-positive bacteria resistant to multiple drugs, including *Staphylococcus aureus* and methicillin-resistant *Staphylococcus aureus* (MRSA), with decreased vulnerability to vancomycin. Daptomycin is four- to eightfold as vigorous as vancomycin against MRSA and methicillin-susceptible *Staphylococcus aureus* (MSSA), and most of these kinds of activities are retained against *S. aureus* with abridged susceptibility to vancomycin (Gómez Casanova et al. 2017).

Daptomycin received approval from the Food and Drug Administration (FDA) of USA to be used for injection in 2003 for the treatment of cSSSI affected patients, and later it was approved for treating patients affected with *Staphylococcus aureus* bacteremia in 2006, including patients with methicillin-susceptible and methicillin-resistant isolates caused right-sided infective endocarditis. In 2016, the FDA approved daptomycin RF, daptomycin's new formulation for injection, to be used for the same signs (Frankenfeld et al. 2018).

Standard monotherapy consisting of vancomycin and daptomycin versus a regimen of ceftaroline with daptomycin in MRSA bacterium has displayed a more decisive clinical trial of its possible benefits against this principle root of infection-associated mortality (Geriak et al. 2019).

The most common side effect of daptomycin is eosinophilic pneumonia. Among the patients who are likely to be seen with eosinophilic pneumonia induced by daptomycin were male (83%) and elder people (mean age 65.4 ± 15 years). The daptomycin doses have ranged from 4 to 10 mg/kg/day but involved multiple numbers of renal dysfunction patients. The mean duration of treatment with daptomycin following the onset of EP symptoms was recorded as 2.8 ± 1.6 weeks. The majority of patients were found to have presented with infiltrates or opacities of CT/CXR (86%), peripheral eosinophilia (77%), fever (57%), and also dyspnea (94%). After daptomycin has been discontinued for about 24 h–1 week, improvement in symptoms has been seen. Most patients have also been prescribed to take corticosteroids (66%) for their treatment (Uppal et al. 2016).

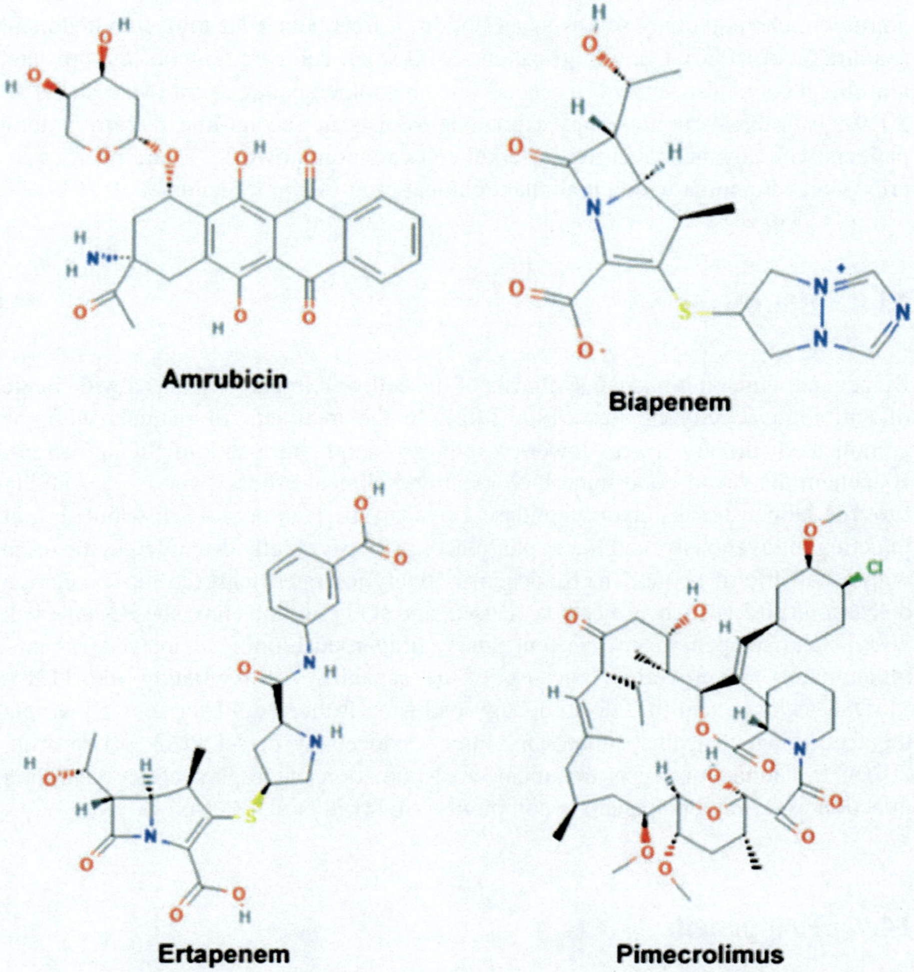

Fig. 14.2 Structure of novel antibiotics produced by actinomycetes (Sources: PubMed Compound Database)

14.7 Amrubicin

Amrubicin hydrochloride, a doxorubicin derivative isolated from *Streptomyces peucetius* var. *caesius*, is an utterly synthetic 9-aminoanthracycline and, upon entering, the body converts itself to its active form (Fig. 14.2). Amrubicin has demonstrated its activity which is similar to that of doxorubicin on tumors of transplant animals, including sarcoma 180, Lewis lung carcinoma and P 388 leukemia, and further compelling antitumor activity against human tumor breast xenografts, gastric and lung cancer (Raja and Prabakaran 2011).

In Japan, for patients who are with non-small cell lung cancer (NSCLC), amrubicin chemotherapy serves as an option of treatment after third-line treatment (Sakurai et al. 2020). Coming to patients with small cell carcinoma of the prostate, amrubicin could also serve as a second-line chemotherapeutic agent (Maesaka et al. 2019). Amrubicin monotherapy's potential assists in second-line therapy among patients with advanced/recurrent Merkel cell carcinoma (MCC), a rare neuroendocrine skin carcinoma with a malignant clinical progression (Sakaida et al. 2017).

14.8 Biapenem

Biapenem, which inhibits the synthesis of the cell wall in bacteria, has a wide range of antibiotic activity in vitro (Fig. 14.2). In the treatment of patients with the complicated urinary tract, lower respiratory, and intra-abdominal infections, Biapenem displayed good microbiological and clinical efficacy due to its stability towards human renal dihydropeptidase I hydrolysis. It is broadly distributed after injecting intravenously, and linear pharmacokinetics is chiefly discarded in the urine with a half-life of about 1 h. Biapenem is finely tolerated with the most common deleterious effects such as diarrhea, nausea, and skin eruptions or rashes (Kennewell 2006). With biapenem, in a recent study, drug repositioning displayed an anti-inflammatory severe effect because of its capability of regulating the TLR4-MyD88 pathway and mTOR-autophagy pathway. It thus could serve as a possible therapeutic agent against pulmonary injury induced by diesel PM2.5 (Lee et al. 2020). In China, for sepsis treatment, a combination of drugs such as Xuebijing injection (XBJ) and biapenem is commonly used (Liu et al. 2020).

14.9 Ertapenem

Ertapenem, a wide-spectrum carbapenem antibiotic, is principally used to treat infections by aerobic gram-negative bacteria (Fig. 14.2). Ertapenem is linked to asymptomatic and transient elevations in serum enzymes similar to other carbapenems. The carbapenems have also been associated with exceptional occasions of clinically obvious and acute cholestatic liver damage (Bethesda 2012). Ertapenem delivers intense activity against various pathogens, which are commonly linked to ventilator-associated and hospital-acquired bacterial pneumonia, which includes many gram-negative pathogens and *Staphylococcus aureus* susceptible to methicillin (Bader et al. 2019). It has to be administered only once a day as it has a higher half-life, contrary to other carbapenems (Joyner et al. 2019). In another study, when 2000 mg of ertapenem administered once a day to patients infected by *Mycobacterium tuberculosis*, it reached the expected value of $f40\%\ T > \mathrm{MIC}$ among almost all the patients, thereby advocated exploration in phase II study (Zuur et al. 2018). Patients with thrombocytosis should be cautious and examined with a wary and thorough history considering the potential side effects of ertapenem, as it may induce thrombocytosis (Docobo et al. 2017).

14.10 Pimecrolimus

Pimecrolimus, a 33-epi-chloro-derivative of ascomycin macrolactam, has an immunosuppressant property (Fig. 14.2). Binding to macrophilin-12 (FKBP-12) receptor, pimecrolimus forms a complex which blocks the calcineurin-mediated signal transduction cascade dependent on calcium. Calcineurin, an enzyme, through dephosphorylation, is responsible for activating the activated T-cells (NF-AT) nuclear factor, a T-cell transcriptional regulatory factor. Subsequently, the inflammatory mediators from mast cells and T-cells, and Th1 (T helper 1) and Th2 (T helper 2) type cytokines production and release are prevented, and the signal expressions necessary for the inflammatory T-lymphocytes activation is inhibited. Pimecrolimus, however, does not affect primary fibroblasts and dendritic cells/Langerhans cells, as its mode of action is cell-selective.

Pimecrolimus, being marketed first by Novartis using Elidel as the trade name, is an immunomodulating agent, which Galderma now endorses since early 2007 in Canada. Presently it is available as a topic body cream that aids in treating eczema, and atopic dermatitis (https://pubchem.ncbi.nlm.nih.gov/compound/ELIDEL). Among topical calcineurin inhibitors (TCIs), this cream is a non-steroidal prescription-based medication approved by the FDA for treating mild to moderate eczema (Ahn and Robinson 2021). Because of the widened clinical usage of pimecrolimus, its efficacy and economic assessment as evidence are essential, which is primely relevant to three medical conditions, including adult psoriasis, adult seborrheic dermatitis, and adult atopic dermatitis (Deonandan and Severn 2017).

14.11 Gemtuzumab

Gemtuzumab ozogamicin, which is a prodrug of calicheamicin, is attached to an anti-CD33 monoclonal antibody. The LL-E3328 antibiotics, also known as the calicheamicins, were discovered from the *Micromonospora echinospora sp. Calichensis* fermentation products. In cells, lysosomes cleave the covalent linkage between calicheamicin and monoclonal antibodies, thereby permitting the release of calicheamicin. Calicheamicin, a hydrophobic member, belongs to the enediyne family of antibiotics, which cleaves DNA and helps to treat patients with acute myeloid lymphoma (Raja and Prabakaran 2011).

When gemtuzumab ozogamicin (2.5–10 mg/m^2) is administered in 1–4 cycles as a sole agent, it exhibited effective linking between children affected by very advanced acute myeloid leukemia (AML) to hematopoietic stem cell transplantation (Niktoreh et al. 2019). Furthermore, the FDA approved Gemtuzumab ozogamicin

(Mylotarg; Pfizer Inc.), on September 1, 2017, together with cytarabine and daunorubicin for treating adult patients with recently spotted CD33-positive AML as monotherapy (Jen et al. 2018).

Treatment with Gemtuzumab ozogamicin is accompanied by an increased risk of hepatic veno-occlusive disease/sinusoidal obstruction syndrome (VOD/SOS) and hepatotoxicity, particularly succeeding hematopoietic stem cell transplantation. Other nonspecific earnest perilous effects linked with the Gemtuzumab ozogamicin treatment are bleeding/thrombocytopenia, myelosuppression, tumor lysis syndrome, and infusion-related reaction (Cortes et al. 2020).

14.12 Summary

Still, a lot of secondary metabolites from actinobacteria are in clinical and preclinical studies. With the development of synthetic biology, more and more secondary metabolites from actinobacteria will be found, and more medicines will be produced cheaply. In the future, actinobacteria will be one of the potential sources of new antibiotics against drug-resistant pathogens, antineoplastic drugs or immunomodulators.

Acknowledgments This work was partially funded by the National Natural Science Foundation of China (NSFC) No. 81903887.

References

Abdel-Razek AS, El-Naggar ME, Allam A, Morsy OM, Othman SI (2020) Microbial natural products in drug discovery. PRO 8:470. https://doi.org/10.3390/pr8040470

Ahn D, Robinson CA (2021) Pimecrolimus. In: StatPearls. StatPearls Publishing, Treasure Island, FL. https://www.ncbi.nlm.nih.gov/books/NBK545135/

Babiker HM, Karass M, Recio-Boiles A, Chandana SR, McBride A, Mahadevan D (2019) Everolimus for the treatment of advanced pancreatic ductal adenocarcinoma (PDAC). Expert Opin Investig Drugs 28:583–592. https://doi.org/10.1080/13543784.2019.1632289

Bader JC, Lakota EA, Dale GE, Sader HS, Rex JH, Ambrose PG, Bhavnani SM (2019) Pharmacokinetic-pharmacodynamic evaluation of ertapenem for patients with hospital-acquired or ventilator-associated bacterial pneumonia. Antimicrob Agents Chemother 63:e00318–e00319. https://doi.org/10.1128/AAC.00318-19

Cassell GH (2001) Development of antimicrobial agents in the Era of new and reemerging infectious diseases and increasing antibiotic resistance. JAMA 285:601. https://doi.org/10.1001/jama.285.5.601

Cortes JE, de Lima M, Dombret H, Estey EH, Giralt SA, Montesinos P, Röllig C, Venditti A, Wang ES (2020) Prevention, recognition, and management of adverse events associated with gemtuzumab ozogamicin use in acute myeloid leukemia. J Hematol OncolJ Hematol Oncol 13:137. https://doi.org/10.1186/s13045-020-00975-2

Deonandan R, Severn M (2017) Pimecrolimus for the treatment of adults with atopic dermatitis, seborrheic dermatitis, or psoriasis: a review of clinical and cost-effectiveness. https://www.ncbi.nlm.nih.gov/books/NBK507656/

Docobo RA, Bukhari S, Qutrio Baloch Z (2017) Ertapenem-induced thrombocytosis. Cureus. https://doi.org/10.7759/cureus.1263

Ebada SS, Edrada RA, Lin W, Proksch P (2008) Methods for isolation, purification and structural elucidation of bioactive secondary metabolites from marine invertebrates. Nat Protoc 3:1820–1831. https://doi.org/10.1038/nprot.2008.182

Frankenfeld C, Mittal S, Melendez Y, Mendez-Vigo L, Lamp KC, Keller KN, Bertolami SR (2018) Daptomycin: a comparison of two intravenous formulations. Drug Des Devel Ther 12:1953–1958. https://doi.org/10.2147/DDDT.S167010

Genilloud O (2017) Actinomycetes: still a source of novel antibiotics. Nat Prod Rep 34:1203–1232. https://doi.org/10.1039/C7NP00026J

Geriak M, Haddad F, Rizvi K, Rose W, Kullar R, LaPlante K, Yu M, Vasina L, Ouellette K, Zervos M, Nizet V, Sakoulas G (2019) Clinical data on Daptomycin plus Ceftaroline versus standard of care monotherapy in the treatment of methicillin-resistant *Staphylococcus aureus* bacteremia. Antimicrob Agents Chemother 63:e02483–e02418. https://doi.org/10.1128/AAC.02483-18

Gómez Casanova N, Siller Ruiz M, Muñoz Bellido JL (2017) Mechanisms of resistance to daptomycin in Staphylococcus aureus. Rev Espanola Quimioter Publicacion of Soc Espanola Quimioter 30:391–396. https://pubchem.ncbi.nlm.nih.gov/compound/ELIDEL

Jen EY, Ko C-W, Lee JE, Del Valle PL, Aydanian A, Jewell C, Norsworthy KJ, Przepiorka D, Nie L, Liu J, Sheth CM, Shapiro M, Farrell AT, Pazdur R (2018) FDA approval: Gemtuzumab ozogamicin for the treatment of adults with newly diagnosed CD33-positive acute myeloid leukemia. Clin Cancer Res 24:3242–3246. https://doi.org/10.1158/1078-0432.CCR-17-3179

Joyner ML, Cole Manning C, Forbes W, Bobola V, Frazier W (2019) Modeling Ertapenem: the impact of body mass index on distribution of the antibiotic in the body. Math Biosci Eng 16: 713–726. https://doi.org/10.3934/mbe.2019034

Kennewell P (2006) Comprehensive medicinal chemistry II. Volume 1: Global perspective. Elsevier Science

Lee L, Ito T, Jensen RT (2018) Everolimus in the treatment of neuroendocrine tumors: efficacy, side-effects, resistance, and factors affecting its place in the treatment sequence. Expert Opin Pharmacother 19:909–928. https://doi.org/10.1080/14656566.2018.1476492

Lee W, Baek M-C, Kim K-M, Bae J-S (2020) Biapenem as a novel insight into drug repositioning against particulate matter-induced lung injury. Int J Mol Sci 21:1462. https://doi.org/10.3390/ijms21041462

Liu L-W, Shi Y-Y, Li Z-L, Zuo L-H, Tang M, Jing Z-W, Zhao H-Y, Xue P, Zhou L, Du Q-Z, Zhang X-J, Sun Z (2020) Metabolomic insights into the synergistic effect of biapenem in combination with Xuebijing injection against sepsis. Front Pharmacol 11:502. https://doi.org/10.3389/fphar.2020.00502

LiverTox (2012) Clinical and research information on drug-induced liver injury [Internet]. National Institute of Diabetes and Digestive and Kidney Diseases, Bethesda, MD

Maesaka F, Nakai Y, Tomizawa M, Owari T, Miyake M, Inoue T, Anai S, Tanaka N, Fujimoto K (2019) Amrubicin is effective against small cell carcinoma of the prostate as a second-line chemotherapeutic agent: a case report. IJU Case Rep 2:133–136. https://doi.org/10.1002/iju5.12058

Manivasagan P, Venkatesan J, Sivakumar K, Kim S-K (2014) Pharmaceutically active secondary metabolites of marine actinobacteria. Microbiol Res 169:262–278. https://doi.org/10.1016/j.micres.2013.07.014

Muralidharan G, Micalizzi M, Speth J, Raible D, Troy S (2005) Pharmacokinetics of tigecycline after single and multiple doses in healthy subjects. Antimicrob Agents Chemother 49:220–229. https://doi.org/10.1128/AAC.49.1.220-229.2005

Niktoreh N, Lerius B, Zimmermann M, Gruhn B, Escherich G, Bourquin J-P, Dworzak M, Sramkova L, Rossig C, Creutzig U, Reinhardt D, Rasche M (2019) Gemtuzumab ozogamicin in children with relapsed or refractory acute myeloid leukemia: a report by Berlin-Frankfurt-Münster study group. Haematologica 104:120–127. https://doi.org/10.3324/haematol.2018.191841

O'Shaughnessy J, Thaddeus Beck J, Royce M (2018) Everolimus-based combination therapies for HR+, HER2− metastatic breast cancer. Cancer Treat Rev 69:204–214. https://doi.org/10.1016/j.ctrv.2018.07.013

On Behalf of NIKE Group, Gallo M, Malandrino P, Fanciulli G, Rota F, Faggiano A, Colao A (2017) Everolimus as first line therapy for pancreatic neuroendocrine tumours: current knowledge and future perspectives. J Cancer Res Clin Oncol 143:1209–1224. https://doi.org/10.1007/s00432-017-2407-5

Overwater I, Rietman A, van Eeghen A, de Wit M (2019) Everolimus for the treatment of refractory seizures associated with tuberous sclerosis complex (TSC): current perspectives. Ther Clin Risk Manag Volume 15:951–955. https://doi.org/10.2147/TCRM.S145630

Patterson MC, Mengel E, Vanier MT, Moneuse P, Rosenberg D, Pineda M (2020) Treatment outcomes following continuous miglustat therapy in patients with Niemann-Pick disease Type C: a final report of the NPC Registry. Orphanet J Rare Dis 15:104. https://doi.org/10.1186/s13023-020-01363-2

Pineda M, Walterfang M, Patterson MC (2018) Miglustat in Niemann-Pick disease type C patients: a review. Orphanet J Rare Dis 13:140. https://doi.org/10.1186/s13023-018-0844-0

Raja A, Prabakaran P (2011) Actinomycetes and drug: an overview. Am J Drug Discov Dev 1:75–84. https://doi.org/10.3923/ajdd.2011.75.84

Raja A, Prabakaran P, Gajalakshm P (2010) Isolation and screening of antibiotic producing psychrophilic actinomycetes and its nature from Rothang Hill soil against Viridans Streptococcus sp. Res J Microbiol 5:44–49. https://doi.org/10.3923/jm.2010.44.49

Sakaida E, Ebata T, Iwasawa S, Kurimoto R, Yonemori S, Ota S, Nakatani Y, Sekine I, Takiguchi Y (2017) Potential activity of amrubicin as a salvage therapy for merkel cell carcinoma. Intern Med 56:567–570. https://doi.org/10.2169/internalmedicine.56.7675

Sakurai R, Kaira K, Miura Y, Sunaga N, Saito R, Oyama T, Hisada T, Yamada M (2020) Clinical significance of topoisomerase-II expression in patients with advanced non-small cell lung cancer treated with amrubicin. Thorac Cancer 11:426–435. https://doi.org/10.1111/1759-7714.13289

Sivaramkrishna H, Girish M (2009) Microbes: an eternal source of innovative drugs. Express Pharma:16–30

Sun J, Chen C, Cui C-Y, Zhang Y, Liu X, Cui Z-H, Ma X-Y, Feng Y, Fang L-X, Lian X-L, Zhang R-M, Tang Y-Z, Zhang K-X, Liu H-M, Zhuang Z-H, Zhou S-D, Lv J-N, Du H, Huang B, Yu F-Y, Mathema B, Kreiswirth BN, Liao X-P, Chen L, Liu Y-H (2019) Plasmid-encoded tet (X) genes that confer high-level tigecycline resistance in *Escherichia coli*. Nat Microbiol 4: 1457–1464. https://doi.org/10.1038/s41564-019-0496-4

Uppal P, LaPlante KL, Gaitanis MM, Jankowich MD, Ward KE (2016) Daptomycin-induced eosinophilic pneumonia – a systematic review. Antimicrob Resist Infect Control 5:55. https://doi.org/10.1186/s13756-016-0158-8

van Gelder T, Fischer L, Shihab F, Shipkova M (2017) Optimizing everolimus exposure when combined with calcineurin inhibitors in solid organ transplantation. Transplant Rev 31:151–157. https://doi.org/10.1016/j.trre.2017.02.007

Wang J, Pan Y, Shen J, Xu Y (2017) The efficacy and safety of tigecycline for the treatment of bloodstream infections: a systematic review and meta-analysis. Ann Clin Microbiol Antimicrob 16:24. https://doi.org/10.1186/s12941-017-0199-8

Xie J, Roberts JA, Alobaid AS, Roger C, Wang Y, Yang Q, Sun J, Dong H, Wang X, Xing J, Lipman J, Dong Y (2017) Population pharmacokinetics of tigecycline in critically Ill patients with severe infections. Antimicrob Agents Chemother 61. https://doi.org/10.1128/AAC.00345-17

Xiong Y, Chen J, Sun X, Xu G, Li P, Deng Q, Yu Z, Chen Z, Zheng J (2021) The antibacterial and antibiofilm activity of telithromycin against *Enterococcus* spp. isolated from patients in China. Front Microbiol 11:616797. https://doi.org/10.3389/fmicb.2020.616797

Zuur MA, Ghimire S, Bolhuis MS, Wessels AMA, van Altena R, de Lange WCM, Kosterink JGW, Touw DJ, van der Werf TS, Akkerman OW, Alffenaar JWC (2018) Pharmacokinetics of 2000 milligram ertapenem in tuberculosis patients. Antimicrob Agents Chemother 62. https://doi.org/10.1128/AAC.02250-17